Process Analysis, Design, and Intensification in Microfluidics and Chemical Engineering

Harrson Silva Santana
University of Campinas, Brazil

João Lameu da Silva Jr
Federal Institute of Education, Science, and Technology of South of Minas Gerais, Brazil

Osvaldir Pereira Taranto
University of Campinas, Brazil

A volume in the Advances in Chemical and Materials Engineering (ACME) Book Series

IGI Global
DISSEMINATOR OF KNOWLEDGE

Published in the United States of America by
IGI Global
Engineering Science Reference (an imprint of IGI Global)
701 E. Chocolate Avenue
Hershey PA, USA 17033
Tel: 717-533-8845
Fax: 717-533-8661
E-mail: cust@igi-global.com
Web site: http://www.igi-global.com

Library of Congress Cataloging-in-Publication Data

Names: Santana, Harrson Silva, 1984- editor. | Silva Jr., Joao Lameu da (Silva Junior), 1986- editor. | Taranto, Osvaldir Pereira, 1963- editor.
Title: Process analysis, design and intensification in microfluidics and chemical engineering / Harrson Silva Santana, Joao Lameu da Silva Jr., and Osvaldir Pereira Taranto, editors.
Description: Hershey, PA : Engineering Science Reference, an imprint of IGI Global, [2019] | Includes bibliographical references.
Identifiers: LCCN 2018016420| ISBN 9781522571384 (hardcover) | ISBN 9781522571391 (ebook)
Subjects: LCSH: Chemical processes. | Microfluidics.
Classification: LCC TP155.7 .P7554 2019 | DDC 532/.05--dc23 LC record available at https://lccn.loc.gov/2018016420

This book is published in the IGI Global book series Advances in Chemical and Materials Engineering (ACME) (ISSN: 2327-5448; eISSN: 2327-5456)

British Cataloguing in Publication Data
A Cataloguing in Publication record for this book is available from the British Library.

All work contributed to this book is new, previously-unpublished material.
The views expressed in this book are those of the authors, but not necessarily of the publisher.

For electronic access to this publication, please contact: eresources@igi-global.com.

Advances in Chemical and Materials Engineering (ACME) Book Series

ISSN:2327-5448
EISSN:2327-5456

Editor-in-Chief: J. Paulo Davim, University of Aveiro, Portugal

MISSION

The cross disciplinary approach of chemical and materials engineering is rapidly growing as it applies to the study of educational, scientific and industrial research activities by solving complex chemical problems using computational techniques and statistical methods.

The **Advances in Chemical and Materials Engineering (ACME) Book Series** provides research on the recent advances throughout computational and statistical methods of analysis and modeling. This series brings together collaboration between chemists, engineers, statisticians, and computer scientists and offers a wealth of knowledge and useful tools to academics, practitioners, and professionals through high quality publications.

COVERAGE

- Statistical methods
- Nanomaterials
- Fatigue and Creep
- Toxicity of Materials
- Thermo-Chemical Treatments
- Numerical Techniques
- Wear of Materials
- Sustainable Materials
- Industrial chemistry
- Heat Treatments

The Advances in Chemical and Materials Engineering (ACME) Book Series (ISSN 2327-5448) is published by IGI Global, 701 E. Chocolate Avenue, Hershey, PA 17033-1240, USA, www.igi-global.com. This series is composed of titles available for purchase individually; each title is edited to be contextually exclusive from any other title within the series. For pricing and ordering information please visit http://www.igi-global.com/book-series/advances-chemical-materials-engineering/73687. Postmaster: Send all address changes to above address.

Titles in this Series

For a list of additional titles in this series, please visit:
https://www.igi-global.com/book-series/advances-chemical-materials-engineering/73687

The Geometry of Higher-Dimensional Polytopes
Gennadiy Vladimirovich Zhizhin (Russian Academy of Sciences, Russia)
Engineering Science Reference • ©2019 • 286pp • H/C (ISBN: 9781522569688) • US $195.00

Composites and Advanced Materials for Industrial Applications
K. Kumar (Birla Institute of Technology, India) and J. Paulo Davim (University of Aveiro, Portugal)
Engineering Science Reference • ©2018 • 402pp • H/C (ISBN: 9781522552161) • US $225.00

Emerging Synthesis Techniques for Luminescent Materials
Ratnesh Tiwari (Bhilai Institute of Technology, India) Vikas Dubey (Bhilai Institute of Technology, India) and Sanjay J. Dhoble (Rashtrasant Tukadoji Maharaj Nagpur University, India)
Engineering Science Reference • ©2018 • 505pp • H/C (ISBN: 9781522551706) • US $215.00

Handbook of Research on Ergonomics and Product Design
Juan Luis Hernández Arellano (Autonomous University of Ciudad Juárez, Mexico) Aide Aracely Maldonado Macías (Autonomous University of Ciudad Juárez, Mexico) Juan Alberto Castillo Martínez (University of Rosario, Colombia) and Porfirio Peinado Coronado (Autonomous University of Ciudad Juárez, Mexico)
Engineering Science Reference • ©2018 • 446pp • H/C (ISBN: 9781522552345) • US $235.00

Mechanical Properties of Natural Fiber Reinforced Polymers Emerging Research and ...
Sarah S. Gebai (International University of Beirut, Lebanon) Ali M. Hallal (International University of Beirut, Lebanon) and Mohammad S. Hammoud (International University of Beirut, Lebanon)
Engineering Science Reference • ©2018 • 228pp • H/C (ISBN: 9781522548379) • US $165.00

For an entire list of titles in this series, please visit:
https://www.igi-global.com/book-series/advances-chemical-materials-engineering/73687

IGI Global
DISSEMINATOR OF KNOWLEDGE

701 East Chocolate Avenue, Hershey, PA 17033, USA
Tel: 717-533-8845 x100 • Fax: 717-533-8661
E-Mail: cust@igi-global.com • www.igi-global.com

Table of Contents

Detailed Table of Contents

Analysis of hydrodynamics and mass transfer Taylor flows in micro channels of both gas-liquid and liquid-liquid systems on the basis of classical theoretical approach with some simplifying assumptions was performed. Results of theoretical analysis for description of hydrodynamic parameters and mass transfer characteristics were confirmed by comparison with the author's own and available in literature experimental data. It was shown that the main parameters of two-phase Taylor flows could be quite precisely described theoretically: mean bubble/droplet velocity, liquid film thickness, real gas holdup (which is always smaller than so-called dynamic holdup), pressure drop. Peculiarities of liquid-liquid flows compared to gas-liquid Taylor flows in capillaries are discussed. Wettability effect on hydrodynamics was examined. Tools of mass transfer intensification of gas-liquid and liquid-liquid Taylor flow in micro channels are analyzed. Three-layer model for heat and mass transfer has been proposed and implemented for the case of solid-liquid mass transfer for gas-liquid Taylor flows; optimal process conditions for this process are found theoretically and discussed from physical point of view.

The chapter aims to introduce the computational fluid dynamics (CFD). A review was provided, outlining its development and applications on chemical engineering and microfluidics. The fundamental points of the CFD, listing the advantages and precautions of this numerical technique were provided. The description of CFD methodology including the three essential stages (pre-processing, solving, and post-processing) was made. The fundamental transport equations—total mass (continuity), momentum, energy, and species mass balances—and the usual boundary conditions used in CFD were explained. The main approaches used in multicomponent single-phase flows, single-phase flow in porous media, and multiphase flows in microscale were detailed, as well as the numerical mesh types and its quality parameters. A brief introduction of finite volume method (FVM) used by most of the available CFD codes was also performed, describing the main numerical solution features. Finally, the conclusions and future prospects of CFD applications are exposed.

In this chapter, a numerical study and assessment of the mixing efficiency of a novel microfluidic device for mixing two fluids are presented. The device under study consists of a two-dimensional straight microchannel with a square pillar centered across the channel. The main fluid flows through the microchannel from the main inlet to the outlet, while the second fluid is injected through the pillar as two small jets at its upstream corners. For different values of the Reynolds number, intensity ratio between the jets and the main channel stream and jets injection angle, the authors have conducted several numerical simulations to characterize both the mixing efficiency and the required input power to make the fluids flow. The optimum configuration has been revealed for high values of the Reynolds number, low intensity ratios, and high injection angles. Thanks to vortex shedding and the corresponding downstream oscillations, a mixing efficiency of around 90% can be reached. The worst mixing efficiency is obtained for a configuration without vortex shedding, having a mixing efficiency of only around 2%.

The chapter explores the dynamics of liquid penetration in a closed end vertical capillary. This model is very important for impedance spectroscopy methodology where oxidized porous silicon provides an in vitro medium, and one important criteria of this methodology is the liquid penetration depth inside the silicon pores as the impedance is greatly affected by this penetration depth. This problem is also important in order to understand how the presence of entrapped air inside a micro pore can influence the dynamics of capillary flow. For this purpose, the model is studied both analytically and numerically. In this study, different pore size (500 nm and 2 μm diameter) with equal pore depth (~10 μm) have been used. Finally, the analytical solution is compared with the numerical results. In addition, the linearization of the system is also investigated and found the critical viscosity of which demarcates the over-damped and under-damped regimes. Further, this study is extended by incorporating the dynamic contact angle effects on the meniscus dynamics.

The immiscible two-phase flow behaves nonlinearly, and it is a challenging task to control and stabilize the liquid-liquid interface. Parallel flow forms under a proper balance between the driving force, the friction resistance, and the interfacial tension. The liquid-solid interaction as well as the liquid-liquid interaction plays an important role in manipulating the liquid-liquid interface. With vacuum-driven flow, long and stable parallel flow is possible to be obtained in oil-water systems and can be used for fabricating micro- and nanomaterials. Ultra-small Cu nanoparticles of 4~10 nm were synthesized continuously through chemical reactions taking place on the interface. This makes it possible for in situ synthesis of conductive nanoink avoiding oxidation. Well-controlled interface reactions can also be used to produce ultra-long sub-micro Cu wires up to 10 mm at room temperature. This method provided new and simple additive fabrication methods for making integrated microfluidic devices.

The use of micro-heat exchangers increased with the advancement of microfluidics. These microdevices present some advantages like elevated surface area-to-volume ratio resulting in high heat transfer rates. Micro-heat exchanger with phase change is a new application of such devices. The simultaneous momentum, heat, and mass transfer at microscale still require investigations due to the inherent complexity. The main goal of the chapter is to demonstrate experimentally and numerically the capability of the micro-heat exchanger use in the continuous process of ethanol excess recovery from the biodiesel. The influence of flow rate, ethanol/biodiesel molar ratio, and temperature on the ethanol evaporation performance was evaluated. The flow rate and the ethanol/biodiesel molar ratio influenced negatively the evaporation. In contrast, the temperature was affected positively. The mathematical model was able to capture the main features of the continuous evaporation; however, further improvements must be performed in order to consider the thermodynamics characteristics.

This chapter provides a guide for microfluidic devices development and optimization focused on chemical analysis applications, which includes medicine, biology, chemistry, and environmental monitoring, showing high-level performance associated with a specific functionality. Examples are chemical analysis, solid phase extraction, chromatography, immunoassay analysis, protein and DNA separation, cell sorting and manipulation, cellular biology, and mass spectrometry. In this chapter, most information is related to microfluidic devices design and fabrication used to perform several steps concerning chemical analysis, process preparation of reagents, samples reaction and detection, regarding water quality monitoring. These steps are especially relevant to lab-on-chip (LOC) and micro-total-analysis-systems (μTAS). μTAS devices are developed in order to simplify analytical chemist work, incorporating several analytical procedures into flow systems. In the case of miniaturized devices, the analysis time is reduced, and small volumes (nL) can be used.

Many real-world applications depend on temperature sensing and/or control. This includes a wide range of industrial processes, chemical reactors, and SCADA systems, in addition to other physical, mechanical, and biological systems. With the advancement of technology, it became possible to produce a new generation of smart and compact temperature sensors, which are capable of providing digital outputs that are more accurate, robust, and easily interfaced and integrated into measurement and control systems. This chapter first surveys traditional analog temperature sensors, such as RTDs and thermocouples, to provide a strong motivation for the need to adopt better and smarter techniques that mainly rely on digital technology (e.g., CMOS designs). Different interfacing techniques that do not need ADCs are introduced, including the programmable Arduino microcontrollers. Different applications will be explored that include automotive accessories, weather forecast, healthcare, industrial processing, firefighting, and consumer electronics. Both wired and wireless technologies, including the IoT, will be investigated as means for transmitting the sensed data for further processing and data logging. A special case study to provide information redundancy in industrial SCADA systems will be analyzed to illustrate the advantages and limitations of smart temperature sensors. The chapter concludes with a summary of the design effort, accuracy, performance, and cost effectiveness of smart temperature sensors while highlighting future trends in this field for different applications.

This chapter presents a numerical investigation of heat and mass transfer characteristics during the evaporation of liquid films in vertical geometries. A two-phase model is developed to simulate laminar film evaporation into laminar gas flow. The liquid film evaporation is evaluated under adiabatic and heated wall conditions for both pure and binary liquid film. The model is based on a finite difference method to solve the governing equations of the two phases. The obtained results concerns two industrial processes. The first part of the chapter is devoted to the analysis of the thermal protection of vertical channel wall, while the second part is devoted to the desalination process by falling liquid film. The simulations results allowed the determination of the optimal operating conditions for both processes.

Preface

Defined as the science and technology applied in structures of dimensions ranging from tens to hundreds of micrometers, the Microfluidics represents a great potential for chemical processes design, development and optimization. On the other hand, the Chemical Engineering comprehends skills of project design of industrial processes, usually resulting in large chemical plants as we already know. Therefore, what is common between these two distinct process scales?

The microfluidic devices provide advantages regarding the traditional macroscopic process. The main advantages of microfluidic devices are: low amount of reactants and samples, short reaction times, low fabrication costs, high surface area-to-volume ratio (about 20000 m^2 m^{-3} against 1000 m^2 m^{-3} of conventional reactors) and high rates of heat and mass transfer.

In addition, the microscale processes can achieve higher production ratcs applying the numbering-up concept. This concept consists in a modular scale-up design, using multiple units of the optimized microdevice operating continuously in parallel, resulting in a higher volumetric rate capacity and flexible operation and maintenance. The numbering-up concept provides a smart, effective and faster time-to-market result, allowing the development of new products and processes. This is fundamental to increase the product variety, an essential key for the modern industries to overcome the challenges outcome from the volatile globalized markets.

In this context, the present book will approach the relationship of these two technology areas and how they influence the industrial process and society. The purpose of this book is to be used as a source of information and references for beginners to expert readers in Microfluidics and its application in Chemical Engineering. Each chapter, written by specialists in their occupation fields, provides concepts and specific information, updating the literature and providing advice and insights about the topics. The main idea of the book is to provide an approach beyond the Microfluidic or Chemical Engineering fields, building new pathways and means to develop processes and products in industrial and academic areas, contributing to the Microfluidics progress, and even, looking forward that the Microfluidics could be integrate regular subject matter in Chemical Engineering courses.

The target audience of this book is consisted by professionals and researchers working in engineering markets, Microfluidics and Chemical Engineering companies, and Universities and Institutes of Technology.

WHAT IS MICROFLUIDICS?

Microfluidics is commonly defined as the science and the technology of systems handling low amount of fluids using structures with dimensions ranging from tens to hundreds of micrometers - the microdevices (Whitesides, 2006).

Another definition was made by Nguyen and Wereley (2006): Microfluidics is the science and the engineering of systems which the fluid behavior is distinct from the conventional flow mostly due to the reduced length scale of the system. This definition is more comprehensive than the previous, once it provides an important insight about the fluid flow in the microdevices. In the conventional macroscale, the fluid flow behavior is described by the classical continuum mechanics developed since the age of Isaac Newton (1643-1747). In contrast, the definition of Microfluidics by Nguyen and Wereley (2006) could lead to the erroneous idea that the classical fluid mechanics could not predict correctly the flow behavior in microdevices. This assumption must be treated carefully, since depending on the case, the classical fluid mechanics is capable to predict the microscale fluid flow properly (Santana, Silva Jr., & Taranto, 2015).

The aforementioned definitions clarified the main aspects of Microfluidics. From the interpretation by Whitesides (2006), the Microfluidics handles low amount of fluids in microscale. Nguyen and Wereley (2006) augmented the definition introducing the fluid flow theory. Now, we will expand the Microfluidics concept.

Lo (2013) defined Microfluidics as the science and the technology about the fluid flow behavior, controlled manipulation of fluids and design of devices and systems capable to execute reliably tasks in microchannels with typical dimensions ranging from tens to hundreds of micrometers. We particularly appreciate this definition, once it combines the previous interpretation with the microdevice design, an essential feature in microscale, mostly for chemical processes, in which a scale down (i.e., the reduction of the scale) is not an easy task, since each scale possesses their own transport phenomena characteristics. Accordingly, at the beginning of chemical process development in microscale, the microdevices are designed considering the Microfluidics characteristics.

The tools developed in Microfluidics allowed the obtaining of advantages regarding macroscale methods. The achievement of high conversion and selectivity associated with short residence times enables the advancement of chemical processes, resulting in the creation of new products and production methods in a faster way for the industry and has been increasing the interest of professionals in the most diverse knowledge fields, including the Chemical and Processing Engineering.

WHAT IS THE CHEMICAL ENGINEERING?

According to the European Federation of Chemical Engineering (EFCE), the Chemical Engineering covers conception, development, design, improvement and application of processes and products, including also, economic analysis, design, manufacturing, operation, control and management of industrial processing plants jointly with research and educational fields about these subjects (Gillet, 2001).

This definition highlights the broad field of Chemical Engineering, involving not only the professionals employed in daily functions inside the industries (the Process Engineers), but also, professionals involved with the conception and implementation of industrial plant projects and professional engineers engaged in Universities and researches and technology institutes. From this viewpoint, we can state that the Chemical Engineer manages the laboratory scale chemistry to industrial scale process, allowing an effective large scale production of the products used daily by all of us.

Another interesting description about the Chemical Engineer is provided by the American Institute of Chemical Engineers (AIChE): "The Chemical Engineering is the career that the knowledge of mathematics, chemistry and other natural sciences, acquired by study, experience and practice is applied in order to develop economic means for the use of materials and energy resources in the benefit of mankind" (AIChE, 2016).

Based on the aforementioned knowledge, the Chemical Engineer can act and collaborate in diversified areas from the large scale production of commodities to research institutes and industries including pharmaceutical, electronics and robotics.

PROCESS INTENSIFICATION: THE LINK BETWEEN CHEMICAL ENGINEERING AND MICROFLUIDICS

There is no general and shared definition about Process Intensification (PI). Along the years, some authors proposed distinct interpretations, aiding the conceptualization of PI.

The first definition was stated by Ramshaw (1995) at the 1st International Conference on Process Intensification in the Chemical Industry. Ramshaw (1995) defined PI as the strategy for size reduction of a chemical plant in order to achieve a determined production goal. The size reduction could be performed by the size reduction of individual process parts or by the decrease in the number of involved unities. By the time, Ramshaw (1995) addressed volume reductions about orders of 100 or more. These orders of magnitude appeared very challenging, once the miniaturization concepts were still not widespread in Chemical Engineering, causing the Chemical Engineers to think over the size reduction to achieve the process intensification.

Stankiewicz and Moulijn (2000) stated that a volume decrease by the order of two already can be classified as PI. They expanded the PI definition exclusively over the reduction of plant size or equipment size. According to Stankiewicz and Moulijn (2000), the PI consists in the development of new equipment and techniques in order to promote significant enhancement in manufacturing and processing, considerably decreasing the ratio between equipment size/production capacity, the energy consumption and the waste generation, and consequently resulting in sustainable and economic technologies.

In brief, any development in the Chemical Engineering leading to a technology substantially small, clear and energetically efficient is a Process Intensification, according to the previous definitions.

Lutze et al. (2010) extended the PI definitions. They declared that the PI can be achieved by addition or reinforcement of the process phenomena by the integration of operations and functions, or alternatively, by the oriented reinforcement of the phenomena in a specific operation, enhancing some attributes as the increment of capacity, productivity, flexibility and safety and the decrease of energy consumption and waste generation.

In this context, Microfluidics appears as an interesting tool aiming for the Process Intensification in the most diverse Chemical Engineering process fields.

BOOK ORGANIZATION

The main idea of the book is to provide an approach beyond the Microfluidics or Chemical Engineering fields, building new pathways and means to develop processes and products in industrial and academic areas. The book contemplates 10 chapters. The description of each one of these chapters is provided following.

Chapter 1 evaluates the hydrodynamics and mass transfer of Taylor flows in microchannels of both Gas-Liquid and Liquid-Liquid systems. The classical theoretical approach with some simplifying assumptions was performed and the results were confirmed by comparison with experimental data. It was shown that the main parameters of two-phase Taylor flows could be quite precisely described theoretically: mean bubble/droplet velocity, liquid film thickness, real gas holdup, pressure drop. Moreover, it was discussed the peculiarities of Liquid-Liquid flows compared to Gas-Liquid Taylor flows in capillaries, as well, the wettability effect on hydrodynamics and mass transfer intensification of Gas-Liquid and Liquid-Liquid Taylor flow in micro channels.

Chapter 2 presents an overview about the development and applications of Computational Fluid Dynamics (CFD) on Chemical Engineering and Microfluidics. The fundamental points of the CFD, listing the advantages and precautions of this numerical technique were provided. The fundamental transport equations – total mass (continuity), momentum, energy and species mass balances – and the usual boundary conditions used in CFD were explained. The main approaches used in multicomponent single-phase flows, single-phase flow in porous media and multiphase flows in microscale were detailed, as well, the numerical mesh types and its quality parameters. A brief introduction of Finite Volume Method (FVM), used by most of the available CFD codes was also performed, describing the main numerical solution features.

Chapter 3 investigates the mixing efficiency of a novel micromixing unit, a passive micromixer based on blowing and vortex shedding mechanisms, which are used to promote an irregular flow downstream. The main goal of the chapter is the determination of the governing parameters that provide the highest mixing efficiency at the exit of the channel and the required input power to make both fluids to flow and mix. The system under study consisted in a two-dimensional straight microchannel with a square pillar centered across the channel. The main fluid flows through the microchannel from the main inlet to the outlet, while the second fluid is injected through the pillar forming two small jets at the upstream corners.

Chapter 4 presents a method for deriving an analytical solution to the momentum balance of a liquid in a capillary tube. The authors introduced an extended solution of the well-established Lucas-Washburn equation. The extended solution included the entrapped gas pressure term and excluded the gravity term. Furthermore, the governing equation is non-dimensionalized and a non-dimensional viscous term was found. Numerical approach was also included in this study where commercial CFD software was used for solving the problem numerically. Comparison between analytical and numerical solutions is also shown.

In Chapter 5, the control of immiscible liquid-liquid two-phase flow has been discussed and tested thoroughly. Oleic acid-water flow in microchannels of 240 μm in width and 8 μm in height was studied as a microfluidic demo system for building blocks of future multifunctional integrated System-on-Chip (SoC) or System-in-a-Package (SiP).

The main goal of Chapter 6 was to demonstrate experimentally and numerically the capability of the micro-heat exchanger use in the continuous process of ethanol excess recovery from the biodiesel. The influence of flow rate, ethanol/biodiesel molar ratio and temperature on the ethanol evaporation performance was evaluated in experimental and numerical runs.

Chapter 7 provides a guide for microfluidic devices development and optimization focused on chemical analysis applications, which includes medicine, biology, chemistry and environmental monitoring, showing high-level performance associated with a specific functionality, exploiting Low Temperature Co-fired Ceramic (LTCC) technology as prototype material used for microfluidic devices fabrication.

In Chapter 8 the topic of smart sensors with emphasis on temperature sensing, one of the most important variable in Microfluidics and in Chemical Engineering, is introduced. This chapter first surveys traditional analog temperature sensors to provide a strong motivation for the need to adopt better and smarter techniques that mainly rely on digital technology, e.g. Complementary-Metal-Oxide-Semiconductor (CMOS) designs. Different interfacing techniques that do not need Analog-to-Digital-Conversion (ADCs) are introduced. Different applications were explored including automotive accessories, weather forecast, health care, industrial processing, firefighting, and consumer electronics. Both wired and wireless technologies, including the Internet-of-Things (IoT), were investigated as means for transmitting the sensed data for further processing and data logging.

Chapter 9 presents a numerical investigation of heat and mass transfer characteristics during the evaporation of liquid films in vertical tube. A two-phase model was developed to simulate laminar film evaporation into laminar gas flow. The liquid film evaporation was evaluated under adiabatic and heated wall conditions for both pure and binary liquid film. The obtained results concerns two industrial processes. The first part of the chapter was devoted to the analysis of the thermal protection of vertical channel wall, while the second part was devoted to the desalination process by falling liquid film.

Chapter 10 presents a validation of the selection criteria for the construction of granulation and drying equipment. Based on the simulation results, various constructions of the apparatus components are offered. The computer simulation allowed offering an algorithm of the optimization calculations of the granulation and drying equipment. New constructive solutions to perform the granulation and the drying processes in the devices with a directional fluidized bed are offered.

The editors are grateful for the support and the patience of our families during this project. We could never have completed this journey without you. The editors thank God for having you as family.

Moreover, the editors would like to immensely thank all the authors that contributed with theirs researches and make possible the accomplishment of this project. The editors also are grateful to the reviewers that contributed to the quality of this book. Finally, the authors acknowledge the book development editor and the IGI Global team.

Harrson Silva Santana
University of Campinas, Brazil

João Lameu da Silva Júnior
Federal Institute of Education, Science, and Technology of South of Minas Gerais, Brazil

Osvaldir Pereira Taranto
University of Campinas, Brazil

REFERENCES

American Institute of Chemical Engineers (AIChE) Constitution. (2016). Available in: https://www.aiche.org/about/governance/constitution

Gillet, J. E. (2001). Chemical Engineering Education in the Next Century. *Chemical Engineering & Technology*, *24*(6), 561–570. doi:10.1002/1521-4125(200106)24:6<561::AID-CEAT561>3.0.CO;2-X

Lo, R. C. (2013). Application of Microfluidics in Chemical Engineering. *Chemical Engineering & Process Techniques*, *1*, 1002.

Lutze, P., Gani, R., & Woodley, J. M. (2010). Process intensification: A perspective on process synthesis. *Chemical Engineering and Processing: Process Intensification.*, *49*(6), 547–558. doi:10.1016/j.cep.2010.05.002

Nguyen, N.-T., & Wereley, S. T. (2006). *Fundamentals and applications of microfluidics*. Boston: Publishing House Artech House.

Ramshaw, C. (1995). *The Incentive for Process Intensification*. International Conference on Process Intensification for the Chemical Industry, London, UK.

Santana, H. S., Silva, J. L. Jr, & Taranto, O. P. (2015). Numerical simulations of biodiesel synthesis in microchannels with circular obstructions. *Chemical Engineering and Processing: Process Intensification*, *98*, 137–146. doi:10.1016/j.cep.2015.10.011

Stankiewicz, A. I., & Moulijn, J. A. (2000). Process Intensification: Transforming Chemical Engineering. *Chemical Engineering Progress*, *96*(1), 22–33.

Whitesides, G. M. (2006). The origins and the future of microfluidics. *Nature*, *442*(7101), 368–373. doi:10.1038/nature05058 PMID:16871203

Chapter 1
Analysis of Hydrodynamics and Mass Transfer of Gas-Liquid and Liquid-Liquid Taylor Flows in Microchannels:
Theoretical Approach and Experimental Proofs

Rufat Abiev
St. Petersburg State Institute of Technology (Technical University), Russia

ABSTRACT

Analysis of hydrodynamics and mass transfer Taylor flows in micro channels of both gas-liquid and liquid-liquid systems on the basis of classical theoretical approach with some simplifying assumptions was performed. Results of theoretical analysis for description of hydrodynamic parameters and mass transfer characteristics were confirmed by comparison with the author's own and available in literature experimental data. It was shown that the main parameters of two-phase Taylor flows could be quite precisely described theoretically: mean bubble/droplet velocity, liquid film thickness, real gas holdup (which is always smaller than so-called dynamic holdup), pressure drop. Peculiarities of liquid-liquid flows compared to gas-liquid Taylor flows in capillaries are discussed. Wettability effect on hydrodynamics was examined. Tools of mass transfer intensification of gas-liquid and liquid-liquid Taylor flow in micro channels are analyzed. Three-layer model for heat and mass transfer has been proposed and implemented for the case of solid-liquid mass transfer for gas-liquid Taylor flows; optimal process conditions for this process are found theoretically and discussed from physical point of view.

DOI: 10.4018/978-1-5225-7138-4.ch001

INTRODUCTION

In the past decades micro and minireactors are actively investigated for their use instead of conventional devices as well as for new processes which are crucially increased due to smaller transversal sizes of the micro and minichannels (capillaries in other word). One understands under the micro channels usually the channels with hydraulic diameter roughly smaller than 1 mm (often 100–200 μm or even smaller, down to 10 μm). The mini channels have larger diameters, approximately up to 2 mm (see Background section for details). The advantages of micro and mini channels are:

1. Smaller diffusional path, e.g. for typical diffusivity in the liquid $D = 10^{-9}$ m^2/s the characteristic time of diffusion could be assessed as $t_{dif} \approx d^2/D$, i.e. for d = 100 mm $t_{dif} \approx 10^7$ s = 2778 hours, for d = 10 mm $t_{dif} \approx 10^5$ s = 28 hours, for d = 1.0 mm $t_{dif} \approx 10^3$ s = 16.7 min, for d = 0.1 mm $t_{dif} \approx 10$ s. Hence, the decrease of the channel diameter results in much shorter time necessary to perform mass transfer due to diffusion alone.
2. The flow in micro and minichannels is close to the plug flow due to absence of turbulent mixing, especially for so-called Taylor (or slug, or chain) flow (Figure 1), where the bubbles/droplets of dispersed phase have the length larger than diameter of the channel.
3. The smaller volume of micro and minireactors is easier to control due to less inertia of hydrodynamics, heat and mass transfer parameters.
4. For explosive chemicals produced in micro and minireactors there is no risk of explosion because of their diameter is smaller than the critical value of diameter.
5. Excellent mixing in each element of Taylor flow structure due to inner local vortices (so-called Taylor vortices), allowing to increase mass transfer several times compared to one-phase laminar flow.

The application of micro and mini technologies can not only improve the quality of end products, reducing energy and resource consumption, but also increase the manufacturing mobility and the possibility of its rapid reset to produce other products. Micro- and minireactors can reduce the space occupied by the plant, making it safer and more environmental friendly.

Micro and minireactors are widely studied attracting significant interest of both researcher and industry. Nonetheless, many of published results have either experimental or numerical nature, and some theoretical approach is needed to provide more sophisticated understanding of phenomena and processes taking place in this promising type of chemical equipment.

Figure 1. Schematic diagram used in formulating the continuity equation for the slug flow of a two-phase mixture

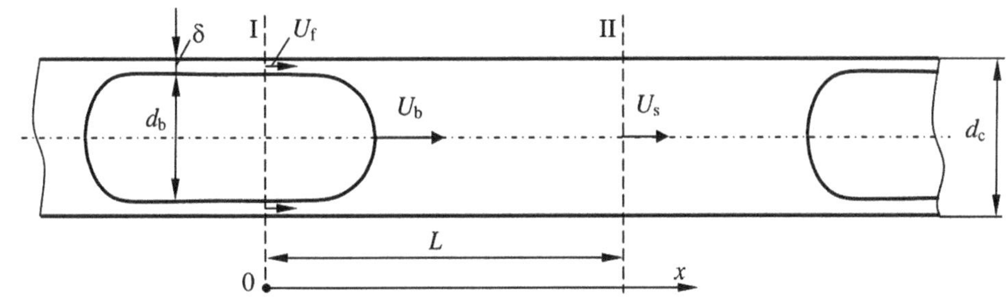

Thus, the aim of this chapter is to build up a general view to the two-phase Taylor flows in capillaries which allows to describe mathematically the structure (length of the bubbles/droplets of dispersed phase - plugs and the droplets of continuous phase - slugs) and hydrodynamics of this flow, and further to have a key for heat and mass transfer description and optimization. Methods of multiphase fluid dynamics are used for this aim with some assumptions along. The proposed mathematical models allow to predict with quite high precision the hydrodynamics and mass transfer characteristics of two-phase Taylor flows and to improve design of industrial and lab scale devices.

BACKGROUND

Mini and micro devices like pumps, mixers, heat exchangers, reactors, extractors and valves have been already developed and produced in small series (Hessel, Müller, & Kolb, 2005; the Catalogue, 2009). Cross section dimension of such equipment is usually in the range between 10 μm and 3 mm. In some papers it was proposed to distinguish between micro and mini scale; boundary between them is considered to be about 100 μm. Upper boundary of miniscale according to several authors is in the range between 1 mm and 5.4 mm. A physically based limit for miniscale size of the pipes could be attributed to the Bond number

$$Bo = \frac{\Delta \rho g d^2}{\sigma} \tag{1}$$

imposing the balance between gravitational and capillary forces, (in some papers Eötvös number is used instead) corresponding to the critical value of Bo = 4, from which the upper limit of minichannels diameter follows:

$$D_{crit} = 2\sqrt{\frac{\sigma}{\Delta \rho g}} \tag{2}$$

Microreactors can be competitive for the fast reactions, when mass transfer limits a reaction rate, as well for increased heat transfer, when it is necessary to remove quickly heat from the reactants. It is possible due to unusually high values of heat and mass transfer rates in microreactors, which could be up to two orders of magnitude higher than in conventional types of reactors (Hessel, Müller, & Kolb, 2005). Another significant advantage of microreactors is a very narrow residence time distribution that can substantially reduce the formation of byproducts in the consecutive reactions (Hessel, Müller, & Kolb, 2005; Kreutzer, Kapteijn, Moulijn, & Heiszwolf, 2005a). Microreactors are used for the reactions in mixtures of gases as well as in gas-liquid, liquid-liquid systems (Kreutzer, Kapteijn, Moulijn, & Heiszwolf, 2005a; Hessel, Angeli, Gavriilidis, & Löwe, 2005), and more recently for the synthesis of ionic liquids (Renken, Hessel, Löb, Uerdingen, & Kiwi-Minsker, 2007). One of the new trends in micro technology is a combination of flow in microchannels with the imposition of additional forces, for instance, microwaves (Gao, Rebrov, Verhoeven, Schouten, Kleismit, Kozlowski, Cetnar, Turgut, & Subramanyam, 2010; He, Haswell, & Fletcher, 2004).

In terms of design microreactors can be divided into single- and multichannel types. One of the types of multichannel microreactors are so-called monolithic reactors with catalyst of honeycomb structure, representing a block of parallel channels with hydraulic diameter in the range between 0.3 mm and 3.1 mm, with an active catalyst covered on their inner surface (Roy, Bauer, Al-Dahhan, Lehner, & Turek, 2004; Bauer, Schubert, Lange, & Abiev, 2006). But so far the main application of such reactors is gas-liquid catalytic reactions.

Most favorable regime for conducting gas-liquid catalytic reactions is a slug flow (Kreutzer, Kapteijn, Moulijn, & Heiszwolf, 2005a; Roy, Bauer, Al-Dahhan, Lehner, & Turek, 2004; Bauer, Schubert, Lange, & Abiev, 2006), which is shown in Figure 1. Gas-liquid slug flow regime is a consecutive flow of gas bubbles separated from each other by liquid slugs. The advantages of this regime are a good mixing inside the liquid slug due to the so-called Taylor vortices, as well as short diffusion path for gas molecules penetrating through the liquid film between the bubble and the wall of the catalyst (Kreutzer, Kapteijn, Moulijn, & Heiszwolf, 2005a; Roy, Bauer, Al-Dahhan, Lehner, & Turek, 2004; Bauer, Schubert, Lange, & Abiev, 2006; Onea, Wörner, & Cacuci, 2009).

Process equipment containing multiple small parallel channels has gained increasing attention within the past decades in industry and academia due to its potential to intensify processes by enhancing heat and mass transfer or by more precisely controlling contact times (Kreutzer, Kapteijn, Moulijn, & Heiszwolf, 2005a; Gascon, van Ommen, Moulijn, & Kapteijn, 2015a). Several processes have been successfully realised such as chemical or biocatalytic reactions (Kreutzer, Kapteijn, Moulijn, & Heiszwolf, 2005a; Watts, & Haswell, 2005; Kiwi-Minsker, & Renken, 2005; Moulijn, Kreutzer, Nijhuis, & Kapteijn, 2011; Burns & Ramshaw, 1999; Bolivar & Nidetzky, 2013; Schubert, Brandner, Fichtner, Linder, Schygulla, & Wenka, 2001), extraction (Kashid, Harshe, & Agar, 2007; Tsaoulidis & Angeli, 2014) and absorption (Ganapathy, Shooshtari, Dessiatoun, Alshehhi, & Ohadi, 2014).

The extraordinary heat and mass transfer capacity results from the high surface-to-volume ratios within these devices as well as from a microfluidic flow regime – Taylor flow. This regime is characterised by a flow of two immiscible fluids in separate compartments and offers large interfacial areas, short diffusion distances, internal circulation vortices, and nearly plug flow behaviour (Kreutzer, Kapteijn, Moulijn, & Heiszwolf, 2005a; Kreutzer, Du, Heiszwolf, Kapteijn, & Moulijn, 2001). Further literature reviews on hydrodynamics in micro and minichannels (Rebrov, 2010; Shao, Gavriilidis, & Angeli, 2009; Angeli & Gavriilidis, 2008).

Another interesting application of Taylor Gas-Liquid (Vapor-Liquid) flows is the closed loop pulsating heat pipes (CLPHP) (Khandekar, Dollinger, & Groll, 2003; Mehta & Khandekar, 2014; Das, Nikolayev, Lefèvre, Pottier, Khandekar, & Bonjour, 2010; Khandekar, Gautam, & Sharma, 2009; Bhagat & Watt, 2015). Interestingly that one of the two main types of flow is slug flow (the other is annular flow), containing liquid slug and gas-vapor bubbles.

Despite such a great interest to this subject, published results in the literature are disjointed, sometimes even contradictory. In this regard, there is a need of reliable experimental data for mathematical models verification, as well as a building of physically based relations for the mass transfer in the case of slug flow.

The aim of this chapter is to represent general view to the mathematical modeling of hydrodynamics and mass transfer of Gas-Liquid and Liquid-Liquid Taylor flows in micro channels by means of theoretical approach using classical equations and modern formulae necessary to close the model.

Quite complete reviews on hydrodynamics and mass transfer studies in micro-reactors were presented recently (Kashid, Renken, & Kiwi-Minsker, 2011; Sobieszuk, Aubin, Pohorecki, 2012; Sotowa, 2014). These papers detail the problems of mass transfer in the systems of gas-liquid, liquid- liquid and gas-liquid-solid phases. However, it should be noted the experimental orientation of the studies considered in (Kashid, Renken, & Kiwi-Minsker, 2011; Sobieszuk, Aubin, Pohorecki, 2012; Sotowa, 2014), the results of which are mostly empirical correlations of mass transfer coefficients on dimensionless numbers and regime parameters in microreactors. Such approach to the description of mass transfer makes it possible to carry out technological calculations of microreactor equipment, but does not permit a detailed study of the mechanisms of the matter transport.

The structure of the chapter is organized as follows: 1) a simplified model of Taylor flow hydrodynamics; 2) the method for calculating slug flow hydrodynamics, including that at high pressure gradients; 3) the peculiarities of Liquid-Liquid flows; 4) correlations for Taylor flow parameters - bubble velocity, gas void fraction (hold-up) and pressure drop; 5) analysis of local pressure gradient inversion and form of bubbles in Taylor flow in microchannels; 6) mass transfer intensification by means of Taylor vortices; 7) three-layer model and its application to the mass transfer modeling, and; 8) case study: optimization of wall-liquid mass transfer process.

HYDRODYNAMICS AND MASS TRANSFER OF GAS-LIQUID AND LIQUID-LIQUID TAYLOR FLOWS IN MICROCHANNELS

A Simplified Model of Taylor Flow Hydrodynamics

One of the first classical works that study the flow of a gas–liquid mixture in a capillary was performed by Taylor (1961). At the same time, Bretherton (1961) published a paper in which the hydrodynamic theory of lubrication was used to solve the problem describing the effect of the flow on the thickness of the liquid film around the bubble. A modified Bretherton equation, which can be used for a wider range of the capillary number, $Ca < 1$, was obtained by Aussillous and Quéré (2000).

The slug flow of a gas–liquid mixture was theoretically described for a pipe with dead ends when the rise of the bubble is caused by gravity and the liquid displaced by the bubble flows downward (Pokusaev, Zaitsev, & Zaitsev, 1999) and a flow through capillary (Reinelt, 1987).

Some regularities derived by the theory were in one or another way supported by experimental results (Thulasidas, Abraham, & Cerro, 1995; Liu, Vandu, & Krishna, 2005) or numerical simulation (Taha & Cui, 2004; Van Baten & Krishna, 2005; Kreutzer, Kapteijn, Moulijn, & Heiszwolf, 2005b). Phenomena observed in experiments have been systematically theoretically described in the papers of (Kreutzer, Du, Heiszwolf, Kapteijn, & Moulijn, 2001; Kapteijn, Moulijn, & Heiszwolf, 2005a; Kapteijn, Moulijn, Kleijn, & Heiszwolf, 2005b), by some other groups: group of R. Krishna (Liu, Vandu, & Krishna, 2005; Van Baten & Krishna, 2005), group of J. Schouten (Warnier, de Croon, Rebrov, & Schouten, 2010), and partly also by our team (Abiev, 2008, 2009, 2010, 2011, 2012, 2013, 2015; Abiev & Lavretsov, 2011a, 2011b, 2012; Abiev & Dymov, 2013; Abiev & Svetlov, 2014; Popova, Abiev, Lappalainen, Svetlov, Andreeva, Trifonov, & Ostrovskii, 2014; Svetlov & Abiev, 2016).

Non-circular channels were in the focus of several papers of Thulasidas, Abraham, and Cerro, (1995, 1997); Triplett, Ghiaasiaan, Abdel-Khalik, and Sadowski (1999); Wörner, 2012, 2013, Kececi, Wörner, Onea, and Soyhan (2009); Wörner, Ghidersa, and Onea (2007); Öztaskin, Wörner, and Soyhan (2009); Onea, Wörner, and Cacuci (2009); Falconi et al. (2016), Gupta, Fletcher, and Haynes (2010); Zhang, Fletcher, and Li (2016).

Using the mass conservation law for the control volume, a general continuity equation was derived:

$$\frac{\partial\left(\rho_i \varepsilon_i\right)}{\partial t} + \operatorname{div}\left(\rho_i \alpha_i u_i\right) = q_{Vi} \tag{3}$$

where the volume fraction ε_i of phase "*i*" is not equal to the fraction of the area α_i occupied by the same phase *i* on the surface of the control volume of the multiphase mixture *V* (that is, the surface fraction of phase *i*):

$$\alpha_i = \frac{S_i}{S} \tag{4}$$

i.e. $\varepsilon_i \neq \alpha_i$.

Let us separate out a control volume bounded by two sections I and II that are immobile relative to the capillary. At a given instant of time, one of these sections passes through the cylindrical portion of the bubble or drop (from here on, the particle of the dispersed phase will be called a bubble for definiteness) and the other section passes through the liquid slug at a sufficient distance from the ends of bubbles (Figure 1).

Let the *x* axis be directed along the flow of the mixture and section I be taken as its origin. As was shown in (Abiev, 2008), the radial distribution of the velocity in the bubble is not uniform, which is attributed to the action of the shear stresses at its surface (Figure 2). It is also known from Taylor (1961), Taha and Cui (2004), Van Baten and Krishna (2005) papers that the flow in the slug can be circulatory.

For the bubble, we have

$$q_2\left(x=0\right) \equiv q_b = \int_{A_b} u_2 dA_b = 2\pi \int_0^{R_b} u_2\left(r\right) r dr \tag{5a}$$

$$U_2\left(x=0\right) \equiv U_b = q_b / A_b \tag{5b}$$

Figure 2. Schematic diagram used in formulating the motion equation for the slug flow of a two-phase mixture: (a), general view; (b), zone of interaction between the film and bubble; 1, liquid film; 2, bubble

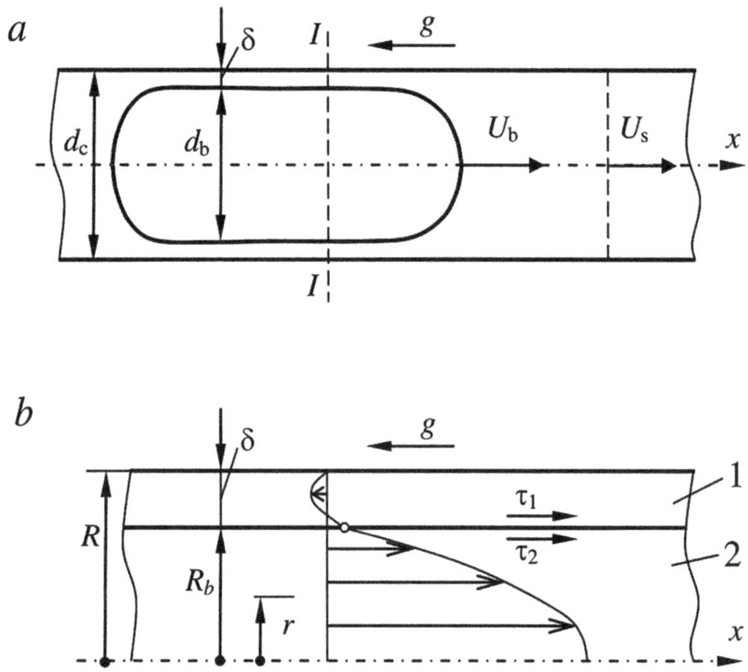

for the liquid film around the bubble, we obtain

$$q_1\left(x=0\right)\equiv q_f=\int_{A_f} u_1 dA_f=2\pi\int_{R_b}^{R} u_1\left(r\right) r dr \quad (6a)$$

$$U_1\left(x=0\right)\equiv U_f=q_f/A_f \quad (6b)$$

for the liquid slug, we can write

$$q_1\left(x=L\right)\equiv q_s=\int_{A_s} u_1 dA_s=2\pi\int_{0}^{R} u_1\left(r\right) r dr \quad (7a)$$

$$U_1\left(x=L\right)\equiv U_s=q_s/A_s \quad (7b)$$

From continuity equation it was derived

$$U_f A_f + U_b A_b = U_s A_c \tag{8}$$

or

$$q_f + q_b = q_s \tag{9}$$

For moderate values of capillary and Weber numbers the experimental data are best fitted by the Aussillous and Quéré (2000) equation:

$$\frac{\delta}{R} = \frac{1.34 \cdot Ca^{2/3}}{1 + 3.35 \cdot Ca^{2/3}} \tag{10}$$

For higher Weber and Reynolds numbers the equation of Han and Shikazono (2009) is available which is applicable for extended range of parameters.

The final expressions for the distributions of velocities are written below:

- The velocity profile in the liquid film is given by

$$u_1(r) = E_1 \left(R^2 - r^2\right) - C_1 \ln\left(\frac{R}{r}\right) \tag{11}$$

- The velocity profile in the dispersed phase (bubble or drop)

$$u_2(r) = E_1 \left[R^2 - R_b^2\right] + E_2 \left[R_b^2 - r^2\right] - C_1 \ln\left(\frac{R}{R_b}\right) \tag{12}$$

The final expressions for the distribution of shear stresses are written below:

- The stress profile in the liquid film around the bubble is given by:

$$\tau_1(r) = \mu_1 \frac{\partial u_1}{\partial r} = \mu_1 \left(-2 E_1 r + \frac{C_1}{r}\right) \tag{13}$$

- The stress profile in the dispersed phase (bubble or drop) is written as:

$$\tau_2(r) = \mu_2 \frac{\partial u_2}{\partial r} = -2\mu_2 E_2 r \tag{14}$$

where

$$E_1 = \frac{g_1'}{4\nu_1} = \frac{1}{4\nu_1}\left(g_x - \frac{1}{\rho_1}\frac{\partial p}{\partial x}\right);$$

$$C_1 = (\rho_1 - \rho_2) g_x \frac{R_b^2}{2\mu_1};$$

$$E_2 = \frac{g_2'}{4\nu_2} = \frac{1}{4\nu_2}\left(g_x - \frac{1}{\rho_2}\frac{\partial p}{\partial x}\right) \tag{15}$$

The instantaneous flow rates of the liquid were found by integrating velocity profile:

- In the film q_f

$$\begin{aligned} q_f &= 2\pi \int_{R_b}^{R} u_1(r)\, r\, dr = \\ &= \pi R^2 \left[E_1 \frac{R^2}{2} - \frac{C_1}{2}\right] - \pi R_b^2 \left[E_1\left(R^2 - \frac{R_b^2}{2}\right) - C_1\left(\ln\left(\frac{R}{R_b}\right) + \frac{1}{2}\right)\right] \\ &= \frac{\pi}{8\nu_1}\left(g_x - \frac{1}{\rho_1}\frac{\partial p}{\partial x}\right)\left(R^2 - R_b^2\right)^2 - \pi C_1\left(\frac{R^2 - R_b^2}{2} - R_b^2 \ln\left(\frac{R}{R_b}\right)\right) \end{aligned} \tag{16}$$

- Of the gas carried by the bubble:

$$q_b = 2\pi \int_0^{R_b} u_2(r) r\, dr =$$
$$= \pi R_b^{\,2} \left[E_1 \left(R^2 - R_b^{\,2} \right) - C_1 \ln \left(\frac{R}{R_b} \right) + E_2 \frac{R_b^{\,2}}{2} \right]$$
$$= \frac{\pi}{4} g_x \left[\frac{R_b^2 \left(R^2 - R_b^{\,2} \right)}{\nu_1} + \frac{R_b^4}{2\nu_2} \right] - \frac{\pi}{4} \frac{\partial p}{\partial x} \left[\frac{R_b^2 \left(R^2 - R_b^{\,2} \right)}{\mu_1} + \frac{R_b^4}{2\mu_2} \right] - \pi C_1 R_b^2 \ln \left(\frac{R}{R_b} \right) \quad (17)$$

After simple algebraic manipulations with Equations (9), (16), and (17), the pressure gradient in the film and, hence, along the bubble axis was derived:

$$\frac{\partial p}{\partial x} = \frac{\left(G_2 - G_3 \right) g_x - q_s}{G_1}, \quad (18)$$

where the symbols are defined by equations:

$$G_1 = \frac{\pi}{8} \left[\frac{R^4 - R_b^4}{\mu_1} + \frac{R_b^4}{\mu_2} \right], \qquad G_2 = \frac{\pi}{8} \left[\frac{R^4 - R_b^4}{\nu_1} + \frac{R_b^4}{\nu_2} \right],$$
$$G_3 = \frac{\pi}{4} \frac{\rho_1 - \rho_2}{\mu_1} R_b^2 \left(R^2 - R_b^{\,2} \right). \quad (19)$$

ALGORITHM FOR CALCULATING SLUG FLOW HYDRODYNAMICS, INCLUDING HIGH PRESSURE GRADIENTS

The simultaneous solution of the continuity and Navier–Stokes equations for a slug flow allows us to determine the pressure gradient along the bubble axis.

Calculations show that this gradient at low liquid velocities in a water–air system is low for capillaries with diameters larger than 1 mm. At the same time, when the liquid velocity in capillaries with diameters less 1 mm comes to 1 m/s, the pressure gradient can be about 1500 Pa/m and its effect on the bubble is many times stronger than the effect of the gravitational force.

In the mathematical model developed in (Abiev, 2008) given by (8), (10), (16)–(19), the velocities of the continuous and dispersed phases can be determined by calculation instead of being arbitrarily specified. Also, this model allows one to calculate the gas volume fraction when the flow rate of the gas phase is known and the bubble-slug length when the size of bubbles is known.

The input data are the capillary radius R; phase physical properties ρ_1, ρ_2, μ_1, μ_2, σ; gravitational acceleration projection g_x, and average liquid flow rate (that is, the flow rate of the liquid carried by the slug which is equal to two-phase flow rate) $q_s = q_{tp} = q_l + q_2$.

The output data of the calculation are the flow rate of the gas carried by the bubble q_b, bubble velocity U_b, flow rate of the liquid in the film around the bubble q_f, liquid film thickness δ, and the pressure gradient along the film $\partial p/\partial x$.

The computational algorithm is described below:

1. The known flow rate of the liquid in the slug qs is divided by the cross-section of the capillary to calculate the liquid velocity:

$$U_s = q_s / A_c. \tag{20}$$

2. Liu et al. (2005) Equation (21) could be used as initial approximation to determine the bubble velocity

$$U_b^0 = U_s (1 - 0.61\ \mathrm{Ca}_s^{0.33})^{-1}, \tag{21}$$

3. The bubble velocity is used to calculate the capillary number Ca.
4. The liquid film thickness δ is determined from Equation (10).
5. Equations (18) and (19) are used to calculate $\partial p/\partial x$ at iteration *n*.
6. Equations (16) and (17) are used to determine q_b, q_f, and then at iteration *n*.
7. Comparison of the value of a chosen parameter, such as the relative error of the flow rate of the liquid in the film.

$$\psi = 2(|qs - qb| - |qf|)/(|qs - qb| + |qf|)$$

Or the ratio of the absolute error of the calculated flow rate of the liquid in the film to the rate itself

$$\psi = |qs - qb - qf|/|qf|,$$

with the specified value of computational error Δ decides whether the calculation should be terminated (when $\psi \le \Delta$) or go to step 3 (when $\psi > \Delta$).

It should be noted that though the described model was developed for round channels (which is clearly seen when finding the flow through the film in the Equation (16) it can be also used in general for the channels with arbitrary cross-section. In rectangular channels, the film has a different thickness along the angular coordinate (Thulasidas, Abraham, & Cerro, 1997; Marschall, Falconi, Lehrenfeld, Abiev, Wörner, Reusken, & Bothe, 2017; Kolb & Cerro, 1991; Hazel & Heil, 2002). Also a significant amount of liquid can move in the back flow at the corners of the channel (Yao, Dong, Zhao, & Chen, 2014a, 2014b; Dong, Yao, Zhang, Chen, Yuan, & Xu, 2015). To find the flux of a continuous phase through a film, integration over the angular coordinate is necessary. This topic was discussed in detail by Falconi et al. (2016) using numerical modeling techniques.

PECULIARITIES OF LIQUID-LIQUID FLOWS

An attempt to use mathematical model (8), (10), (16)–(19) to describe the hydrodynamics of the slug flow in a liquid–liquid system was done by Abiev and Dymov (2013). Figure 3 presents the ratio of the velocities of the dispersed and continuous phases U_d/U_c as a function of the capillary number Ca.

The actual velocities of droplets at respective flow rates of the continuous and dispersed phases were obtained during experiments. Using the previously developed mathematical model (8), (16)–(19), which showed good performance when calculating the parameters of a gas–liquid system, the inverse problem of finding the thickness of the film of the continuous phase around a liquid slug was solved. The obtained values were approximated by the following relationship similar in a structure to the Aussillous and Quéré (2000) equation (Abiev & Dymov, 2013):

$$\frac{\delta}{R} = \frac{0.38Ca^{1/3}}{1 + 0.34Ca^{2/3}} \tag{22}$$

The adequacy of mathematical model (8), (16)–(19) that including Equation (22) for describing the thickness of the film was tested. The results are presented in Figure 4 as the ratio of the velocities of the dispersed and continuous phases U_d/U_c versus the capillary number Ca.

As can be seen from Figure 3, calculated line *4* is in good agreement with experimental points *1* and *2*. It should be noted that the mathematical model shows a good agreement with experimental data obtained both in a polyvinylchloride tube and in a glass tube. Figure 4 presents the same dependence as in Fig. 3, only for a water–benzyl alcohol system. Experimental points are also in good agreement with the calculated line.

Figure 3. Ratio of velocities of dispersed and continuous phases U_d/U_c as a function of capillary number Ca in water–sunflower oil system: (1) experiment with a polyvinylchloride tube with an inner diameter of 2 mm, (2) experiment with a glass tube with an inner diameter of 3.04 mm, (3) calculation by mathematical model (Abiev, 2008) with the Aussillous & Quéré (2000) equation (10), and (4) calculation by mathematical model of Abiev & Dymov (2013) with Eq. (22)

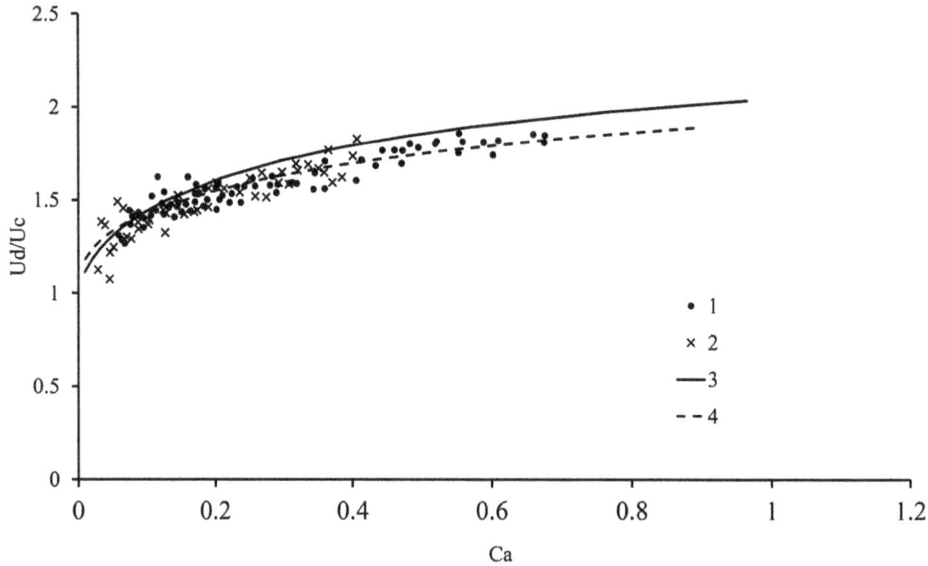

The previously constructed mathematical model (Abiev, 2008) makes it possible to describe the hydrodynamics of a slug flow for a liquid–liquid system over a wide range of capillary numbers (Ca = 0.05–0.7). As a result of solving the inverse problem, relationship (22) is derived for calculating the thickness of a film around the droplets of the dispersed phase. The use of procedure (Abiev, 2008) and Equation (22) makes it possible to calculate the true volume fraction of the dispersed phase with an accuracy of ±20%.

Figure 4. Ratio of the velocities of the dispersed and continuous phases U_d/U_c as a function of the capillary number Ca in a water–benzyl alcohol system. The points represent experiment with a polyvinylchloride tube with an inner diameter of 2 mm; the curve represents calculations by mathematical model with Equation (22).

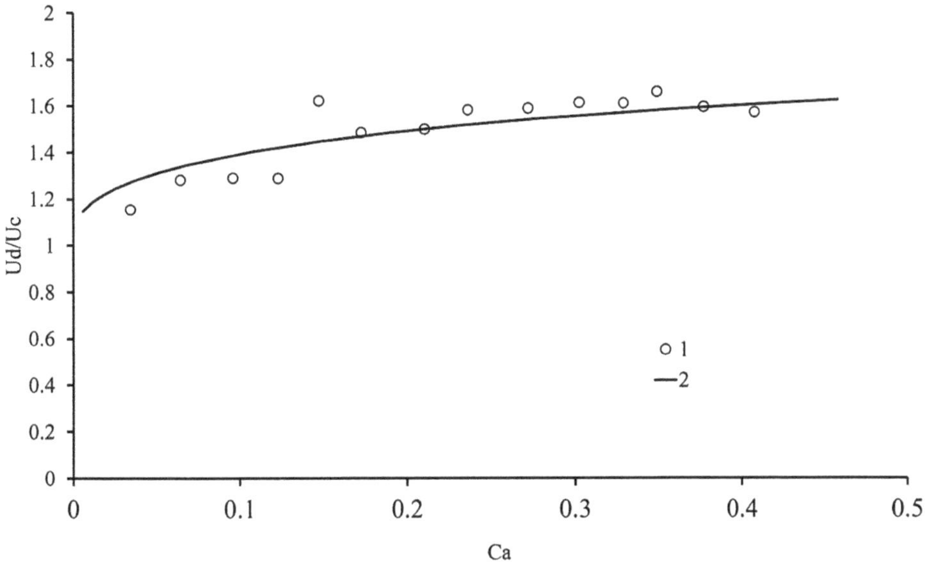

CORRELATIONS OF TAYLOR FLOW PARAMETERS: BUBBLE VELOCITY, GAS VOID FRACTION (HOLD-UP) AND PRESSURE DROP

Bubble Velocity Approximations

Since the micro channels due to their small sizes have much more expressed capillary and viscous forces compared to the inertia and gravitational forces, therefore the correlation found earlier for the velocities of bubble moving in the pipes with the sizes more than critical diameter calculated by Equation (2) (approximately 5.5 mm for water-air system and 3 mm for organic liquids) are not applicable.

Interestingly that the ratio of the bubble velocity related to the velocity of the two-phase flow (which is equal to the slug flow) according to the experimental observations of Thulasidas et al. (1995) could be much higher than unity (up to $\eta(Ca) = U_b/U_s = 2.5–2.6$), especially for high level of capillary number ($Ca > 0.01$).

This effect was explained by kinematical reasons in (Abiev, 2008; Kececi, Wörner, Onea, & Soyhan, 2009): the higher the velocity of the bubble, the thicker is the film around the bubble according to the Bretherton's theory. It has been shown that average film velocity is at least 100 times lower than the velocity of the bubble. In the continuity equation (8) first term ($U_f A_f$) is negligible, $A_b < A_c$, and therefore $U_b > U_s$.

Liu et al. (2005) have proposed to calculate slip ratio with correlation (21), obtained from the experimental data for capillary numbers Ca_s in the range between 0.0002 and 0.39 in an upward flow of gas–liquid mixtures in capillaries with square and circular cross-section with hydraulic diameters from 0.91 to 3.2 mm and viscosity of the liquid phase 0.95–15.9 mPa s.

From the condition $\eta = 2.5$, it follows that this model is limited to the value of Ca_s=0.952. In addition, it has been obtained only on their experimental data (Liu, Vandu, & Krishna, 2005), and has not been verified by the data of other authors.

Relatively simple approximation taking into account up and down directions flow was found by Abiev & Dymov (2013) and Abiev (2013) which provides high accuracy in a wide range of capillary numbers (between $Ca = 3 \cdot 10^{-4}$ and $Ca = 30$). Relation of gas bubble velocity and the capillary number was calculated by the model (Abiev, 2008) and approximated with a following function:

$$\eta(Ca) = 1 + k\,\{1 - \exp\,[-F(Ca)]\}, \qquad (23)$$

The exponent index of F(Ca) was approximated by four types of functions, e.g. by:

$$F(Ca) = \exp\,[b + c\ln(Ca) + g\ln(Ca)^2], \qquad (24)$$

which has maximum standard deviation S from the data calculated by the model (Abiev, 2008) as low as 0.033. Values of coefficients in Equations (23)-(24) for downward flow and horizontal orientation of capillary are shown in Table 1, determination coefficient is $R^2 = 0.999$.

Table 1. Values of coefficients in equations (23)-(24) for upward and downward flows and horizontal orientation of capillary

Flow Direction	*b*	*c*	*g*	*k*	*S*
upward flow	0.0155	0.458	–0.021	1.717	0.03221
downward flow	–0.0665	0.4776	–0.0192	1.714	0.03301
horizontal	–0.0204	0.4714	–0.0211	1.716	—

(Abiev & Lavretsov, 2012; Abiev, 2013)

Gas Hold-Up

In a number of studies (Bercic & Pintar, 1997; Thulasidas, Abraham, & Cerro, 1997; Tsoligkas, Simmons, Wood, & Frost, 2007a), the considerable influence of the length of bubbles and slugs on the intensity of mass transfer in capillaries from a gas to a liquid and further to the surface of a capillary is proven. In particular, the mass transfer from the gas to the liquid (from the quantity of methane that passes from the bubbles to the water) and from the liquid to the surface of the capillaries with diameters of 1.5, 2.5, and 3.1 mm (from the quantity of benzoic acid dissolved by water) was determined by Bercic and Pintar (1997). It was also shown by Bercic and Pintar (1997) that the main role in mass transfer between a liquid with a dissolved gas and the surface of a capillary is played by liquid slugs, rather than the film around the bubbles.

The influence of the length of the slugs on the mass transfer in both cases is explained by the enhancement of the rate of circulation in the slugs when their length is decreased (Thulasidas, Abraham, & Cerro, 1997; Tsoligkas, Simmons, Wood, & Frost, 2007a; Tsoligkas, Simmons, & Wood, 2007b).

A method for calculating the void fraction and relative size of bubbles at the known flow rates of phases is constructed by Abiev (2010) using the mathematical model of a gas–liquid slug flow in capillaries that was developed earlier in (Abiev, 2008).

The results of the calculations are in good agreement with the experimental data of other authors. The boundedness of linear approximations such as the Armand equation by the small values of the capillary numbers is revealed.

Relative area of a bubble in the cross section of a capillary:

$$\varepsilon_A = A_b / A_c \tag{25}$$

and the average relative length of a bubble:

$$\varepsilon_L = \sum_{i=1}^{n} L_{bi} \Big/ \sum_{i=1}^{n} L_{UCi} = \langle L_b \rangle / \langle L_{UC} \rangle \tag{26}$$

were introduced.

A simple relation of the geometrical parameters of a gas–liquid flow in the capillary:

$$\varepsilon_V = \varepsilon_A \varepsilon_L \tag{27}$$

It is important to note that formula (27) makes it possible to determine the average relative length of a bubble ε_L at the known value of the void fraction ε_V.

A general relation between real volume fraction of gas ε_V and dynamic hold-up:

$$\beta = \frac{Q_2}{Q_1 + Q_2} \tag{28}$$

is derived in the forms of (23)

$$\beta = \frac{U_b A_b}{U_s A_c} \varepsilon_L = \frac{U_b}{U_s} \varepsilon_A \varepsilon_L = \frac{U_b}{U_s} \varepsilon_V = \eta \varepsilon_V \tag{29}$$

It was shown (Abiev, 2010) that the void fraction depends not only on the dynamic gas holdup, but also on the capillary number (and at high two-phase flow velocities also from the Weber number), as well as on the direction of the flow, because all these parameters have their impact on bubble to slug velocities ratio η(Ca). It is found that the ratio of the dynamic gas holdup to the void fraction varies from 1 to 2.5 as the capillary number increases. A tenfold error in the experimental determination of the length of liquid slugs by use of simplified calculation procedure is revealed.

Pressure Drop

The problem of pressure drop calculation and modeling was discussed in several papers, e.g. Mishima and Hibiki (1996); Liu et al. (2005); Kreutzer et al. (2005), Chalfi and Ghiaasiaan (2008); Warnier et al. (2010); Jovanovic et al. (2011), though till now there are quite high scattering of experimental and theoretical data. This could be attributed to the complexity of the two-phase flow and numerous factors having impact onto pressure drop taken into account by different research groups.

Abiev (2011) has analyzed the contribution of different types of pressure drop in the capillaries and the mathematical model taking into account the restructuring of the velocity profile in liquid slugs Δp_{trans} and energy losses due to the formation and renovation of the interface in the motion of bubbles $\Delta p_{\Delta F}$ was introduced:

$$\Delta p_{trans} = k_1 N_{uc} L_s \left\{ 1 + \frac{1}{2b_1 L_s} \left[1 - \exp\left(-b_1 L_s\right) \right] \right\} \tag{30}$$

where $k_1 = 8\mu_1 U_s / R^2$, $b_1 = \nu_1 \gamma_m^2 / U_s R^2$, and

$$\Delta p_{\Delta F} = \frac{2R_b}{R^2}\frac{w}{U_s}\frac{L_c}{L_{uc}}\sigma \tag{31}$$

However, pressure drop model (Abiev, 2011), which includes only the terms (30) and (31), gave results of approximately 5-20% lower than those found in experiments. This indicates the presence of additional pressure drop, which primarily occurs along cylindrical part of the bubble, since the energy is lost in the liquid film around the bubble (Abiev, 2008), and the pressure is not constant as previously thought. The values of pressure drop in all the bubbles determined by the equation:

$$\Delta p_b = \left(\frac{dp}{dx}\right)_b \sum L_b = \left(\frac{dp}{dx}\right)_b L_c \varepsilon_L \tag{32}$$

The total pressure drop is the sum of the terms (30) - (32):

$$\Delta p_{Tot} = \Delta p_{\text{trans}} + \Delta p_{\Delta F} + \Delta p_b \tag{33}$$

Correlation of experimental and calculated data for the pressure drop obtained in (Abiev, 2011), were found very good ($R^2 = 0.916$).

The role of wetting hysteresis in creating the additional pressure drop in Taylor flows in minichannels and microchannels was apparently first reported in Jovanovic et al. (2011), where only the case of a continuous medium that readily wets the material of the capillary walls was considered. In a number of studies like Kawahara, Chung, and Kawaji (2002), Liu et al. (2005), Zhao and Bi (2001), Mishima and Hibiki (1996), H. Lee and S. Lee (2001) this form of the pressure drop was also not taken into account.

The estimation of wettability difference for advanced and receding contact angles was recently reported by Abiev (2015). It was found that the calculation of the advancing contact angles using the Bracke formulas leads to the excessive values of the theoretically predicted pressure drop across the microchannel compared with the experimental values; therefore, further studies should be aimed at correcting formulas for calculating the dynamic advancing and receding contact angles.

ANALYSIS OF LOCAL PRESSURE GRADIENT INVERSION AND FORM OF BUBBLES IN TAYLOR FLOW IN MICROCHANNELS

Description of the Problem

Kreutzer et al. (2005b) have found a uniform pressure distribution along the length of Taylor bubble, the pressure in the bubble being substantially higher than in the surrounding liquid slug section (Figure 5), likewise Duran Martinez, Billet, Julcour-Lebigue, and Larachi (2015) obtained a longitudinal pressure profile which has even a positive slope, hence in a bubble zone along the film the pressure is not falling, but rising, i.e. $\left(\frac{\partial p}{\partial z}\right)_f > 0$. Thus, there is an inversion of the pressure gradient in the liquid film, while in the liquid slug the velocity profile obeys Poiseuille-Hagen law (Figure 6).

Figure 5. The form of bubble and the axial pressure distribution along the microchannel according to Kreutzer et al. (2005b) with our comments (see the text for details). T – zone of thinning of the film around the bubble. Axial pressure distribution in the film is constant.

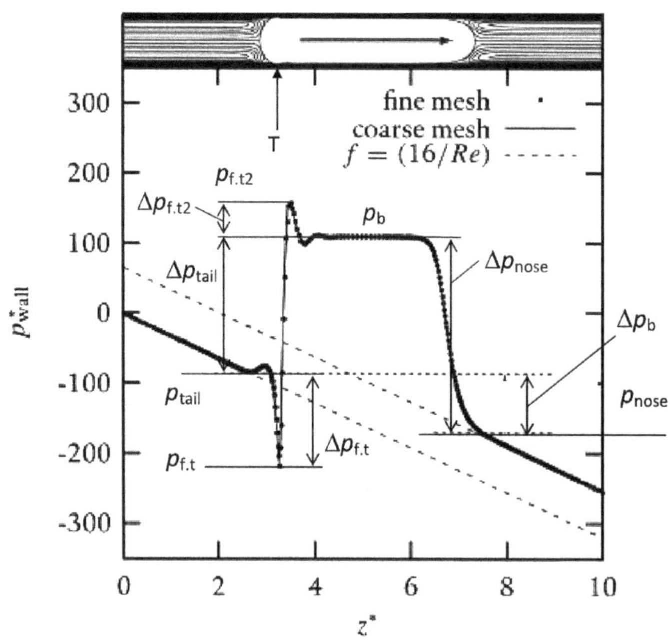

Figure 6. Scheme of Taylor gas-liquid flow in micro channel and calculated pressure profile in the liquid phase. Axial pressure distribution has positive slope in the film, whereas the pressure distribution slope is negative in the liquid slugs.

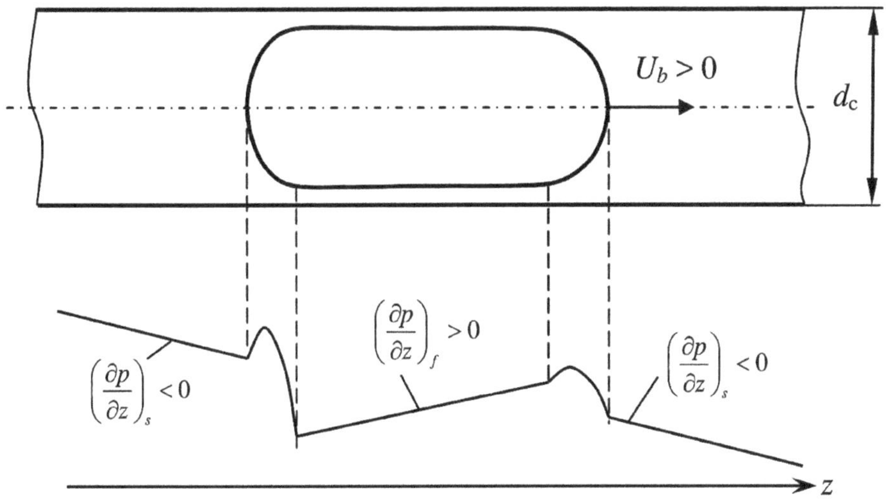

The first of these two observations can be easily attributed to the almost uniform pressure distribution in the gas phase. As shown by Abiev (2011), the pressure gradient in the bubble can be of the order of 1000 Pa/m at two-phase superficial velocity of about 1 m/s for the liquid with properties similar to the properties of water. The jump in Kreutzer et al. (2005b) (see Figure 5) was attributed to the action of capillary pressure.

In the second case, the increasing pressure may cause doubts in the correctness of the calculation. The mathematical model of hydrodynamic characteristics of the Taylor flow built by Abiev (2011) is in a good agreement with numerical simulations of Duran Martinez et al. (2015), including the velocity distribution in the film, in the liquid slug, and in the bubble (far from the ends of the bubble). At first glance the increasing pressure in the film around the bubble looks at least unusual. Since we could not find in the literature any physically plausible explanation, in this chapter we attempt to find the reasons for this inversion of the pressure drop.

In particular, the issues of this chapter are formulated as follows: How a situation when pressure gradient in the film is larger than that in the overall two-phase flow could be feasible? What is the origin of the convex form of bubble nose and the planar or even concave form of its tail? How the wavy form near the tail of the bubble is linked with the axial pressure distribution in the liquid film and the interfacial tension in the liquid and the gas phases?

All these questions are important for better understanding of the process driving forces, thus giving a tool to engineers for better design of microreactors and micro heat exchangers. The bubble's nose elongation results in a significant change of the liquid film thickness in this area, influencing thus mass transfer characteristics by the change of the gas-liquid interface and the velocity in the liquid film.

The observed deformation of the bubbles could be used also in experimental praxis as a detector of pressure drop and reduced two-phase flow rate in the Taylor flow.

The Reasons of the Negative Sign in the Film Velocity Around the Taylor Bubble

Here a more detailed discussion of the "kinematic" reasons of the negative sign in the film velocity around the Taylor bubble shown in Figure 7 is performed.

By integrating the continuity equation for the two-phase Taylor flow in the capillary (on the assumption of steady-state process and incompressible liquids, although gas compressibility can become apparent in practice in a sufficiently long microchannels but within the same bubble, having a length of about 2–30 diameters of the capillary, i.e. no more than approx. 30 mm for $d \approx 1$ mm, the compressibility can be ignored) the Equation (8) has been obtained by Abiev (2008), characterizing the mass balance between the sections I and II in Figure 1.

Subtracting from the left and right sides of the Equation (8) expression $U_b\left(A_b + A_f\right) = U_b A_c$, taking into account the equality of cross-sectional area of the liquid slug A_c to the sum of cross-sectional area of the bubble A_b and the film A_f: $A_c = A_b + A_f$ we get the continuity equation in the coordinates moving at an average velocity of the bubble U_b (Falconi, Lehrenfeld, Marschall, Meyer, Abiev, Bothe, Reusken, Schlüter, & Wörner, 2016):

$$(U_s - U_b)A_A = \left[U_f(z_b) - U_b\right]A_f(z_b) \quad (34)$$

Here $U_f(z_b)$ is the average axial velocity of the liquid in the film in a section of the axial coordinate z_b, measured from the tail of the moving bubble, U_s being two-phase flow rate (equal to the velocity of the liquid slug, see (Abiev, 20008), A_c being the cross-sectional area of the capillary; $A_f(z_b) = A_c - A_b(z_b)$ being the cross-sectional area of the film in the cross-section with the coordinate z_b.

Abiev (2008) has shown to follow from Equation (34) that U_f became negative in the case of

Figure 7. Velocity in the film for different cross-sections in gas-liquid Taylor flow

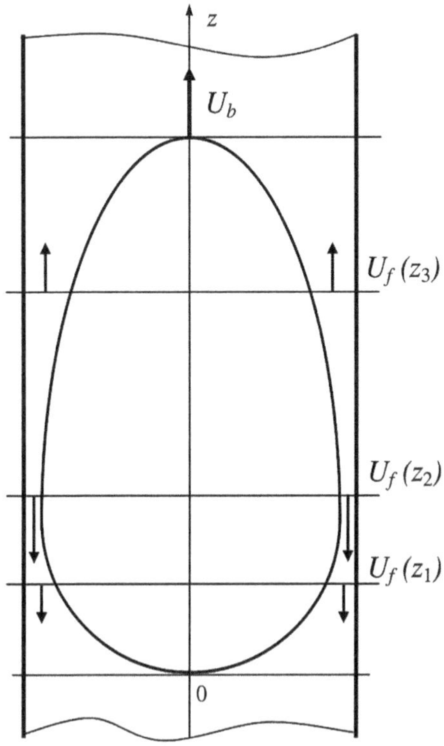

$$A_b(z_b) / A_c > U_s / U_b \tag{35}$$

which corresponded to a local inversion of the flow in the liquid film.

Indeed, for an axisymmetric bubble in the cylindrical capillary $A_b / A_c = d_b^2 / d_c^2$ and from the equation (34) obtained in (Falconi, Lehrenfeld, Marschall, Meyer, Abiev, Bothe, Reusken, Schlüter, & Wörner, 2016) Equation (36) was revealed:

$$U_f(z_b) = \frac{U_s - U_b \dfrac{A_b(z_b)}{A_c}}{1 - \dfrac{A_b(z_b)}{A_c}} = \frac{U_s - U_b \dfrac{d_b^2(z_b)}{d_c^2}}{1 - \dfrac{d_b^2(z_b)}{d_c^2}} \tag{36}$$

which implied the inequality (35).

This expression can be used to calculate $U_f(z_b)$ if the dependence $d_b(z_b)$ and a value of the average bubble velocity U_b are known. Otherwise, the iterative procedure Abiev (2008) can be used.

It follows from Equation (34) that $U_b A_b = U_s A_c - U_f A_f$ or $U_b A_c - U_b A_f = U_s A_c - U_f A_f$. Finely one could write:

$$\left(U_b - U_s\right) A_c = \left(U_b - U_f\right) A_f \tag{37}$$

Film velocity $U_f(z_b)$ is expressed from Eq. (34) and the derivative by z_b is found:

$$\frac{\partial U_f(z_b)}{\partial z_b} = \frac{\partial U_f(z_b)}{\partial A_f}\frac{\partial A_f(z_b)}{\partial z_b} = -\frac{(U_s - U_b)A_c}{A_f^2}\frac{\partial A_f(z_b)}{\partial z_b}$$

Taking into account Equations (37) and (38) was found in (Falconi, 2016).

$$\frac{\partial U_f(z_b)}{\partial z_b} = \underbrace{\frac{U_b - U_f(z_b)}{A_f(z_b)}}_{>0}\frac{\partial A_f(z_b)}{\partial z_b} \tag{38}$$

From Equation (38) follows (Falconi, Lehrenfeld, Marschall, Meyer, Abiev, Bothe, Reusken, Schlüter, & Wörner, 2016) that owing to $\left(U_b - U_f(z_b)\right)/A_f(z_b) > 0$ the cross-section with coordinate z_b, at which value U_f has a minimum corresponds to the position of the cross-section with a minimum value of A_f. For axisymmetric bubble in a cylindrical capillary the film thickness does not depend on an azimuthal coordinate φ, therefore, the axial position where velocity U_f has the minimum value is independent of φ [please note that axial symmetry is disrupted, for example, in micro channels with square cross-section, see (Falconi, Lehrenfeld, Marschall, Meyer, Abiev, Bothe, Reusken, Schlüter, & Wörner, 2016).

For practical applications to Taylor flow it is useful to preliminary assess whether the locally recurrent (inverted) flow in the film is possible or not. From Equation (35) it follows that the area with a local inversion of flow in the film is possible for those sections where the condition (39) is true:

$$\frac{A_b}{A_c}\frac{U_b}{U_s} > 1 \tag{39}$$

The first fraction at the left side of inequality (39) is always less than unity and the second is greater than unity, and both fractions are dependent on the capillary number. In (Liu, Vandu, & Krishna, 2005) experimental investigations of round and square cross-section of the capillaries for a variety of liquids was performed and correlation for U_b / U_s was obtained. Abiev (2013) the correlation based on a hydrodynamic model (Abiev, 2008; Abiev, 2011) is derived, which is valid for a wide range of capillary numbers. The ratio A_b / A_c in this case can be found from the Aussilous and Quéré (2000) equation or other sources (Howard & Walsh, 2013; Schwartz, Princen, & Kiss, 1986; Pohorecki & Kula, 2008). All of this allows us a calculation of the size of the area where the condition (39) for cylindrical capillaries is valid. In (Falconi, Lehrenfeld, Marschall, Meyer, Abiev, Bothe, Reusken, Schlüter, & Wörner, 2016) the existence of the liquid film velocity inversion in microchannels with square section was revealed and it was shown that it was of the same nature as in cylindrical capillaries.

It is important to mention that the size of the zone of inversion of the liquid velocity in the film is large enough, and accounts for about 70-80% of the length of the bubble, which corresponds to the steady-state film thickness far from the nose and the tail of the bubble. A more detailed analysis of the impact on the size of the velocity inversion zones in the film of various factors (the physicochemical properties of the media, the two-phase flow rate, and the geometry of the microchannel) goes beyond the scope of this chapter and will be discussed (elsewhere) in a separate article or book.

The flow almost along the entire length of the film is directed backward, and the analysis of Navier-Stokes equations performed in showed that the only reason of the flow in the film is the pressure gradient. Hence, it can be argued that since the motion direction of the bubble and liquid slug on the one hand, and the film, on the other hand, are opposite, the pressure gradients in these objects will have different signs, i.e.,

$$U_f < 0, \left(\frac{\partial p}{\partial z}\right)_f > 0 \tag{40}$$

$$U_b > 0, \left(\frac{\partial p}{\partial z}\right)_b < 0 \tag{41}$$

Thus, according to the concept proposed in this chapter (Figure 8):

1. The distribution of pressure in the liquid slugs should follow Hagen-Poiseuille equation, while:

$$U_s > 0, \left(\frac{\partial p}{\partial z}\right)_s < 0 \tag{42}$$

2. The pressure gradient in the bubbles (gaseous part of the two-phase flow) is also negative, and the velocity is positive (see (41).);
3. As the film moves backward ($U_f < 0$), the pressure gradient in the film has a positive sign in accordance with (40), and as a result the pressure gradient in the liquid film around the bubble is not equal to the average pressure gradient in the bubble.

We emphasize that the movement of gas in the bubble has a circulating character (see Figure 8), due to the presence of the shear stresses between gas and liquid film (Abiev, 2008), therefore, the longitudinal profile of the pressure distribution in the bubble depends on the radius. Though all the curves with respect to pressure gradient in Figure 8 are qualitative (the concept, main idea is presented here), they represent in general the pressure plots obtained in the works published, e.g., by Kreutzer et al. (2005b), Abadie, Xuereb, Legendre, and Aubin (2013), and are found numerically by CFD modeling. It is expedient to highlight that the pressure gradients in the liquid film and the corresponding average velocities have different signs: $U_b > 0$, $U_f < 0$.

Since the exact calculation of the axial pressure distribution profiles for our best knowledge is not published in the free-for-all literature, it is possible to put forward the assumption that near the axis of the bubble the local pressure gradient in the bubble corresponds to Equations (41), while near its interface with the film the local pressure gradient is also positive.

Thus, in the framework designated in Figure 8 by the double-dashed line, there is an area with an alternative pressure gradient [$\left(\partial p/\partial z\right)_b < 0$, $\left(\partial p/\partial z\right)_f > 0$] and alternative average velocities ($U_b > 0$, $U_f < 0$).

On the Bubble Form of Taylor Flow in Microchannels

From versatile studies, both experimental and numerical (Kreutzer, Kapteijn, Moulijn, Kleijn, & Heiszwolf, 2005b) it is known that when Ca > 0.01 the nose of the bubble has an elongated elliptical shape, and the tail looks like flattened ellipsoid. At the same time near the transition from the ellipsoid part of the tail to the cylindrical side the diameter of the bubble has the highest value (expanded), i.e. the liquid film around the bubble in the rear zone is thinner than in the other part (Figure 5).

Figure 8. The concept of an axial pressure distribution along the liquid and bubble. Frame, contoured by double-dashed line is the area of alternative pressure gradient and alternative average velocities of phases ($U_b > 0$, $U_f < 0$)

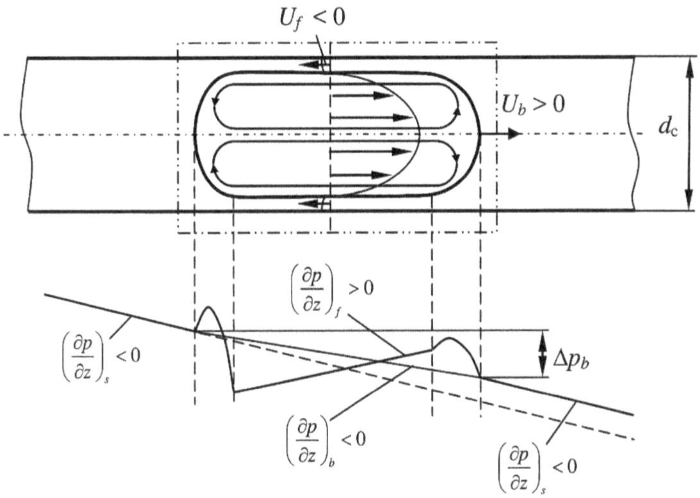

As seen from the graph in Figure 5 (dots show the results of numerical solution on a fine grid, solid lines, in the rough grid), the average pressure in the bubble is constant along the axis of the bubble at a sufficiently large distance and is equal to p_b.

Both pressures on the nose p_{nose} and on the tail p_{tail} projections of the bubble are lower than the average pressure in the bubble. Note that the weak dependence of the pressure along the longitudinal coordinates of the bubble can occur only for very low values of the capillary number (Ca<0.001–0.01).

For relatively large bubble velocities (Ca>0.1, Re>50) rather larger pressure gradients along the bubble (Abiev, 2008, 2011) (*dp/dz* ≈ 1000 Pa/m for water-air system) were revealed, which should be taken into account in the calculations.

Obviously, to let the bubble move some forces should be applied to its ends (to the nose and the tail), i.e., forces due to the pressure (linked with normal stresses) and/or shearing forces on the lateral surface (with the corresponding shear stresses).

As the average velocity of the film is extremely low (approximately two orders of magnitude smaller than those of the bubble and of the liquid slug, see (Abiev, 2011), the shear stresses play a minor role in the movement of the bubble, at least for moderate flow rates (typical for the laminar flow). Furthermore, according to CFD simulations (Kreutzer, Kapteijn, Moulijn, Kleijn, & Heiszwolf, 2005b; Duran Martinez, Billet, Julcour-Lebigue, & Larachi, 2015) and analytical descriptions (Abiev, 2011; Falconi Lehrenfeld, Marschall, Meyer, Abiev, Bothe, Reusken, Schlüter, & Wörner, 2016) (see also the analysis above) the direction of the flow in the film has a negative sign, i.e. backward (inverse) direction of flow. Therefore, shearing stresses can exert a braking effect only, and active driving force is the pressure difference between the ends of the bubble.

This pressure difference is shown in Figure 5 and is equal to:

$$\Delta p_{\mathrm{b}} = p_{\mathrm{tail}} - p_{\mathrm{nose}} > 0. \tag{43}$$

We would like to emphasize the obvious effect: the higher the velocity of the bubble, the greater the pressure difference Δp_{b}. This is due to the fact that with the growth of bubble velocity U_b (and Ca, Re values) the shear stresses are also increasing, and to overcome them a greater pressure drop Δp_{b} between the ends of the bubble is required.

As follows from the results of calculations obtained by Kreutzer et al. (2005b) (Figure 5), the average pressure in the bubble p_{b} is greater than the pressure in the liquid, and we believe that jumps at the ends of the bubble are caused mainly by the capillary pressure. An additional contribution to the increase of pressure can be attributed to the inertial effects associated with the deceleration of liquid during its U-turn at the ends of the bubble.

They can have a significant contribution at local velocities in slugs about $U_{\mathrm{s}} =$ 0.3 m/s (corresponding to a Ca = 0.004 for water-air system), and more: the pressure caused by the liquid U-turn will be at these conditions $\Delta p_{\mathrm{in}} = 2\rho\, U_{\mathrm{s}}^2/2 = 90$ Pa, which corresponds to the order of the pressure difference on the bubble presented in (Kreutzer, Kapteijn, Moulijn, Kleijn, & Heiszwolf, 2005b).

As a first approximation, it could be assumed that the contribution of the inertial pressure is insignificant and is about the same in the nose and the tail of the bubble, then a pressure jump in the tail and the nose part are respectively

$$\Delta p_{\mathrm{tail}} = p_{\mathrm{b}} - p_{\mathrm{tail}} = p_{\mathrm{c.tail}} \tag{44}$$

$$\Delta p_{\mathrm{nose}} = p_{\mathrm{b}} - p_{\mathrm{nose}} = p_{\mathrm{c.nose}} \tag{45}$$

Here $p_{\mathrm{c.tail}}$, $p_{\mathrm{c.nose}}$ are capillary pressures in the tail and the nose part of the bubble. Note that Δp_{tail} is exactly a jump in pressure, i.e., the difference between the pressure in the bubble and outside (see Fig. 5), but not the pressure drop along the z axis.

According to Laplace equation capillary pressure is represented by:

$$p_c = \sigma\left(\frac{1}{R_1} + \frac{1}{R_2}\right) \tag{46}$$

For bodies with axial symmetry (in the studied case, for all ends of bubbles with ellipsoid form) $R_1 = R_2 = R$, and Equation (46) takes the form:

$$p_c = \frac{2\sigma}{R} \tag{47}$$

It follows from inequality (43) $p_{tail} > p_{nose}$, then from Equations (44) and (45) it is consequently:

$$\Delta p_{tail} < \Delta p_{nose} \tag{48}$$

This means that considering Equations (47) and (44), (45):

$$\Delta p_{tail} = p_{c.tail} = \frac{2\sigma}{R_{tail}} < \Delta p_{nose} = p_{c.nose} = \frac{2\sigma}{R_{nose}} \tag{49}$$

In case of $R_{tail} \geq 0$, $R_{nose} > 0$ (the second condition is always present, as the results of experiments show), the inequality (49) is valid at a ratio between the radii of curvature of the nose and tail of the bubble:

$$R_{tail} > R_{nose} \tag{50}$$

observed in many experimental studies (Liu, Vandu, & Krishna, 2005; Van Baten & Krishna, 2005; Kreutzer, Kapteijn, Moulijn, Kleijn, & Heiszwolf, 2005b; Mishima & Hibiki, 1996; Kawahara, Chung, & Kawaji, 2002). If $R_{tail} < 0$, $R_{nose} > 0$, inequality (49) is fulfilled without any additional conditions.

Thus, the greater the rate of bubble U_b (at specified d_c, μ_l, σ values), the more values Δp_{tail} and Δp_{nose} differ from one another, and therefore significantly higher is a difference of radii of curvature of bubble's nose and tail parts. It should be borne in mind that, according to Equation (46) a surface curvature $\kappa = 1/R$ is positive, if the center of curvature is located within the bubble, but in the case of location of the latter out of the bubble the curvature of the bubble becomes negative.

With increasing viscosity μ_l of liquid and/or velocity of bubbles U_b (i.e. with increasing capillary number Ca) the pressure in the tail of the bubble p_{tail} increases so that it becomes first equal and then even higher than the average pressure in the bubble p_b. As follows from Equation (44), with growing Ca number $p_{c.tail}$ value first becomes zero and then negative, i.e. when $p_{tail} \geq p_b$ the pressure in the tail of the bubble is larger than the average pressure level, i.e. $p_{tail} \geq p_b$.

In the literature as a first approximation it is assumed that the shape of the ends of the bubble is hemispherical, considering the ellipsoidal shape as the second approach. In Fig. 9 the shape of the ends of bubbles is shown simplified in the form of an ellipsoid: an oblate for the tail section and elongated for the nose part.

In the first case a radius that determines the capillary pressure is a larger radius (corresponding to R_{tail} in Figure 9) for an oblate ellipsoid and a smaller radius for a prolate ellipsoid (corresponding to R_{nose} in Figure 9), in this case:

$$\Delta p_{tail} = p_b - p_{tail} = p_{c.tail} = 2\sigma/R_{tail} \leq 0 \quad (51)$$

In other words, under the influence of excess pressure at the rear surface the convex shape of the bubble (Figure 9a, $R_{tail} > 0$, $p_{tail} < p_b$) first becomes flat (Figure 9b, $R_{tail} \to \infty$, the pressure is equal to the pressure outside of the bubble: $p_{tail} = p_b$), and then it is completely pressed inwards and becomes concave (Figure 9c, $R_{tail} < 0$, $p_{tail} >$

Figure 9. The bubble deformation with increasing bubble velocity and differential pressure across the length of the bubble: (a) – convex tail part, $R_{tail} > 0$; (b) – the form of the tail section is close to the planar, $R_{tail} \to \infty$; (c) – a concave tail section, $R_{tail} < 0$

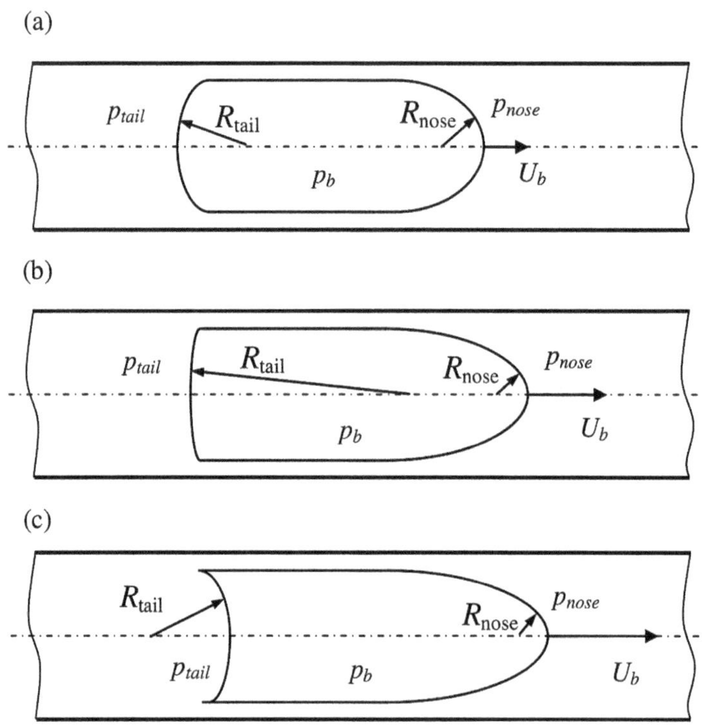

p_b). As can be seen from Figure 9, at low velocities of the bubble and, respectively, a slight pressure drop on it, the radii of curvature of the nose and tail parts of the bubble do not differ significantly from the capillary radius: $R_{tail} \approx R_c \approx R_{nose}$.

The situation with the change of the radius of curvature in the nose part is much easier than for the rear: with increasing bubble velocity U_b and consequently with the growth both of the total pressure drop on the bubble Δp_b and the pressure difference at the nose Δp_{nose}, it follows from Equation (47) that R_{nose} at these conditions decreases (as shown in Figure 9a-c).

MASS TRANSFER INTENSIFICATION BY MEANS OF TAYLOR VORTICES

Experimental corroboration of mass transfer intensification in mini and micro channels is described in this section. An experimental study confirming mass transfer intensification by means of Taylor vortices was performed in (Abiev & Lavretsov, 2012). Experiments were carried out in the glycerol–air system for the determination of the reduction of mass transfer efficiency by the transition from circulation flow to bypass flow mode in the capillary number range 0.03–1.00.

When capillary number runs up to the critical value then film thickness increases so that the liquid begins to flow along gas bubble (Abiev, 2010). The critical value of the Ca predicted by Taylor for the horizontal capillaries is 0.707, calculated value from equation derived in (Abiev, 2010) is 0.730.

For estimation the intensification of mass transfer from liquid to the capillary wall was applied following value, first proposed by Gruber and Melin (2003):

$$E = \frac{Sh_{TP}}{Sh_{SP}} \tag{52}$$

Three experimental points with Ca > 0.73 for both single phase and two-phase flows were obtained to determine supposed jump of E after passing the critical value of Ca.

It can be seen from Figure 10, when Ca < 0.2 the value of E was found as high as 3.5, at Ca ≈ 0.5 it is decreased to 1.3 and after passing the critical value E ≈ 1. High values of E at small capillary numbers can be explained by the presence of circulation inside the slug, which contribute to the renovation of the interface. The transition from circulation regime to bypass regime decreases the value of E to 1.

Figure 10. Dependence of mass transfer efficiency coefficient on the capillary number for system liquid – capillary wall
(Abiev & Lavretsov, 2012)

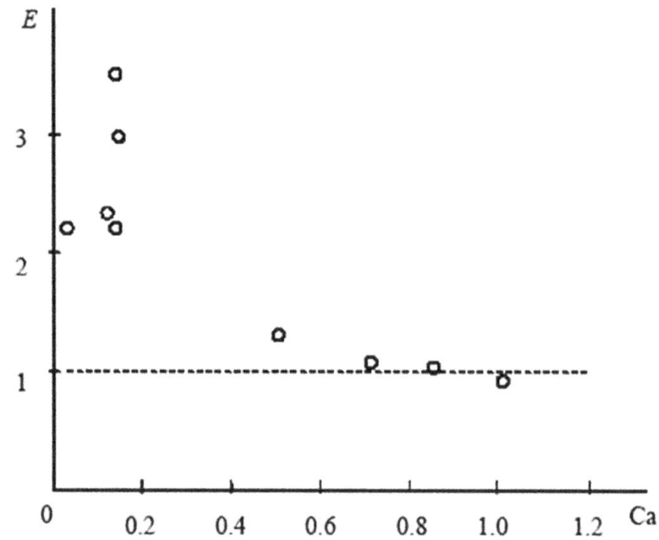

Thus the intensification of mass transfer in slug flow regime at the expense of the Taylor vortices was demonstrated experimentally. It should be noted that the simulation of mass transfer from the channel wall to the liquid slug proved their presence (Gruber & Melin, 2003).

THREE-LAYER MODEL AND ITS APPLICATION TO THE MASS TRANSFER MODELING

Three-Layer Mass Transfer Model for Mini and Micro Channels

Three-layer mass transfer model has been proposed by Abiev (2013). The main idea of three-layer mass transfer model is based on the structure of the toroidal Taylor vortex existing in the slugs. Due to convection dominating over diffusion in the Taylor vortex the main mass flux is defined by the circulation velocity in this vortex. The name 'three-layer' comes from number of layers in this structure: first layer is inner layer (having cylinder shape) of the toroidal vortex, the second layer is outer annular layer of the toroidal vortex and the third layer is the slow (almost stagnant) film around the toroidal vortex which is not involved in the circulation. In order to crucially simplify the mathematical modeling the structure of Taylor vortex was presented as quasi-1D: the vortex was 'unwrapped' to the linear structure one part of which corresponds to the first layer and the second part mimics the second layer of the original vortex. This second part has diffusional contact with third layer, i.e. with the slow film, the latter is contacting with the capillary wall.

The main mass transfer within the first and second layers takes place due to convection and could be described with transient convection equation. In the third layer mass transfer is described by diffusion only.

The task of mass transfer calculation consists from following steps: 1) determination of three-layer geometry: radii of vortex' layers, thickness of the slow film, and; 2) solution of coupled diffusion and convection equations in all the linear layers simultaneously.

Geometry of Three-Layer Mass Transfer Model

Main geometrical characteristics of the slug flow are calculated (vortex center radius R_0 and radius R_1 of surface separating the transition film flow from the circulation flow), equations for dimensional and dimensionless circulation times as well as average velocities in all three layers are derived. Dependence on the length of the slug L_s related to the diameter d_c of capillary is taken into account.

Generalized equation of R_0/R ratio is proposed in (Abiev, 2013) to calculate as

$$R_0/R = \sqrt{\frac{2-\eta\left(Ca\right)}{2}} + a\exp\left(-b\frac{L_s}{d}\right) \tag{53}$$

where $a \approx 0.34$, $b \approx 4.86$ and $\eta = \eta(\text{Ca})$ could be calculated either by Liu et al. (2005) Equation (21) or Equation (23) or more precisely by the method described in section 2.2 (see Abiev, 2008) for details).

The radius R_1 of film-vortex interface is determined by equation obtained by Thulasidas et al. (1997):

$$R_1 = R\sqrt{2-\eta} \tag{54}$$

It is worthwhile to mention that only for $\text{Ca} < \text{Ca}_{cr}$ there are Taylor vortices, whereas for $\text{Ca} > \text{Ca}_{cr}$ the flow changes to by-pass around the bubbles. It was shown in (Abiev, 2009) that critical value of capillary number depends on modified Bond number:

$$\mathrm{Bo} = \frac{\rho_1 d^2 g \cos\gamma}{\sigma} \tag{55}$$

Geometrical properties of slug flow and mass fluxes are shown on Figure 11, general scheme of three layer mass transfer model is presented on Figure 12.

Mass Transfer Equations of Three-Layer Model

Mass transfer equations of three-layer model were first proposed in (Abiev, 2013) and then used for the mass transfer modeling and optimization in (Svetlov & Abiev, 2016) for the case of gas-liquid two-phase flow (see also section Case Study: Wall-Liquid Mass Transfer Process). In the formulation presented in section Geometry of three-layer mass transfer model, the mass transfer is described by the equation of convective diffusion:

$$\frac{\partial C}{\partial t} = -div\left(j_{conv} + j_{diff}\right), 0 < x < L_s, 0 < r < R \tag{56}$$

For the axially symmetric form reduced then to the quasi-1D formulation we have obtained in (Svetlov & Abiev, 2016) for two inner layers:

Figure 11. Geometrical properties of slug flow and mass fluxes for three-layer mass transfer model
(Abiev, 2013)

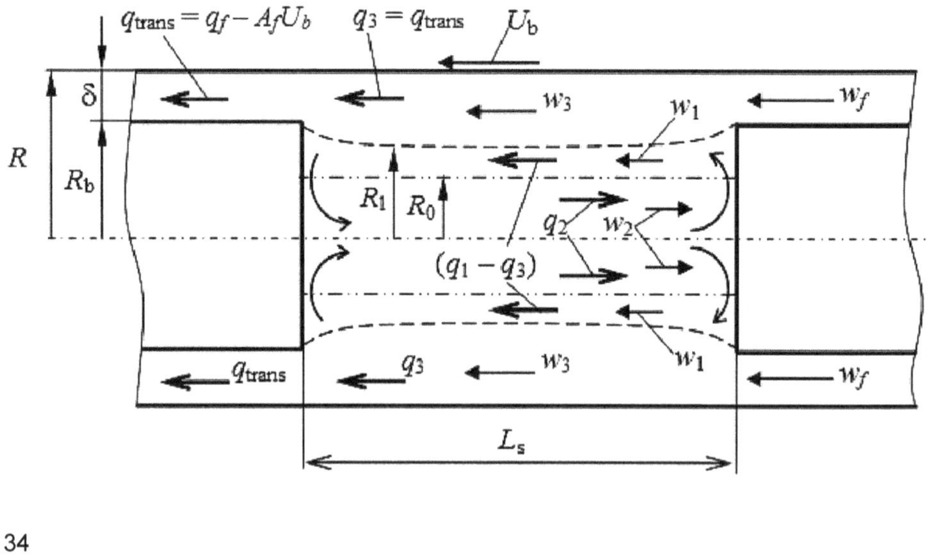

Figure 12. The scheme of three-layer mass transfer model
(Abiev, 2013)

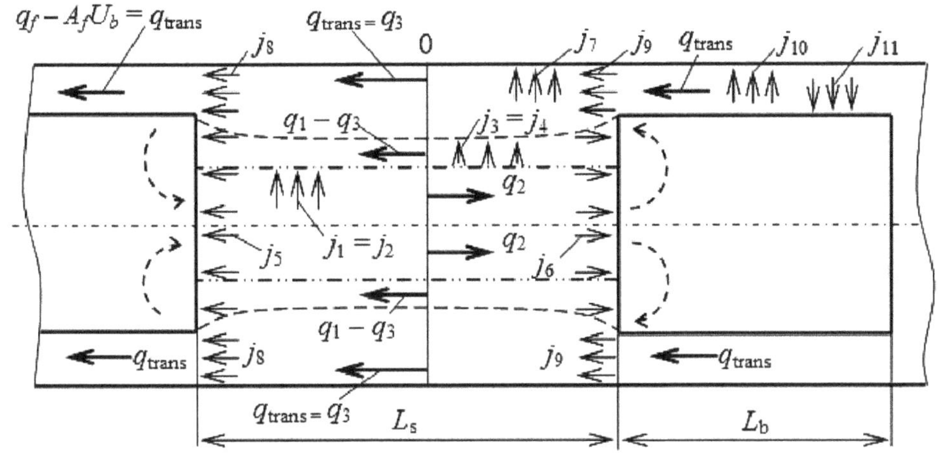

$$\frac{\partial C}{\partial t} + v_x \frac{\partial C}{\partial x} = D\frac{\partial^2 C}{\partial x^2} \tag{57}$$

Mass transfer between transition film flow and the outer ring-shaped layer of circulation flow is described by use of radial diffusion. To implement this, a source term was used, i.e. for elementary volume ΔV of the moving liquid a material balance equation is:

$$\frac{\Delta V \partial C}{\partial t} = \left[\left(j_{conv1} - j_{conv2}\right) + \left(j_{diff1} - j_{diff2}\right)\right] S_2 + q_r S_r \tag{58}$$

where $S_2 = \pi\left(R_1^2 - R_0^2\right)$ is the cross-section area of the outer ring-shape layer (which is equal to that for the inner layer), $S_r = 2\pi \cdot R_1 \cdot \Delta x$ is the lateral interface area between transition film and the outer ring-shaped layer, q_r – the radial flux from the transition film to the bulk liquid in the circulation area. For further details of boundary conditions used in the model and the approximate solution of the PDE the reader could address to (Svetlov & Abiev, 2016).

A case study of three-layer mass transfer model application is described in section below.

CASE STUDY: WALL-LIQUID MASS TRANSFER PROCESS

Further progress in mass transfer simulation by means of three-layer mass transfer model was achieved in (Svetlov & Abiev, 2016) where slug-wall mass transfer is studied. The process of mass transfer between the wall covered with the film of solid substance A (benzoic acid) reacting with the molecules B (NaOH) dissolved in the liquid (water) was used as a model which mimics the transfer of reacting molecules dissolved in the liquid towards the wall on which the active catalyst (Pt or Pd) is deposited.

The process of convection and diffusion in a Taylor regime of gas–liquid flow through microchannels is modeled based on a three-layer mass transfer model in the axisymmetric formulation.

The calculation was performed using the following initial data: $L_s = 2$ mm, $\rho = 1000$ kg/m^3, $\mu = 10^{-3}$ Pa s, $\sigma = 0.0726$ N/m; $U_s = 0.05$ m/s, 0.1 m/s, $d = 0.65$ mm, 1 mm, 1.5 mm.

The circulation circuit of Taylor vortices is described in the form of outer and inner layers surrounded by a thin film that does not participate in circulation (Figure 13). Due to the assumption about the predominance of convection over radial diffusion in the inner layer (the Peclet number order of magnitude is nearly 10^5) and a uniform concentration distribution over the cross sections of each layer, the problem is confined to a quasi-one-dimensional problem with boundary conditions of special type. The obtained numerical solutions allowed to determine the kinetics of mass transfer from a liquid to a channel wall in detail, to calculate the average mass-transfer coefficient, and to reveal optima in the dependences of the mass-transfer coefficient on the two-phase flow velocity (Figure 14) and the capillary diameter (Figure 15); furthermore, there is also an optimum for the length of a liquid slug. The obtained results allowed to clearly recognize the pattern of Taylor convection and diffusion fluxes and to reveal the reasons for a decrease in mass fluxes under non-optimal conditions.

It has been established that the mass-transfer coefficient descends with an increase in the continuous medium viscosity, and a growth in the surface tension leads to an increase in the mass transfer coefficient. Both effects are explained by an increase in the thickness of the film, through which a compound diffuses to the wall of a microchannel.

At first glance, it can be concluded that the mass transfer coefficient must grow with increasing capillary diameter. However, the transition film thickness grows with increasing diameter, which leads to a decrease in the flux of substance from the wall to the outer circulation circuit. An increase in the liquid film resistance leads to a decrease in the intensity of mass transfer so that the difference [$C(t +$

$\Delta t) - C(t)]$ descends and the difference $[C_s - C(t)]$ ascends. This leads to a decrease in the mass transfer coefficient at larger capillary diameters.

A numerical analysis of Figure 14 allows to reveal that the mass-transfer coefficient β grows with increasing slug velocity U_S at a channel diameter d of less than 0.6 mm, but it is much lower than its optimal value (e.g., $\beta < 10^{-4}$ m/s at $U_S = 0.6$ m/s and d < 0.6 mm, whereas the optimal mass transfer coefficient $\beta_{opt} \approx 10^{-4}$ m/s is attained at d = 0.65 mm).

When the slug velocity U_S descends, the difference between the minimum and maximum mass-transfer coefficients (βmin and βmax) increases at a fixed U_S; i.e., the ratio between the maximum and minimum mass transfer coefficients βmax/βmin = 1.6 at $U_S = 0.6$ m/s, while βmax/βmin = 5 at $U_S = 0.2$ m/s. It follows from this that the effect of the diameter on the mass-transfer coefficient is weaker at high liquid flow velocities U_S.

The effect of the slug velocity on β also is not univocal (Figure 15). The calculation was performed at $L_S = 2$ mm, $\rho = 1000$ kg/m^3, $\mu = 10^{-3}$ Pa s, and $\sigma = 0.0726$ N/m. On one hand, the growth in velocity leads to an increase in the thickness of the transition film according to Equations (10) and (22), which produces a negative effect on the mass transfer from the wall of a capillary.

Figure 13. Velocity distribution in the circulation flow in a slug: (a) real velocity profile, (b) three-layer step-wise velocity profile
(Svetlov & Abiev, 2016)

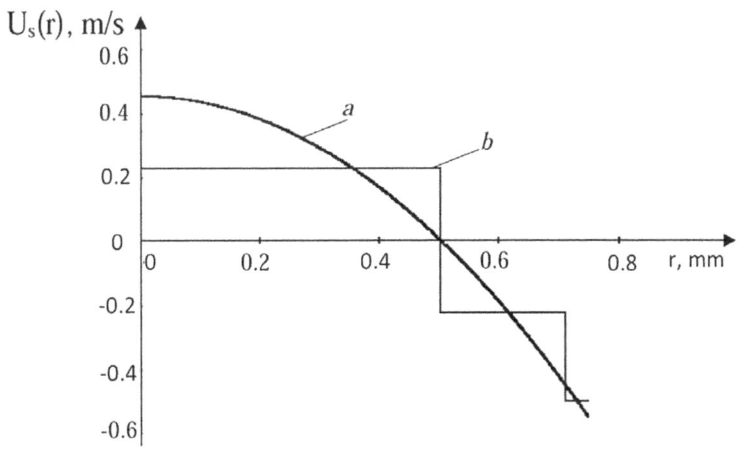

Figure 14. Effect of the channel diameter and the slug velocity on the mass-transfer coefficient at a slug flow velocity U_s of (1) 0.05, (2) 0.1, (3) 0.2, (4) 0.3, (5) 0.4, (6) 0.5, and (7) 0.6 m/s
(Svetlov & Abiev, 2016)

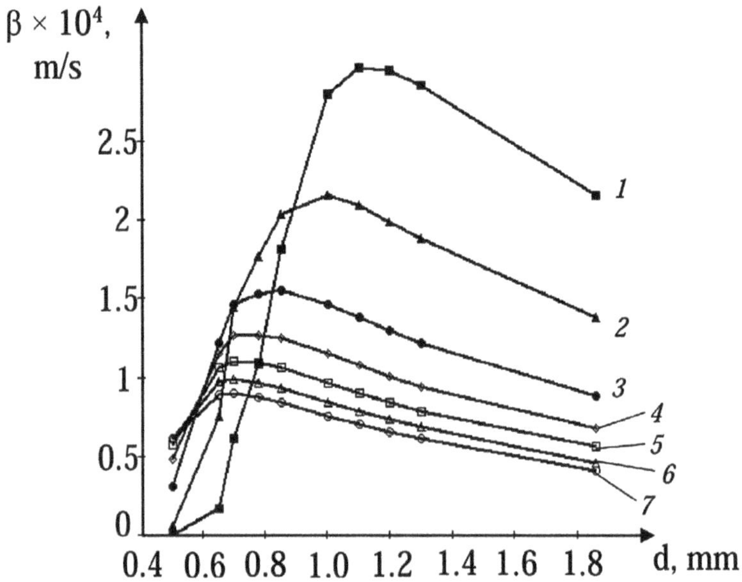

Figure 15. Mass-transfer coefficient versus slug velocity at different capillary diameters d of (1) 0.5, (2) 0.65, (3) 0.85, (4) 1, (5) 1.86, and (6) 2.5 mm
(Svetlov & Abiev, 2016)

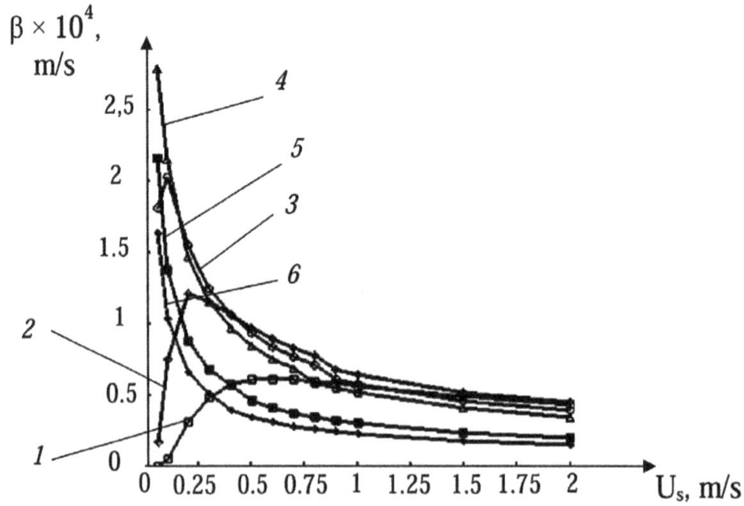

FUTURE RESEARCH DIRECTIONS

The current state of the research in the field of Gas-Liquid and Liquid-Liquid Taylor flows in micro channels already includes quite wide results of experimental investigations and theoretical studies. Nonetheless, there are still many unsolved questions, among them the problem of two-phase flows distribution through several hundreds of even thousands of microchannels, e.g. in monolith catalysts. Besides, the micro sized channels with nonlinear axis: meandering micro channels (Yang, 2017), coiled micro channels (with helical coil) and so-called coiled flow inverters (Gürsel, Kockmann & Hessel, 2017) are still waiting for the creation of general theoretical approach of their fluid dynamics, heat and mass transfer.

Evidently, some elements of theoretical understanding of the two-phase Taylor flows discussed in this chapter could be applied to these more complex systems in the nearest future.

CONCLUSION

An attempt to analyze the state of art of process intensification in mini and micro channels on the base of deep understanding of the hydrodynamics and mass transfer mechanisms is performed in this chapter. It is clear now that researchers possess reliable theoretical (semi-analytical) tools for mathematical modeling and prediction of main characteristics of Taylor flow like bubble velocity, film thickness, velocity profiles in all the parts of the multiphase flow, pressure drop and void fraction, length of the bubbles and the slugs etc. These dynamical and geometrical parameters allow to calculate the mass transfer coefficients and performance of microreactors or other micro apparatuses. In general, the three-layer model could be applied to the heat transfer modeling as well.

Nevertheless there are some challenges to be resolved in coming years like maldistribution of phases among the channels of monoliths or other multichannel devices, control of bubble sizes during their generation and the change of bubble sizes during their movement throughout the micro channels due to coalescence and/or mass transfer/heating, as well as hydrodynamics and mass transfer of micro channels with complex geometry like meandering micro channels and different types of coiled micro channels.

ACKNOWLEDGMENT

This research was supported by the Russian Foundation of Basic Research RFBR [grant number (research project N°) 18-03-00206 A].

REFERENCES

Abadie, T., Xuereb, C., Legendre, D., & Aubin, J. (2013). Mixing and recirculation characteristics of gas-liquid Taylor flow in microreactors. *Chemical Engineering Research & Design*, *91*(11), 2225–2234. doi:10.1016/j.cherd.2013.03.003

Abiev, R. S. (2008). Simulation of the Slug Flow of a Gas–Liquid System in Capillaries. *Theoretical Foundations of Chemical Engineering*, *42*(2), 105–117. doi:10.1134/S0040579508020012

Abiev, R. S. (2009). Circulation and Bypass Modes of the Slug Flow of a Gas–Liquid Mixture in Capillaries. *Theoretical Foundations of Chemical Engineering*, *43*(3), 298–306. doi:10.1134/S0040579509030099

Abiev, R. S. (2010). Method for Calculating the Void Fraction and Relative Length of Bubbles under Slug Flow Conditions in Capillaries. *Theoretical Foundations of Chemical Engineering*, *44*(1), 86–101. doi:10.1134/S0040579510010112

Abiev, R. S. (2011). Modeling of Pressure Losses for the Slug Flow of a Gas–Liquid Mixture in Mini- and Microchannels. *Theoretical Foundations of Chemical Engineering*, *45*(2), 156–163. doi:10.1134/S0040579511020011

Abiev, R. S. (2012). Modern State and Perspectives of Micro Technique Application in Chemical Industry. *Russian Journal of General Chemistry*, *82*(12), 2019–2024. doi:10.1134/S1070363212120237

Abiev, R. S. (2013). Bubbles velocity, Taylor circulation rate and mass transfer model for slug flow in milli and micro channels. *Chemical Engineering Journal*, *227*, 66–79. doi:10.1016/j.cej.2012.10.009

Abiev, R. S. (2015). Effect of Contact Angle Hysteresis on the Pressure Drop under Slug Flow Conditions in Minichannels and Microchannels. *Theoretical Foundations of Chemical Engineering*, *49*(4), 414–421. doi:10.1134/S0040579515040223

Abiev, R. S., & Dymov, A. V. (2013). Modeling the Hydrodynamics of Slug Flow in a Minichannel for Liquid–Liquid Two Phase System. *Theoretical Foundations of Chemical Engineering*, *47*(4), 299–305. doi:10.1134/S0040579513040180

Abiev, R. S., & Lavretsov, I. V. (2011). Hydrodynamics and Mass transfer of Taylor flow of Gas-Liquid Systems in Micro Channels: Theory and Experiment. *Chemical Engineering Journal, 176-177*, 57–64. doi:10.1016/j.cej.2011.03.107

Abiev, R. S., & Lavretsov, I. V. (2011). Hydrodynamics of Gas–Liquid Slug Flow in Capillaries: Comparing Theory and Experiment. *Theoretical Foundations of Chemical Engineering*, *45*(3), 235–247. doi:10.1134/S0040579511030018

Abiev, R. S., & Lavretsov, I. V. (2012). Intensification of mass transfer from liquid to capillary wall by Taylor vortices in minichannels, bubble velocity and pressure drop. *Chemical Engineering Science*, *74*, 59–68. doi:10.1016/j.ces.2012.02.024

Abiev, R.S., & Svetlov, S.D. (2014). Two-phase flow regimes in the channels of microfluidic devices by the use of coaxial and coaxial-spherical micro dispersers. *Scientific Instrument Engineering (Nauchnoye Priborostroeniye)*, *4*, 59–66. (in Russian).

Angeli, P., & Gavriilidis, A. (2008). Hydrodynamics of Taylor flow in small channels: A review. Proc IMechE, Part C. *Journal of Mechanical Engineering Science*, *222*(5), 737–751. doi:10.1243/09544062JMES776

Aussillous, P., & Quéré, D. (2000). Deposition of a Fluid on the Wall of a Tube. *Physics of Fluids*, *15*(10), 2367–2371. doi:10.1063/1.1289396

Bauer, T., Schubert, M., Lange, R., & Abiev, R. (2006). Intensification of Heterogeneous Catalytic Gas–Fluid Interactions in Reactors with a Multichannel Monolithic Catalyst. *Russian Journal of Applied Chemistry*, *79*(7), 1047–1056. doi:10.1134/S107042720607019

Bercic, G., & Pintar, A. (1997). The role of gas bubbles and liquid slug lengths on mass transport in the Taylor flow through capillaries. *Chemical Engineering Science*, *52*(21-22), 3709–3719. doi:10.1016/S0009-2509(97)00217-0

Bhagat, R. D., & Watt, K. M. (2015). Effect Of Bond Number On Thermophysical Properties Of Working Fluid Used In Closed Loop. Pulsating Heat Pipe: A Review. *International Journal Of Science And Research*, *4*(9), 544–551.

Bolivar, J. M., & Nidetzky, B. (2013). Multiphase biotransformations in microstructured reactors: Opportunities for biocatalytic process intensification and smart flow processing. *Green Processes Synthesis*, *2*, 541–559.

Bretherton, F. P. (1961). The Motion of Long Bubbles in Tubes. *Journal of Fluid Mechanics*, *10*(02), 166–188. doi:10.1017/S0022112061000160

Burns, J. R., & Ramshaw, C. (1999). Development of a microreactor for chemical production. *Chemical Engineering Research & Design*, *77*(3), 206–211. doi:10.1205/026387699526106

Chalfi, T. Y., & Ghiaasiaan, S. M. (2008). Pressure drop caused by flow area changes in capillaries under low flow conditions. *International Journal of Multiphase Flow*, *34*(1), 2–12. doi:10.1016/j.ijmultiphaseflow.2007.09.004

Das, S. P., Nikolayev, V. S., Lefèvre, F., Pottier, B., Khandekar, S., & Bonjour, J. (2010). Thermally Induced Two-phase Oscillating Flow inside a Capillary Tube. *International Journal of Heat and Mass Transfer*, *53*(19-20), 3905–3913. doi:10.1016/j.ijheatmasstransfer.2010.05.009

Dong, Z., Yao, C., Zhang, Y., Chen, G., Yuan, Q., & Xu, J. (2015). Hydrodynamics and Mass Transfer of Oscillating Gas-Liquid Flow in Ultrasonic Microreactors. *American Institute of Chemical Engineers Journal*, *61*, 3964–3972.

Duran Martinez, F. L., Billet, A.-M., Julcour-Lebigue, C., & Larachi, F. (2015). Hydrodynamics And Mass Transfer In Taylor Flow. COMSOL conference, Grenoble, France.

Falconi, C. J., Lehrenfeld, C., Marschall, H., Meyer, C., Abiev, R., Bothe, D., ... Wörner, M. (2016). Numerical and experimental analysis of local flow phenomena in laminar Taylor flow in a square mini-channel. *Physics of Fluids*, *28*(1), 012109. doi:10.1063/1.4939498

Ganapathy, H., Shooshtari, A., Dessiatoun, S., Alshehhi, M., & Ohadi, M. (2014). Fluid flow and mass transfer characteristics of enhanced CO2 capture in a minichannel reactor. *Applied Energy*, *119*, 43–56. doi:10.1016/j.apenergy.2013.12.047

Gao, P. Z., Rebrov, E. V., Verhoeven, T. M., Schouten, J. C., Kleismit, R., Kozlowski, G., ... Subramanyam, G. (2010). Structural Investigations and Magnetic Properties of Sol-Gel Ni0.5Zn0.5Fe2O4 Thin Films for Microwave Heating. *Journal of Applied Physics*, *107*(4), 44317. doi:10.1063/1.3309767

Gascon, J., van Ommen, J. R., Moulijn, J. A., & Kapteijn, F. (2015). Structuring catalyst and reactor – an inviting avenue to process intensification. *Catalysis Science & Technology*, *5*(2), 807–817. doi:10.1039/C4CY01406E

Gruber, R., & Melin, T. (2003). Radial mass-transfer enhancement in bubble-train flow. *International Journal of Heat and Mass Transfer*, *46*(15), 2799–2808. doi:10.1016/S0017-9310(03)00053-X

Gupta, R., Fletcher, D. F., & Haynes, B. S. (2010). CFD modelling of flow and heat transfer in the Taylor flow regime. *Chemical Engineering Science*, *65*(6), 2094–2107. doi:10.1016/j.ces.2009.12.008

Gürsel, I. V., Kockmann, N., & Hessel, V. (2017). Fluidic separation in microstructured devices – Concepts and their Integration into process flow networks. *Chemical Engineering Science*, *169*, 3–17. doi:10.1016/j.ces.2017.03.023

Han, Y., & Shikazono, N. (2009). The effect of bubble acceleration on the liquid film thickness in micro tubes. *International Journal of Heat and Fluid Flow*, *30*(5), 842–853. doi:10.1016/j.ijheatfluidflow.2009.02.019

Hazel, A. L., & Heil, M. (2002). The steady propagation of a semi-infinite bubble into a tube of elliptical or rectangular cross-section. *Journal of Fluid Mechanics*, *470*, 91. doi:10.1017/S0022112002001830

He, P., Haswell, S. J., & Fletcher, P. D. I. (2004). Microwave Heating of Heterogeneously Catalysed Suzuki Reactions in a Micro Reactor. *Lab on a Chip*, *4*(1), 38–41. doi:10.1039/b313057f PMID:15007438

Hessel, V., Angeli, P., Gavriilidis, A., & Löwe, H. (2005). Gas−Liquid and Gas−Liquid−Solid Microstructured Reactors: Contacting Principles and Applications. *Industrial & Engineering Chemistry Research*, *44*(25), 9750–9769. doi:10.1021/ie0503139

Hessel, V., Müller, A., & Kolb, G. (2005). *Chemical Micro Process Engineering. Processing and Plants*. Weinheim, Germany: Wiley–VCH Verlag. doi:10.1002/3527603581

Howard, J. A., & Walsh, P. A. (2013). Review and extensions to film thickness and relative bubble drift velocity prediction methods in laminar Taylor or slug flows. *International Journal of Multiphase Flow*, *55*, 32–42. doi:10.1016/j.ijmultiphaseflow.2013.04.005

Jovanović, J., Zhou, W., Rebrov, E. V., Nijhuis, T., Hessel, V., & Schouten, J. C. (2011). Liquid- liquid slug flow: Hydrodynamics and pressure drop. *Chemical Engineering Science*, *66*(1), 42–54. doi:10.1016/j.ces.2010.09.040

Kashid, M. N., Harshe, Y. M., & Agar, D. W. (2007). Liquid-Liquid Slug Flow in a Capillary: An Alternative to Suspended Drop or Film Contactors. *Industrial & Engineering Chemistry Research*, *46*(25), 8420–8430. doi:10.1021/ie070077x

Kashid, M. N., Renken, A., & Kiwi-Minsker, L. (2011). Gas–liquid and liquid–liquid mass transfer in microstructured reactors. *Chemical Engineering Science*, *66*(17), 3876–3897. doi:10.1016/j.ces.2011.05.015

Kawahara, A., Chung, P. M.-Y., & Kawaji, M. (2002). Investigation of two-phase flow pattern, void fraction and pressure drop in a microchannel. *International Journal of Multiphase Flow*, *28*(9), 1411–1435. doi:10.1016/S0301-9322(02)00037-X

Kececi, S., Wörner, M., Onea, A., & Soyhan, H. S. (2009). Recirculation time and liquid slug mass transfer in co-current upward and downward Taylor flow. *Catalysis Today*, *147*, S125–S131. doi:10.1016/j.cattod.2009.07.054

Khandekar, S., Dollinger, N., & Groll, M. (2003). Understanding Operational Regimes of Pulsating Heat Pipes: An Experimental Study. *Applied Thermal Engineering*, *23*(6), 707–719. doi:10.1016/S1359-4311(02)00237-5

Khandekar, S., Gautam, A. P., & Sharma, P. K. (2009). Multiple quasi-steady states in a closed loop pulsating heat pipe. *International Journal of Thermal Sciences*, *48*(3), 535–546. doi:10.1016/j.ijthermalsci.2008.04.004

Kiwi-Minsker, L., & Renken, A. (2005). Microstructured reactors for catalytic reactions. *Catalysis Today*, *110*(1-2), 2–14. doi:10.1016/j.cattod.2005.09.011

Kolb, W. B., & Cerro, R. L. (1991). Coating the inside of a capillary of square cross section. *Chemical Engineering Science*, *46*(9), 2181–2195. doi:10.1016/0009-2509(91)85119-I

Kreutzer, M. T., Du, P., Heiszwolf, J. J., Kapteijn, F., & Moulijn, J. A. (2001). Mass transfer characteristics of three-phase monolith reactors. *Chemical Engineering Science*, *56*(21-22), 6015–6023. doi:10.1016/S0009-2509(01)00271-8

Kreutzer, M. T., Kapteijn, F., Moulijn, J. A., & Heiszwolf, J. J. (2005a). Multiphase monolith reactors: Chemical reaction engineering of segmented flow in microchannels. *Chemical Engineering Science*, *60*(22), 5895–5916. doi:10.1016/j.ces.2005.03.022

Kreutzer, M. T., Kapteijn, F., Moulijn, J. A., Kleijn, C. R., & Heiszwolf, J. J. (2005b). Inertial and interfacial effects on pressure drop of Taylor flow in capillaries. *AIChE Journal. American Institute of Chemical Engineers*, *51*(9), 2428–2440. doi:10.1002/aic.10495

Lee, H. J., & Lee, S. Y. (2001). Pressure Drop Correlations for Two-Phase Flow within Horizontal Rectangular Channels with Small Height. *International Journal of Multiphase Flow*, *27*(5), 783–796. doi:10.1016/S0301-9322(00)00050-1

Liu, H., Vandu, C. O., & Krishna, R. (2005). Hydrodynamics of Taylor Flow in Vertical Capillaries: Flow Regimes, Bubble Rise Velocity, Liquid Slug Length, and Pressure Drop. *Industrial & Engineering Chemistry Research*, *44*(14), 4884–4897. doi:10.1021/ie049307n

Marschall, H., Falconi, C. J., Lehrenfeld, C., Abiev, R., Wörner, M., Reusken, A., & Bothe, D. (2017). *Direct Numerical Simulations of Taylor bubbles in a square mini-channel: detailed shape and flow analysis with experimental validation. In Transport Processes at Fluidic Interfaces* (pp. 663–679). Berlin, Germany: Springer-Verlag.

Mehta, B., & Khandekar, S. (2014). Taylor Bubble-train Flow and Heat Transfer in the Context of Pulsating Heat Pipes. *International Journal of Heat and Mass Transfer*, *79*, 279–290. doi:10.1016/j.ijheatmasstransfer.2014.08.005

Mishima, K., & Hibiki, T. (1996). Some characteristics of air–water two-phase flow in small diameter vertical tubes. *International Journal of Multiphase Flow*, *22*(4), 703–712. doi:10.1016/0301-9322(96)00010-9

Moulijn, J. A., Kreutzer, M. T., Nijhuis, T. A., & Kapteijn, F. (2011). Monolithic Catalysts. and Reactors: High Precision with Low Energy Consumption. In B. C. Gates & H. Knozinger (Eds.), *Advances in Catalysis* (Vol. 54, pp. 249–327). San Diego, CA: Elsevier. Academic Press Inc.

Onea, A., Wörner, M., & Cacuci, D. G. (2009). A Qualitative Computational Study of Mass Transfer in Upward Bubble Train Flow through Square and Rectangular Mini-Channels. *Chemical Engineering Science*, *64*(7), 1416–1435. doi:10.1016/j.ces.2008.11.008

Öztaskin, M. C., Wörner, M., & Soyhan, H. S. (2009). Numerical investigation of the stability of bubble train flow in a square minichannel. *Physics of Fluids*, *21*(4), 042108. doi:10.1063/1.3101146

Pangarkar, K., Schildhauer, T. J., van Ommen, J. R., Nijenhuis, J., Kapteijn, F., & Moulijn, J. A. (2008). Structured Packings for Multiphase Catalytic Reactors. *Industrial & Engineering Chemistry Research*, *47*(10), 3720–3751. doi:10.1021/ie800067r

Pohorecki, R., & Kula, K. (2008). A simple mechanism of bubble and slug formation in Taylor flow in Microchannels. *Chemical Engineering Research & Design*, *86*(9), 997–1001. doi:10.1016/j.cherd.2008.03.013

Pokusaev, B. G., Zaitsev, A. A., & Zaitsev, V. A. (1999). Transfer Processes under Slug Flow Conditions in Three-Phase Media. *Theoretical Foundations of Chemical Engineering*. *33*(6), 539 -549.

Popova, E. A., Abiev, R. Sh., Lappalainen, L. A., Svetlov, S. D., Andreeva, T. V., Trifonov, R. E., & Ostrovskii, V. A. (2014). Synthesis of 5-phenyltetrazole and its N-methyl Derivatives in a Microreactor. *Chemical and Biochemical Engineering Quarterly*, *28*(2), 241–246. doi:10.15255/CABEQ.2013.1936

Rebrov, E. V. (2010). TwoPhase Flow Regimes in Microchannels. *Theoretical Foundations of Chemical Engineering*, *44*(4), 355–367. doi:10.1134/S0040579510040019

Reinelt, D. A. (1987). The Rate at Which a Long Bubble Rises in a Vertical Tube. *Journal of Fluid Mechanics*, *175*(-1), 557–565. doi:10.1017/S0022112087000521

Renken, A., Hessel, V., Löb, P., Uerdingen, M., & Kiwi-Minsker, L. (2007). Chemical Engineering and Processing. *Process Intensification*, *46*, 840–845. doi:10.1016/j.cep.2007.05.020

Roy, S., Bauer, T., Al-Dahhan, M., Lehner, P., & Turek, T. (2004). Monoliths as Multiphase Reactors: A Review. *AIChE Journal. American Institute of Chemical Engineers*, *50*(11), 2918–2938. doi:10.1002/aic.10268

Schubert, K., Brandner, J., Fichtner, M., Linder, G., Schygulla, U., & Wenka, A. (2001). Microstructure devices for applications in thermal and chemical process engineering. *Microscale Thermophysics Engineering*, *5*(1), 17–39. doi:10.1080/108939501300005358

Schwartz, L. W., Princen, H. M., & Kiss, A. D. (1986). On the motion of bubbles in capillary tubes. *Journal of Fluid Mechanics*, *172*(-1), 259–275. doi:10.1017/S0022112086001738

Shao, N., Gavriilidis, A., & Angeli, P. (2009). Flow regimes for adiabatic gas-liquid flow in microchannels. *Chemical Engineering Science*, *64*(11), 2749–2761. doi:10.1016/j.ces.2009.01.067

Sobieszuk, P., Aubin, J., & Pohorecki, R. (2012). Hydrodynamics and Mass Transfer in GasLiquid Flows in Micro-reactors. *Chemical Engineering & Technology*, *35*(8), 1346–1358. doi:10.1002/ceat.201100643

Sotowa, K. I. (2014). Fluid Behavior and Mass Transport Characteristics of Gas^|^#8211;Liquid and Liquid^|^#8211;Liquid Flows in Microchannels. *Journal of Chemical Engineering of Japan*, *47*(3), 213–224. doi:10.1252/jcej.13we141

Svetlov, S. D., & Abiev, R. S. (2016). Modeling Mass Transfer in a Taylor Flow Regime through Microchannels Using a Three-Layer Model. *Theoretical Foundations of Chemical Engineering*, *50*(6), 975–989. doi:10.1134/S0040579516060166

Taha, T., & Cui, Z. F. (2004). Hydrodynamics of Slug Flow Inside Capillaries. *Chemical Engineering Science*, *59*(6), 1181–1190. doi:10.1016/j.ces.2003.10.025

Taylor, G. I. (1961). Deposition of a Viscous Fluid on the Wall of a Tube. *Journal of Fluid Mechanics*, *10*(02), 161–165. doi:10.1017/S0022112061000159

Thulasidas, T. C., Abraham, M. A., & Cerro, R. L. (1995). Bubble Train Flow in Capillaries of Circular and Square Cross Section. *Chemical Engineering Science*, *50*(2), 183–199. doi:10.1016/0009-2509(94)00225-G

Thulasidas, T. C., Abraham, M. A., & Cerro, R. L. (1997). Flow Patterns in Liquid Slugs during Bubble Train Flow Inside Capillaries. *Chemical Engineering Science*, *52*(17), 2947–2962. doi:10.1016/S0009-2509(97)00114-0

Triplett, K. A., Ghiaasiaan, S. M., Abdel-Khalik, S. I., & Sadowski, D. L. (1999). Gas–Liquid Two-Phase Flow in Microchannels. Part I: Two-Phase Flow Patterns. *International Journal of Multiphase Flow*, *25*(3), 377–394. doi:10.1016/S0301-9322(98)00054-8

Tsaoulidis, D., & Angeli, P. (2014). Effect of channel size on mass transfer during liquid-liquid plug flow in small scale extractors. *Chemical Engineering Journal*, *262*, 785–793. doi:10.1016/j.cej.2014.10.012

Tsoligkas, A. N., Simmons, M. J. H., Wood, J., & Frost, C. G. (2007). Kinetic and selectivity studies of gas-liquid reaction under Taylor flow in a circular capillary. *Catalysis Today*, *128*(1-2), 36–46. doi:10.1016/j.cattod.2007.07.001

Tsoligkas, A. N., Simmons, M. J. H., & Wood, J. (2007). The Effect of Hydrodynamics on Reaction Rates in Capillary Reactor. *Proc. 6th Int. Conf. on Multiphase Flow*, 1.

Van Baten, J. M., & Krishna, R. (2005). CFD Simulations of Wall Mass Transfer for Taylor Flow in Circular Capillaries. *Chemical Engineering Science*, *60*(4), 1117–1126. doi:10.1016/j.ces.2004.10.001

Warnier, M. J. F., de Croon, M. H. J. M., Rebrov, E. V., & Schouten, J. C. (2010). Pressure Drop of Gas–Liquid Taylor Flow in Round Micro Capillaries for Low to Intermediate Reynolds Numbers. *Microfluidics and Nanofluidics*, *8*(1), 33–45. doi:10.100710404-009-0448-z

Watts, P., & Haswell, S. J. (2005). The application of micro reactors for organic synthesis. *Chemical Society Reviews*, *34*(3), 235–246. doi:10.1039/b313866f PMID:15726160

Wörner, M. (2012). Numerical modeling of multiphase flows in microfluidics and micro process engineering: A review of methods and applications. *Microfluidics and Nanofluidics*, *12*(6), 841–886. doi:10.100710404-012-0940-8

Wörner, M. (2013). Computational study on scaling of co-current downward Taylor flow in small square channels of three different sizes. *Journal of Chemical Engineering of Japan*, *46*, 335–341. doi:10.1252/jcej.12we068

Wörner, M., Ghidersa, B., & Onea, A. (2007). A Model for the Residence Time Distribution of Bubble Train Flow in a Square Mini Channel Based on Direct Numerical Simulation Results. *International Journal of Heat and Fluid Flow*, *28*(1), 83–94. doi:10.1016/j.ijheatfluidflow.2006.04.011

Yang, L., Loubière, K., Dietrich, N., Le Men, C., Gourdon, C., & Hébrard, G. (2017). Local investigations on the gas-liquid mass transfer around Taylor bubbles flowing in a meandering millimetric square channel. *Chemical Engineering Science*, *165*, 192–203. doi:10.1016/j.ces.2017.03.007

Yao, C. Q., Dong, Z. Y., Zhao, Y. C., & Chen, G. W. (2014a). An online method to measure mass transfer of slug flow in a microchannel. *Chemical Engineering Science*, *112*, 15–24. doi:10.1016/j.ces.2014.03.016

Yao, C. Q., Dong, Z. Y., Zhao, Y. C., & Chen, G. W. (2014b). The effect of system pressure on gas-liquid slug flow in a microchannel. *American Institute of Chemical Engineering Journal*, *60*(3), 1132–1142. doi:10.1002/aic.14306

Zhang, J., Fletcher, D. F., & Li, W. (2016). Heat transfer and pressure drop characteristics of gas–liquid Taylor flow in mini ducts of square and rectangular cross-sections. *International Journal of Heat and Mass Transfer*, *103*, 45–56.

Zhao, T. S., & Bi, Q. C. (2001). Pressure drop characteristics of gas-liquid two-phase flow in vertical miniature triangular channels. *International Journal of Heat and Mass Transfer*, *44*(13), 2523–2534. doi:10.1016/S0017-9310(00)00282-9

ADDITIONAL READING

Born, S., O'Neal, E., & Jensen, K. F. (2014). Organic Synthesis in Small Scale Continuous Flow: Flow Chemistry. Comprehensive Organic Synthesis: Second Edition. 9, 54-93.

Gavriilidis, A., & Angeli, P. (2009) Mixing and Contacting of Heterogeneous Systems. Micro Process Engineering: A *Comprehensive Handbook*. 1, 205-251. Wiley-VCH Verlag GmbH & Co. KGaA.

Günther, A., & Jensen, K. F. (2006). Multiphase microfluidics: From flow characteristics to chemical and materials synthesis. *Lab on a Chip*, *6*(12), 1487–1503. doi:10.1039/B609851G PMID:17203152

Marre, S., & Jensen, K. F. (2010). Synthesis of micro and nanostructures in microfluidic systems. *Chemical Society Reviews*, *39*(3), 1183–1202. doi:10.1039/b821324k PMID:20179831

Zhang, J., Wang, K., Teixeira, A. R., Jensen, K. F., & Luo, G. (2017). Design and scaling up of microchemical systems: A review. *Annual Review of Chemical and Biomolecular Engineering*, *8*(1), 285–305. doi:10.1146/annurev-chembioeng-060816-101443 PMID:28375772

KEY TERMS AND DEFINITIONS

CFD: Computational fluid dynamics.

Dynamic Gas Hold-Up: Ratio of gas flow rate to the total flow rate (see Equation 28).

Liquid Slug: An element of two-phase flow – long droplet of continuous phase.

Micro Channels: Channels with hydraulic diameter roughly smaller than 1 mm (often 100–200 μm or even smaller, down to 10 μm).

Mini Channels: Channels with hydraulic diameter roughly larger than 1 mm.

Real Volume Fraction of Gas: Ratio of gas volume to the unit cell volume (see Equation 27).

Taylor Bubble (Plug): An element of two-phase flow – long bubble or droplet of dispersed liquid.

Taylor Flow: Two-phase flow with bubbles (drops) having size larger than the channel diameter. Contains liquid slugs (continuous phase) and plugs – gas bubbles or drops of dispersed liquid.

Taylor Vortices: Closed loops of circulation within liquid slugs appearing due to parabolic profile of velocity and due to velocity difference between bubbles and liquid slugs.

Unit Cell: Element of Taylor flow structure, containing one bubble (droplet) and one liquid slug.

Chapter 2
An Introduction to Computational Fluid Dynamics and Its Application in Microfluidics

João Lameu da Silva Júnior
Federal Institute of Education, Science, and Technology of South of Minas Gerais, Brazil

ABSTRACT

The chapter aims to introduce the computational fluid dynamics (CFD). A review was provided, outlining its development and applications on chemical engineering and microfluidics. The fundamental points of the CFD, listing the advantages and precautions of this numerical technique were provided. The description of CFD methodology including the three essential stages (pre-processing, solving, and post-processing) was made. The fundamental transport equations—total mass (continuity), momentum, energy, and species mass balances—and the usual boundary conditions used in CFD were explained. The main approaches used in multicomponent single-phase flows, single-phase flow in porous media, and multiphase flows in microscale were detailed, as well as the numerical mesh types and its quality parameters. A brief introduction of finite volume method (FVM) used by most of the available CFD codes was also performed, describing the main numerical solution features. Finally, the conclusions and future prospects of CFD applications are exposed.

DOI: 10.4018/978-1-5225-7138-4.ch002

INTRODUCTION

The Computational Fluid Dynamics or CFD is the numerical technique used in the analysis of flowing systems and other associated phenomena. CFD studies also include mass and heat transfer and chemical reactions (Versteeg & Malalasekera, 2007). The CFD could be understood as the combination of physics, mathematics and computational sciences aiming the fluid flow behavior description (Blazek, 2001).

With the advancement of computational processing power, CFD has become an essential design and optimization tool for engineers, including products and processes development, as well a substantial research tool in physical sciences (Blazek, 2001; Versteeg & Malalasekera, 2007). Engineering applications include reactor development, mixer and agitator design, contaminant dispersion studies, biomedical studies, among others.

In general, CFD solves numerically the transport equations for the case-relevant variables, necessarily including mass conservation (continuity) and momentum balance. From the classical continuum hypothesis, the fluid flow and heat and mass transport phenomena can be described by partial differential equations (PDE) in the most of engineer interest cases. Analytical solutions are possible only for a few simple special cases, as for example, the Hagen-Poiseuille flow.

Cases of industrial and academic interest are often more complex, involving other phenomena simultaneously with momentum transfer, as for example, chemical species mass transfer. For these cases, approximated numerical solution can be obtained from CFD simulations. Initially, the PDE are discretized, resulting in a system of algebraic equations, which can be computationally solved. The discretization procedure results in space and/or time (transient cases) approximations. Therefore, the numerical prediction accuracy depends on the discretization quality (Ferziger & Peric, 2002).

In this context, the CFD versatility can be highlighted, however, some precautions must be considered regarding the predicted results, keeping physical consistency. Wilcox (1994) highlighted three key-factors for a successful CFD modeling: numerical mesh generation, solution algorithm and turbulence modeling. In Microfluidics, interfacial and surface phenomena can play a key role, depending on the specific case. The first two points listed by Wilcox (1994) are discussed forward. Turbulence is difficult to be generated in microstructured devices unless the pressure head is high enough (Kirby, 2010; Tabeling, 2010), or the microflow experience of additional field forces, e.g. electrokinetic force, as demonstrated by Wang et al. (2014). Since generally the microfluidic devices operate under laminar flow regime, turbulence modeling is out of the scope of this Chapter, however it can be found in Versteeg and Malalasekera (2007), Ferziger and Peric (2002) and Wilcox (1994), which approach the turbulence linking it to CFD. Further reference on turbulence is listed in the Additional Reading.

Among the advantages of the CFD, it can be highlighted: time and costs reduction in projects – the design and analysis of several geometries can be performed before the construction of physical prototypes; prediction of fluid flow field in extreme operating conditions – where experimental measurements would be complex or even impracticable. However, in the other hand, as the CFD is a numerical technique, the numerical limitations and mathematical modeling considerations should be taken into account. The CFD predictions must be, whenever possible, validated with experimental data and/or empirical correlations, in order to ensure the prediction validity. In this context, the present chapter aims to describe the main features of the CFD and its application in the Microfluidics field.

CFD METHODOLOGY

Independently of the computational code, there are three essential stages in a CFD study: 1.) Pre-processing; 2.) Solver, and; 3.) Post-processing.

Pre-Processing

This stage starts by defining the region of interest (geometry), called computational domain. This step is often performed using a Computer-Aided Design (CAD) tool. The domain, and subsequently the mathematical model, could be two- or tri-dimensional, or even one-dimensional when using symmetric boundary conditions. The computational domain is discretized in space in a finite number of control volumes or cells, resulting in the numerical mesh or grid.

In general, the higher number of cells (greater grid refinement) the superior solution accuracy. Both, solution accuracy and computational cost (computer hardware and calculation time) depends on mesh refinement and must be analyzed in a CFD project. The pre-processing follows by definition of simulation and system type, including physical properties of the fluid(s), flow regimes and other associated phenomena. The initial (transient cases) and boundary conditions are then defined, generating the input file to be solved in the next stage.

Solver

A common solver algorithm includes three main steps, as described by Versteeg and Malalasekera (2007):

1. Integration of the governing equations of fluid flow over all the (finite) control volumes of the domain;
2. Conversion of the resulting integral equations into a system of algebraic equations (discretization step);
3. Solution of the algebraic equations by an iterative method.

The pre and solving stages are summarized in Figure 1. The solver result file provides flow field information for all transported variables in the entire computational domain.

Post-Processing

CFD results can be graphically visualized and analyzed in this step. With the development of Graphical Process Units (GPU), detailed results including vector fields, streamlines, isovolumes, particle tracking, animations, among others, can be obtained from the numerical solution. These visualization outputs are effective and important for the interpretation of results. However, it is worth to outline the

Figure 1. Main steps of pre-processing and solving stages of a CFD simulation

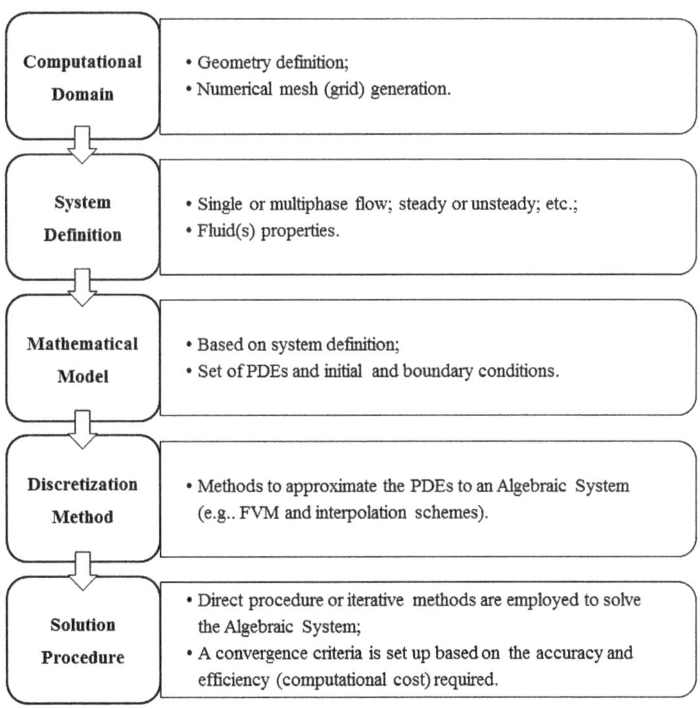

importance of a critical examination of the results before consider they reliable. Also, GPU can be used in the simulation solving process.

MICROFLUIDICS AND COMPUTATIONAL FLUID DYNAMICS

Microfluidics is the science and technology that handles fluid flow in micrometric scale structures. The main advantages of Microfluidics include easiness of temperature control, subproduct minimization, low amount of reactants and samples, short residence times required to achieve the desired performance, high surface area-to-volume ratio, high heat and mass transfer rates, greater safety in flammable, toxic and explosive process operation (Pohar e Plazl, 2009; Whitesides, 2006; Tonkovich Kuhlmann, Rogers, McDanie, Fitzgerald, & Arora, 2005). In this context, the Microfluidics is considered a Process Intensification technique aiming the energy consumption, environmental impact and capital costs reduction. In order to propose optimal microdevices, Computational Fluid Dynamics can be used as a design and analysis tool. Table 1 presents some Microfluidics researches developed using CFD simulations for design and optimization, including microreactors, micro heat exchangers and micromixers.

Table 1. Developed researches using CFD simulations in Microfluidics

Authors	Goals and Main Conclusions
Al-Neama et al. (2017)	• Design and performance analysis of single-phase water flow serpentine microchannels for microelectronic cooling applications. • The conjugated heat-transfer CFD model provided results with good agreement with experimental data for pressure drop and Nusselt numbers. The proposed designs resulted in the minimization of heat transfer resistance, improving the heat transfer, however, at an expense pressure drop increment. Therefore, the authors stand out for the compromise between pressure drop and heat transfer enhancement in the design procedure.
Galletti et al. (2017)	• Mixing investigation in arrow-shape microchannels using Finite Volume and Spectral Element Methods. • Arrow-shaped devices demonstrated superior mixing performance over traditional T-shape microchannels. The increase of Reynolds number up to about 140-150 improved the mixing performance in the arrow-mixer. Above this range, the mixing performance decrease due to the presence of a strong vortical structure at the channel center. Therefore, the noticed optimal operation range of the arrow-mixer microchannels was narrower than for the T-mixer.
Maaref et al. (2017)	• Numerical modeling of water-oil microflow through a porous media using the Cahn-Hilliard Phase Field Method (PFM), in order to evaluate the effects of wettability, heterogeneity and viscosity. • The model was validated with data from visual flooding experiments and then applied to a sensitivity analysis of different wettability states, heterogeneity and viscosity on the water-oil displacement. The authors highlighted the reliable use of the PFM to study the micro and macro-scale mechanisms of the two-phase flow through the porous media.
Santana et al. (2017)	• Microchannel design optimization using static elements for Biodiesel synthesis process intensification. • Firstly, the proposed design was evaluated using CFD simulations that assisted the operating conditions optimization. The design was experimental tested using a PDMS-made microreactor. The microdevice promoted a biodiesel yield of 99.53% at a residence time of about 12 s. The authors emphasized the CFD potential in process intensification studies.

continued on following page

Table 1. Continued

Authors	Goals and Main Conclusions
Silva Jr. et al. (2017)	• The ethanol recovery by evaporation from an ethanol-biodiesel mixture was modeled in a micro-heat exchanger. • A mathematical model considering the momentum, heat and mass transfer was proposed in order to predict the ethanol vaporization phenomena in a parallel microchannel heat exchanger. The flow rate, temperature and ethanol-biodiesel molar ratio influence in the evaporation efficiency was evaluated. Flow rate and temperature presented the major effects on the evaporation process and the superior evaporation efficiency was observed for low flow rates (longer mixture residence times).
Yang et al. (2017)	• Design and performance analysis of microchannel heat sinks (pin-fin configurations). • The geometry design with superior performance in the CFD simulations was manufactured by Ultra Violet-Lithographie Galvanoformung and Abformung (UV-LIGA) and experimentally tested. The CFD and experimental results presented good agreement. The CFD was used as a design tool, predicting the relationship between pressure drop and heat transfer rate, in order to achieve superior cooling performance.
Yuan and Isaac (2017)	• Fluid mixing enhancement by introducing of Lorentz force in order to create a controlled chaotic advection. • The Magneto-Hydro-Dynamics (MHD) model accounted for the introduction of an electric current in the fluid. The current interacted with an external magnetic field generating the Lorentz force that acted as the driving force for fluid flow. The authors highlighted that the CFD-MHD model is a robust tool for development of chaotic advection micromixers, as well as to design controlled-flow devices for targeted delivery.
Zhou et al. (2017)	• CFD application in microfluidic network optimization. • Detailed topology and shape were included. The CFD predictions aided the flow resistance minimization and provided a flexible network set up in order to achieve the desired project requirements.
Argüelles-Vivas and Babadagli (2016)	• Temperature influence in capillary displacement of heavy-oil by gas and residual oil saturation in reservoir micropores. • The increase in gas temperature above 60 ºC swept the oil completely. The authors stands out the CFD approach as a promising tool for the analysis of oil retention in reservoir pores, providing reliable models to study the influence of contact angle, pressure and temperature that are very difficult experimental analysis.
Pradhan and Panigrahi (2016)	• Analysis of liquid slug evaporation by internal convection in a microcapillary. • The CFD simulations exhibited good agreement with Micro-PIV measurements. Both approaches successfully represented the physical behavior of the boiling flow. The channel length presented influence in the slug evaporation performance.
Santana et al. (2016)	• Biodiesel synthesis via vegetable oil transesterification with ethanolic solution of NaOH modeling and simulation in a microchannel reactor. • A multicomponent model assisted the process operating conditions optimization (temperature, catalyst concentration and alcohol/oil molar ratio). The predicted values of biodiesel yield and selectivity presented good agreement with experimental data. The authors highlighted the use of computational simulation in order to accomplish the process design and optimization.
Shojaeian and Kosar (2016)	• Study of convective heat transfer in Non-Newtonian fluids in parallel plates and circular microchannels with constant thermal flux. • The CFD simulations were used to verify the analytical results. The temperature-dependent fluid properties behavior was studied in the viscous dissipation analysis. The authors stated that the fluid properties variations significantly influenced the thermal-hydrodynamics behavior and should be included in the design and optimization procedure of the microdevices.
Abed et al. (2015)	• Forced heat transfer convection in laminar regime in a wavy serpentine microchannel. • Secondary flow vortices improved the fluid mixing, leading to convective heat transfer enhancement. The pressure drop increment was modest relative to the significant heat transfer improvement for the evaluated flow range.
Burkle-Vitzthum et al. (2015)	• Annular microreactor flow modeling for kinetic studies of thermal gas phase reactions with short residence times. • The CFD model was validated with experimental results of residence time distribution and temperature profile. The hydrodynamic behavior of the annular microreactor approached a Plug Flow Reactor condition. This condition was not observed for the tubular microreactor in very short residence times (10-100 ms).
Derksen (2015)	• Liquid-solid mass transfer in a microfluidized bed composed by monosize spherical particles. • The Lattice-Boltzmann Method (LBM) was employed in the numerical simulations. The LBM model provided data about overall adsorption performance, allowing the quantification of local mass transfer rates in the particles surfaces.
Mikaelian et al. (2015)	• Gas-liquid mass transfer investigation in squared and circular microchannels operating under bubbly flow regime. • Correlations for bubble velocity and interfacial mass transfer rate were proposed based on the CFD results. An intrinsic relationship among flow regime pattern, species concentration field and local interfacial mass transfer was observed.

continued on following page

Table 1. Continued

Authors	Goals and Main Conclusions
Pistoresi et al. (2015)	• Flow uniformity evaluation of a Z-type ladder distributor design for mini-channel networks. • A Z-type ladder fluid network composed by 10 channels with square cross section was optimized by two methods: a discrete stairway shape and a continuous tapered shape varying angle from 0° to 30°. Both methods resulted in good flow uniformity. Larger angles and fewer channels resulted in superior flow distribution.
Santana et al. (2015a)	• Micromixer design analysis for application in Biodiesel synthesis by vegetable oil transesterification. • The fluids mixing degree and residence time effects on microdevice performance were evaluated in three micromixer inlet designs. The Reynolds number did not influence the oil conversion in the evaluated range (Re = 10-100), in contrast, the residence time affected the reaction yield, verifying the essential role of the mass diffusion process. Oil conversion of 95% was observed for residence times longer than 10 s.
Santana et al. (2015b)	• Microchannel design optimization using circular obstructions in order to induce split and recombination of the fluid streams. • Two new designs with circular obstructions were proposed and evaluated using a multicomponent flow CFD model. A geometric parameter optimization was carried out alternating the circular obstacles diameter and spacing and the channel width. The fluid mixing and biodiesel synthesis performance were studied. The proposed designs exhibited superior performance over the traditional T-shape microchannel.
Malecha and Malecha (2014)	• Mixing enhancement in complex geometries using pulsation (0-50 Hz) in fluid inlet in 2D and 3D microdevices. • High frequencies promoted mixing enhancement. Low frequencies resulted in time-dependent disturbance on fluids. Low frequencies (up to 25 Hz) provided superior performance over device vibration (more efficient for frequencies above 30 Hz).
Sobieszuk et al. (2014)	• Gas- and liquid-side mass transfer resistance evaluation in microreactors operating in Taylor regime flow. • The gas-side mass transfer, often neglected, could play important role in the overall mass transfer coefficient of gas-liquid systems.
Albadawi et al. (2013)	• Air bubble growth and detachment study in air-water system, using interface capturing methods (Volume of Fluid) in single orifice capillary tubes. • The bubble geometrical properties (center of gravity, contact angle and aspect ratio) influence in the formation and detachment processes was studied and the numerical models were assessed with experimental data. Also, the relationship of inertial and surface forces with the air flow rate on the accuracy of the numerical predictions was evaluated.
Jóźwik et al. (2011)	• 3D modeling and simulation of thermal catalytic • treatment of contaminated air with acetone and isooctane in microreactors. • The model was validated with experimental data. Extrapolated results from mathematical model were used to assess the microreactor length required for a safe effluent composition.
Liu et al. (2010)	• Flow uniformity evaluation of distributors for microchannel networks. • The channel dimensions and bifurcation curvature were investigated. The optimal distributor design was based on the CFD predictions of flow uniformity and pressure drop.
Rahmat and Hubert (2010)	• Two-phase flow in a micro heat pipe (MHP) evaluation considering phase change phenomena. • The MHP consisted in a triangular cross section microchannel. The effective thermal conductivity, pressure and velocity fields were obtained for liquid and vapor phases and compared with literature. The effective length for heat transfer was determined and the proposed mathematical model was employed to simulate a microchannel network in order to predict the two-phase flow recirculation.
Delsman et al. (2004)	• Design, modeling and simulation of an integrated reactor/heat-exchanger in microscale for preferential oxidation of CO in H_2-rich reformate gas, using 2D heat transfer and 3D fluid dynamic models. • The proposed design resulted in high heat recovery efficiency of 90% and a reduction of CO concentration to 10 ppm.

The common aspect of the abovementioned researches was the application of CFD as a design tool for geometry optimization and studies of the physical phenomena in microscale. Moreover, the micromixer design optimization emerge with great prominence as demonstrated by several performed researches (Santana, Tortola, Reis, Silva Jr., & Taranto, 2017; Santana, Silva Jr., & Taranto, 2015b); Ortega-Casanova (2016); Shamloo et al. (2016); Woldemariam et al. (2016); Chen et al. (2014); Hsiao et al. (2014); Solehati et al. (2014); Wang et al. (2014); Cheri et al. (2013); Soleymani et al. (2008); Chung and Shih (2007); Kim et al. (2005); Hong et al. (2004), and; Wang et al. (2002), mostly approaching the mixing enhancement dependence on Reynolds number and analysis of pressure drop and number of mixing units, geometry simplification in order to provide easiness in the device manufacturing and microdevices development for ready integration in lab-on-a-chip systems (miniaturized devices developed to perform laboratory operations on a small scale using miniaturized - Daw and Finkelstein, 2006).

MATHEMATICAL MODELING

In the present section, the main aspects of the mathematical modeling using Computational Fluid Dynamics are discussed, including physical and numerical features.

Physical Behavior of a Fluid Flow

The physical behavior of a fluid flow can be classified in two main categories (Versteeg & Malalasekera, 2007): equilibrium problems and marching problems. Equilibrium problems include steady state cases. These types of problems are modeled by elliptic equations that present unique solutions when boundary conditions are specified over the entire domain. The equilibrium problems are also called boundary-value problems. Unsteady flows, classified as marching problems, are governed by parabolic equations. These equations require initial conditions in the entire domain and boundary conditions for all times ($t > 0$). The problems are then called initial-boundary-value problems and reaches steady state conditions when time tends to infinity. Furthermore, physical problems involving wave phenomena, vibrations other transient processes with negligible energy dissipation rates are described hyperbolic equations.

Verification and Validation in Computational Fluid Dynamics

Since errors and uncertainties are intrinsic to numerical methods, Verification and Validation (V&V) procedures are primordial to provide accuracy and reliability of the simulations.

Versteeg and Malalasekera (2002) discussed the widely accepted V&V terminology, given by AIAA (1998) and Oberkampf and Trucano (2002). Verification is the process of determining that a model implementation accurately represents the conceptual description of the model and the solution to the model. Accordingly, the verification process estimates of errors, including discretization roundoff and iterative convergence errors. Discretization errors are related to the numerical grid and can be quantified comparing results from different refinement levels, as described in the following section. Roundoff errors are quantified comparing the results provided from different computer-machine precision (single or double precision). Iterative convergence errors are inherent from the interpolation scheme truncation and can be quantitatively analyzed by the difference between a specified flow field property predicted with distinct truncation criteria.

Validation is defined by AIAA (1998) and Oberkampf and Trucano (2002) as the process of determining the degree to which a model is an accurate representation of the real world from the perspective of the intended uses of the model. Oberkampf and Trucano (2002) established that quantitative evaluation of the physical modeling uncertainty requires comparison of simulation results with high-quality experimental data. However, there is no a single method to perform this comparison. Some authors (Tu, Yeoh, & Liu, 2008; Andreani, Haller, Heitsch, Hemstromc, Karppinen, Maceke, Schmid, Paillere, & Toth, 2008; Oberkampf & Trucano, 2002; Stern, Wilson, Coleman, & Paterson, 2001a; 2001b; 1999; Iudicello, 1998) have been proposing Best Practice Guidelines (BPGs) and V&V Procedures for CFD simulations.

Numerical Mesh Generation

The numerical mesh or grid generation procedure is one of the key-factors for a successful CFD study. The grid generation is related to the discretization errors, and a poor quality mesh could influence negatively all the numerical predictions. There are two main types of numerical meshes: structured and unstructured grids.

In the structured meshes the grid points are located at coordinate lines intersections. The element location in the grid structure and in the solution matrix is given by indexes: i and j indices (2D) and i, j and k (3D). The structured grid mesh is made using blocks to fit the fluid domain to the solid body (e.g. a cylinder in an external flow or the pipe wall for internal flow studies). The main geometric element types are rectangles (2D) and rectangular parallelepipeds or hexahedrons (3D) (Figure 2a). The main advantage of this grid type is the element orthogonality. However, the structured grid generation becomes hard for very complex geometry. A sub-division of the geometry and then prosecuting the mesh generation in sub-blocks can be suitable for complex geometries.

The unstructured mesh is characterized for the irregular connectivity of the neighbor elements. This approach provides higher flexibility for meshing procedure, allowing a more efficient use of the computational resources. The grid location is not imposed, as in the structured type, favoring the mesh storage with less computational effort. The 2D meshes consist in triangles or quadrilaterals elements. The 3D meshes are constructed with triangular prisms, tetrahedral, hexahedral or even polyhedral elements. Figure 2 presents meshes composed by different element types, including a structured grid with parallelepipeds (Figure 2a), an unstructured mesh with tetrahedral elements (Figure 2b) and a unstructured grid with hybrid elements (tetra and hexahedral – Figure 2c).

Independently of the mesh type and the computational code used in the solution, some common quality parameters must be checked in the numerical mesh, in order to benefit the numerical convergence in the solving procedure. The element orthogonality is an important parameter, and measures the grid distortion by the real element angle between adjacent faces to the optimal angle, i.e., 90° for quadrilateral and 60° for triangular faces. Elements with orthogonal angles below 45° or above 135° are considered high distorted cells, and could lead to numerical solution instabilities and numerical prediction spoiling (Tu, Yeoh, & Liu, 2008). Another important feature of the mesh generation step is the element size ratio. This parameter provides the scaling size between the neighbor elements. Usually, an element size ratio between 1.0 and 1.2 provides good spatial discretization. Several other mesh generation and quality parameters are found in the meshing codes. Particular recommended values for the mesh parameters should be consulted and used according to the different CFD packages employed in the discretization procedure and numerical solution. Some guidelines in mesh generation are presented by Tu et al. (2008), covering grid design and quality features and local refinement and solution adaptation.

Grid Independence in Spatial Discretization

The solution dependence on the grid spatial discretization should be examined. This step consists in the testing of different refinement levels of spatial discretization.

Figure 2. Grid types: (a) structured hexahedral element mesh; (b) unstructured tetrahedral mesh, and; (c) unstructured hybrid elements grid

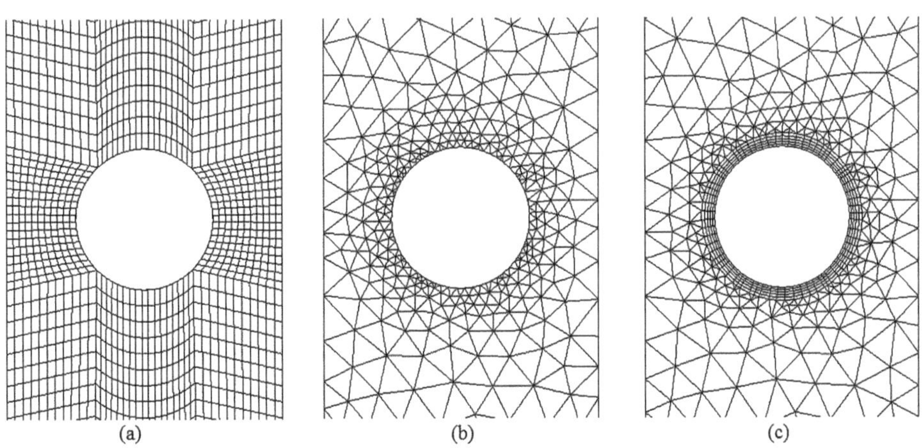

The numerical results provided by these grids must be compared, as for example, the velocity profile and the pressure drop. In general, the independence mesh test could be accomplished in two forms: the first one consists in the increment of the refinement level until the results predicted by these meshes tend to be similar. The definition of the grid refinement level is made by the mesh containing less control volumes that provided similar results to the most refined, in order to maintain the discretization accuracy level with less computational effort. The second way is based on the Grid Convergence Index (GCI) methodology, presented by Roache (1998) and Celik et al. (2008). In this method, the discretization error is estimated using three refinement levels. Extrapolated values for the relevant flow properties are obtained from Richardson extrapolation and the relative errors and the fine-grid convergence index are estimated.

Stability in Solution of Transient Cases

For transient simulations, the time step is an essential parameter linked to the spatial discretization quality and the solution stability. In these cases, the subsequent convergence progress depends on the nondimensional parameter called Courant-Friedlich-Lewy number, or simply Courant number, *CFL*, defined by Equation 1:

$$CFL = u\frac{\Delta t}{\Delta x} < 1 \tag{1}$$

where u is the fluid velocity, Δt is the time step and Δx is the mesh element size. The restriction of $CFL < 1$ was firstly proposed by Courant, Friedlich and Lewys, in order to provide solution stability. Small *CFL* numbers are also required to accurately solve transient details.

The Finite Volume Method (FVM)

The Finite Volume Method (FVM) was introduced by the Imperial College in 1980s and it was based on weighted residual methods, as finite difference and finite element methods (Mohamad, 2011). This method was developed in order to overcome the Finite Difference Method (FDM) problem to deal with complex grids and non-homogenous boundary conditions. Generally, any transported quantity φ, can be represented by the following generic PDE:

$$\frac{\partial}{\partial t}\left(\rho\phi\right) + \nabla\cdot\left(\rho u\phi\right) = \nabla\cdot\left(\Gamma\nabla\phi\right) + S_{\phi} \tag{2}$$

where the first and second left-hand side terms represent the temporal variation of φ (accumulation in the control volume) and the convective transport of φ, respectively. The first and second right-hand side terms represent the diffusive transport of φ (Γ represents the diffusivity coefficient) and additional sources of φ.

The FVM method employs discretization schemes over the finite number of control volumes (CVs) provided in the grid generation. Algebraic equations for each CV are obtained using interpolation schemes to evaluate surface and volume integrals. The number of equations and unknowns are equal to the number of CVs (Ferziger & Peric, 2002). At spatial and time limits tending to zero, the FVM is conservative at discrete level presenting a clear physical interpretation of the fundamental conservation laws. The algebraic system obtained presents always a sparse matrix form independently of the grid type used. A sparse matrix is characterized by the most of its elements being zero and the nonzero terms are closer to the diagonal. The main aspects of the solution methods for the algebraic system are detailed following.

Aspects of Numerical Solution

Versteeg and Malalasekera (2007) related the three main characteristics of a numerical solution method being conservativeness, boundedness and transportiveness. Moreover, Ferziger and Peric (2002) stands out for stability, convergence and accuracy of the solution methods.

Conservativeness, is related to the capability of the numerical solution method to represent the fundamental conservation laws (continuity, momentum and energy balances) in local and global scales. The FVM ensures this property if used with the conservation form of the transport equations (Ferziger & Peric, 2002).

The boundedness property concerns to the physical meaning of fluid and field properties. The numerical solution must respect a bounded range. As example, some physical properties, as for example specific mass, are always positive. Volume, molar or mass fractions must lie between 0 to 1. According to Ferziger and Peric (2002), the boundedness is guaranteed only by first-order discretization schemes, however, unbounded solutions provided by high-order schemes can only happen when very coarse grid are employed.

Transportiveness is associated to the ability of the solution method to represent the real transport phenomena in local and, consequently, global scales. The discretization scheme must represent the real direction and the correct magnitude of the transported property flux. The transportiveness is evaluated by the non-dimensional number of Peclet that represents the ratio of convective transport by diffusive transport (Versteeg & Malalasekera, 2007).

The stability property concerns about the capacity of the numerical method to do not increase the error with the solution procedure progress. According to Ferziger and Peric (2002), a numerical method presents convergence when the exact solution is obtained from the discretized equations at the limit of grid spacing tending to zero.

The accuracy of the numerical solution method is coupled to the inherent errors. Since CFD is a numerical method to solve transport phenomena problems, it always includes implicit errors of solution algorithm. Furthermore, numerical methods also incorporate three types of systematic errors:

1. **Modeling Errors:** Due to the difference between the real flow problem and the simplification performed in the mathematical model. The modeling errors can be neglected for Newtonian fluid flowing in laminar regime, since the complete form of Navier-Stokes equations represent accurately the fluid flow behavior. However, for complex fluid flows, as for example, turbulent flows or multiphase flows, additional models are required to close the equations system. Thus, for these cases, the accuracy of the predicted results also depends of the additional models. Ferziger and Peric (2002) highlighted that is crucial to control and estimate convergence and discretization errors before the evaluation of physical phenomena modeling.
2. **Discretization Errors:** Intrinsic from the procedure to transform the differential to algebraic equations. In general, discretization errors decrease with the grid spacing refinement, however, this represents an increment in computational cost. Furthermore, high-order interpolation schemes could diminish these errors, although they also increase the computational cost.
3. **Convergence Errors:** Inherent from the difference between iterative and exact solutions. These types of errors are easily controlled and quantified due to the different resources available for direct or iterative solution methods.

In summary, an Engineer should use CFD as an efficient engineering tool, i.e., obtaining the required accuracy with the smallest computational effort, or even, obtaining the maximum accuracy with the available computational resource (Ferziger & Peric, 2002).

Algebraic System Solution Procedures

There are two distinct categories of algebraic system solution procedure: direct methods and iterative methods. Examples of the direct methods are the Gaussian elimination and the Thomas algorithm (tri-diagonal matrix algorithm, TDMA). The TDMA is a direct method for one dimensional cases, and can be applied iteratively line-by-line to solve multi-dimensional CFD cases. The TDMA is computationally inexpensive requiring a minimum amount of storage (Tu, Yeoh, & Liu, 2008; Versteed & Malalasekera, 2007). However, in general, the direct methods require high computational effort and huge amount of computer memory regarding the iterative methods (Blazek, 2001).

In the iterative methods, an initial guess for solution are stated, and then, the algorithm systematically improves the solution until the convergence is achieved. Commonly, iterative methods are more economical than direct methods, since for mostly of the CFD cases, the number of required iterations usually decreases with the solution progress (Ferziger & Peric, 2002) and the memory storage requirements are lower, becoming specially interesting for large cases (Versteeg & Malalasekera, 2007). Examples of iterative methods are Jacobi iteration method and Gauss-Seidel iteration method, detailed in Versteeg and Malalasekera (2007).

In order to improve the simulation efficiency Multigrid methods have been developed and implemented in the CFD codes. These methods merge the advantages of using different mesh sizes in the solution procedure. As discussed, finer meshes provides superior results accuracy. However, intermediate solutions do not need high accuracy levels, once the iterative solution will eventually achieve the final solution. Briefly, a Multigrid method employs a refinement cycle during the iterative solution procedure, i.e., the iterations are carried out in meshes of different size. Versteeg and Malalasekera (2007) highlighted the Multigrid methods advantages as, although the greater computational effort required in the finer grids, this additional processing cost is counterbalanced by the benefit of improved convergence rate experienced on the coarser meshes. Accordingly, a more efficient solution procedure is obtained from Multigrid methods.

Fundamental Transport Equations

Most of the CFD solvers are based on the differential-form equations of mass conservation, momentum (Navier-Stokes equations for Newtonian fluids) and energy balances (for non-isothermal cases):

$$\frac{\partial \rho}{\partial t} + \nabla \cdot \left(\rho u\right) = 0 \tag{3}$$

$$\frac{\partial}{\partial t}\left(\rho u\right) + \nabla \cdot \left(\rho u u\right) = -\nabla P + \mu \nabla^2 u + \rho g \tag{4}$$

$$\frac{\partial}{\partial t}\left(\rho C_p T\right) + \nabla \cdot \left(\rho u T\right) = k \nabla^2 T + \mu \Phi_v \tag{5}$$

where ρ is the specific mass, u is the velocity, P is the pressure, μ is the dynamic viscosity, g is the gravity acceleration, C_p is the specific heat capacity, T is the temperature, k is the thermal conductivity, Φ_v is the viscous dissipation term. Additional source terms can be added in these equations, as for example, heat generation due to chemical reactions or secondary momentum transfer. For special cases like multicomponent mixtures, individual species mass balances are also included in the solution, as detailed forward.

Recently, Lattice-Boltzmann Methods (LBM) has received great attention in the numerical simulation of flowing systems. The LBM-based CFD codes employ the Boltzmann transport equation that provides a statistical description of the system as a distribution function. The LBM method was not covered in this Chapter, however, details of this approach is given by Chen and Doolen (1998) and Mohamad (2011).

Usual Initial and Boundary Conditions

The initial and boundary conditions should be analyzed for each specific case; however, there are some usual specifications. Dirichlet boundary conditions are very in CFD cases, which prescribed values or conditions are defined at the boundaries. As for example, the velocity or the mass flow at an inlet. The Neumann boundary condition prescribes the property gradient at the boundary condition. In such case, the property value is estimated from the cell. The initial field must include set of specifications for all flow variables. For inlets, a complete specification has to be performed except for the pressure. Specifically, the pressure reference is often specified at one location inside the domain. For outlets, a common boundary condition used is the zero gradient of the transported variable in the normal direction of the boundary surface. At surface walls, the velocity of a fluid presents the no-slip condition, however in special cases, as for example, particulate systems, the solid phase should be present a free slip condition, or even, the shear stresses could be prescribed at the wall. Symmetry and periodic/cyclic conditions are also employed as boundary conditions, in order to reduce the computational time required and simplify the computational domain geometry.

Mathematical Modeling of Multicomponent Single-Phase Flow

Single-phase flows containing different chemical species can be modeled using the multicomponent approach. The fundamental PDE for this simulation approach is given by Equations 3, 4 and 5 (for non-isothermal). For a system composed by n species, $n - 1$ individual mass balances are solved from Equation 6:

$$\frac{\partial}{\partial t}\left(\rho Y_i\right)+\nabla\cdot\left(\rho u Y_i\right)=\rho D_i\nabla^2 Y_i+S_{Y_i} \tag{6}$$

Where Y_i is the mass fraction ($Y_i = \rho_i/\rho$), D_i is the mass diffusivity coefficient and S_i is the mass source of chemical species i. In order to keep physical consistency, the last mass balance is constrained by mass fraction summation of 1, as given by Equation 7:

$$\sum_{i=1}^{n} Y_i = 1 \tag{7}$$

This approach presented good predictions in Microfluidics, allowing to capture mixing interface of species presenting low miscibility as demonstrated by Santana et al. (2017, 2015b).

Mathematical Modeling of Single-Phase Flow in Porous Media

Ranade (2002) relates two modeling approaches for fluid flow through porous media: a lumped parameter approach that accounts the irregular bed voidage as isotropic or anisotropic porous zone, and a rigorous approach modeling the geometric interstices, which requires a higher computational processing demand. The latter one has achieved progress using LBM-based CFD solvers (Chukwudozie & Tyagi, 2013); Inamuro et al. (1999), Yoshino and Inamuro (2003).

The lumped parameter approach is used in most of CFD solvers. It starts from the Eulerian framework accounting the bed interstices by the void fraction or bed porosity, ε. Basically, an additional source term is included in the momentum equation based on the Ergun (1952) formulation for pressure drop in packed beds. The model takes into account two momentum exchange contribution: friction loss (predominant in viscous laminar flows); inertial loss (prevalent for flow with higher Reynolds numbers). In general, the bed interstices are accounted by the void fraction or bed porosity, ε, defined as the ratio between the actual volume available for fluid flow, V_f, to the total bed volume, V:

$$\varepsilon = \frac{V_f}{V} \tag{8}$$

The convective and diffusive flux terms of the transport equations are weighted by the area porosity tensor, **K**, a second rank tensor that relates the available area vector $\boldsymbol{A}'$ to the infinitesimal planar surface area A:

$$\boldsymbol{A}' = \boldsymbol{K} \cdot A \tag{9}$$

Also, the volumetric terms of the transport equations (local accumulation and source terms) are weighted by the bed porosity, ε. The area porosity tensor, **K**, can be treated as isotropic or anisotropic. The fundamental transport equations are then written as:

$$\frac{\partial}{\partial t}\left(\varepsilon\rho\right) + \nabla\left(\rho\boldsymbol{K} \cdot u\right) = 0 \tag{10}$$

$$\frac{\partial}{\partial t}\left(\varepsilon\rho u\right) + \nabla \cdot \left(\rho\boldsymbol{K} \cdot u \cdot u\right) = -\varepsilon\nabla p + \mu\boldsymbol{K} \cdot \left(\nabla^2 \cdot u\right) + \varepsilon\rho g + \varepsilon S_M \tag{11}$$

$$\frac{\partial}{\partial t}\left(\varepsilon\rho C_p T\right) + \nabla \cdot \left(\rho\boldsymbol{K} \cdot uT\right) = k\boldsymbol{K} \cdot \left(\nabla^2 T\right) + \mu\boldsymbol{K} \cdot \Phi_v \tag{12}$$

Where S_M is the momentum source due to the additional pressure drop in the porous media, modeled in accordance with Equation 13:

$$S_M = -\left(\frac{\mu}{K_{perm}} u + K_{loss}\frac{\rho}{2}\left|u\right| u\right) \tag{13}$$

The first and second terms of the right-hand side of Equation 13 represent the momentum loss due to the bed permeability and the fluid-bed friction inertial losses, respectively. The K_{perm} and K_{loss} terms depends on particle diameter and bed voidage, and are obtained from Ergun equation as:

$$K_{perm} = \frac{d_p^{\ 2}}{150}\frac{\varepsilon^3}{\left(1-\varepsilon\right)^2} \tag{14}$$

$$K_{loss} = \frac{3.5}{d_P} \frac{(1-\varepsilon)}{\varepsilon^3} \tag{15}$$

where d_p is the average solid particle diameter.

Mathematical Modeling of Multiphase Flows

The mathematical modeling of multiphase flows can be performed using two main approaches: Eulerian Multi-Fluid and Eulerian-Lagrangian. The former one can be coupled to level-set methods in order to capture interface behavior of two or more immiscible fluids, resulting in the Volume of Fluid (VOF) method. The main features of these approaches are provided in this section.

Multi-Fluid Eulerian Approach

This approach can be employed for systems composed by gases, liquids and particulate solids. Each phase is treated as continuum and interpenetrates each other. The transport equations are weighted by the volume fraction of each *k-th* phase, α_k, defined as Equation 16:

$$\alpha_k = \frac{V_k}{\sum_{i=1}^{k} V_k}$$

with

$$\sum_{i=1}^{k} \alpha_k = 1 \tag{16}$$

The morphology of the phases must be defined, and a reference density, ρ_{ref}, is defined to take into account the buoyancy force. The fundamental transport equations are given by:

$$\frac{\partial}{\partial t}(\alpha_k \rho_k) + \nabla \cdot (\alpha_k \rho_k u_k) = S_k \tag{17}$$

$$\frac{\partial}{\partial t}\left(\alpha_k \rho_k u_k\right) + \nabla \cdot \left(\alpha_k \rho_k u_k u_k\right) = -\alpha_k \nabla P + \alpha_k \mu_k \nabla^2 u_k + \alpha_k \left(\rho_k - \rho_{ref}\right) g + S_k^{mom} \tag{18}$$

$$\frac{\partial}{\partial t}\left(\alpha_k \rho_k C_{P,k} T_k\right) + \nabla \cdot \left(\alpha_k \rho_k u_k T_k\right) = \alpha_k k_k \nabla^2 T_k + \alpha_k \mu_k \Phi_{v,k} + S_k^{heat} \tag{19}$$

The S_k, S_k^{mom} and S_k^{heat} represent possible sources of mass (e.g. mass transfer among phases), momentum (e.g. interphase momentum exchange due to drag force) and heat (e.g. interphase heat tranfer), respectively. This approach requires additional closure models due to the interaction among the phases.

Volume of Fluid Approach

The Volume of Fluid (VOF) approach is a multiphase model used for systems containing two or more immiscible phases. A moving boundary condition is accounted at the fluid-fluid interface. The Eulerian description and the inherent complexity of the fluid interactions represent the main challenge for the numerical solution. The moving interface location experiences high gradients of fluid and flow field properties. This implies in a requirement of high spatial and temporal refinement (large number of CVs and very small time steps, usually below 10^{-5} s) in order to obtain numerical convergence. Therefore, the main disadvantage of the VOF is the high computational processing required (Ranade, 2002). In contrast, the VOF approach provides details of the interfacial phenomena, being very interesting for microcapillary and gas-liquid Taylor flow investigations, since it furnishes a rigorous modeling of wall adhesion and surface tension effects.

For a system composed by N immiscible phases, the VOF-based models introduce $N - 1$ additional transport equations for the volume fraction of the *k-th* phase:

$$\frac{\partial \alpha_k}{\partial t} + \nabla \cdot \left(\alpha_k u\right) = 0 \tag{20}$$

Also, a single set of shared mass and momentum balances complete the PDE system:

$$\frac{\partial \rho}{\partial t} + \nabla \cdot \left(\rho u\right) = 0 \tag{21}$$

$$\frac{\partial}{\partial t}(\rho u) + \nabla \cdot (\rho u u) = -\nabla P + \mu \nabla^2 u + \rho g + F_s \tag{22}$$

where F_s represents additional momentum sources, as for example, the continuous surface force, F_{CSF}, modeled by Brackbill et al. (1992) for a two-phase flow, as given by Equation 23 (Ranade, 2002):

$$F_{CSF} = 2\sigma\alpha_2\kappa\nabla\alpha_2$$

with

$$\kappa = \frac{1}{|n|}\left[\left(\frac{n}{|n|}\cdot\nabla\right)|n| - (\nabla \cdot n)\right] \tag{23}$$

where σ is the surface or interfacial tension coefficient, α_2 is the volume fraction of the secondary phase, n is the surface normal and κ is the local surface curvature.

Based on the Eulerian framework, the fluid properties ρ and μ are weighted by the volume fraction of the phase in the control volume.

Euler-Lagrange Approach

The Euler-Lagrange (EL) approach is used to simulate the motion of dispersed particles in continuous fluids. The continuous phase is treated in the Euler framework, using the conservation laws given by Equations 3-5. The discrete phase motion is predicted by solving force balances for each dispersed particle. The Euler-Lagrange modeling could couple the phase interactions in a two or in a four way fashion. The called Euler-Lagrange one-way coupling predicts the continuous flow field independent of the discrete phase motion. For the other coupling methods, additional models to take into account the continuous phase/particulate phase interactions (two-way) and additionally, the particle-particle interactions (four-way) are required. The equation of motion of a single dispersed particle is given by Equation 24:

$$m_P \frac{du_P}{dt} = \sum F_P \tag{24}$$

where m_p is the particle mass and u_p is the particle velocity. The right-hand side of the Equation 24 represents the vector summation of all forces acting in the particle, usually including gravitational field and buoyancy force, drag force, among others. Therefore, additional closure models and correlations could be required, e.g. for the nondimensional drag coefficient. The main advantage of Lagrangian approach is that particle-level phenomena can be rigorously predicted (Ranade, 2002). The EL approach is less usual in Microfluidics, however with the progress of micro and nanoparticle technology, this approach can be used to model the fluid system containing these particulate solids.

CONCLUSION

The Computational Fluid Dynamics represents an interesting tool for applications in Microfluidics in two fundamental levels of investigation: transport phenomena and design and optimization tool. Some successful researches employing CFD in Microfluidics were detailed, highlighting the achieved goals. The main steps of a CFD study were depicted. A description of numerical mesh generation listing fundamental quality parameters and independence grid analysis was performed. The primary numerical solution features were outlined, including a brief introduction of the solution methods available in the CFD codes. Details about the main features of the mathematical modeling were provided for the classical Navier-Stokes based CFD solvers. The fundamental transport equations that describe the main flow systems found in Microfluidics were presented, including: single-phase multicomponent flow, single-phase flow in porous media and multiphase flow. The main characteristics of each modeling approach were highlighted.

Considering the aforementioned peculiarities of the numerical techniques, the Engineers should employ the CFD as an efficient and powerful tool for physics investigation of Microfluidics, as well as for Process Intensification using these microstructured devices. The rapid progress in the computer hardware and costs reduction benefits the use of CFD in R&D. All these aspects result in a prospective future with joint advances between Computational Fluid Dynamics and Microfluidics.

ACKNOWLEDGMENT

The author acknowledges the support by the Coordenação de Aperfeiçoamento de Pessoal de Nível Superior - Brasil (CAPES) - Finance Code 001, and the Fundação de Amparo à Pesquisa de Minas Gerais (FAPEMIG) [grant number TEC-APQ-02144-17].

REFERENCES

Abed, W. M., Whalley, R. D., Dennis, D. J. C., & Poole, R. J. (2015). Numerical and experimental investigation of heat transfer and fluid flow characteristics in a micro-scale serpentine channel. *International Journal of Heat and Mass Transfer*, *88*, 790–802. doi:10.1016/j.ijheatmasstransfer.2015.04.062

AIAA. (1998). *Guide for the Verification and Validation of Computational Fluid Dynamics Simulations*. AIAA Guide G-077-1998.

Al-Neama, A. F., Kapur, N., Summers, J., & Thompson, H. M. (2017). An experimental and numerical investigation of the use of liquid flow in serpentine microchannels for microelectronics cooling. *Applied Thermal Engineering*, *116*, 709–723. doi:10.1016/j.applthermaleng.2017.02.001

Albadawi, A., Donoghue, D. B., Robinson, A. J., Murray, D. B., & Delauré, Y. M. C. (2013). On the analysis of bubble growth and detachment at low Capillary and Bond numbers using Volume of Fluid and Level Set methods. *Chemical Engineering Science*, *90*, 77–91. doi:10.1016/j.ces.2012.12.004

Andreani, M., Haller, K., Heitsch, M., Hemstromc, B., Karppinen, I., Maceke, J., ... Toth, I. (2008). A benchmark exercise on the use of CFD codes for containment issues using best practice guidelines: A computational challenge. *Nuclear Engineering and Design*, *238*(3), 502–513. doi:10.1016/j.nucengdes.2007.01.021

Argüelles-Vivas, F. J., & Babadagli, T. (2016). Gas (air)–heavy oil displacement in capillary media at high temperatures: A CFD approach to model microfluidics experiments. *Chemical Engineering Science*, *140*, 299–308. doi:10.1016/j.ces.2015.10.028

Blazek, J. (2001). *Computational Fluid Dynamics: Principles and Applications*. Oxford, UK: Elsevier Science Ltd.

Brackbill, J. U., Kothe, D. B., & Zemach, C. (1992). A continuum method for modelling surface tension. *Journal of Computational Physics*, *100*(2), 335–354. doi:10.1016/0021-9991(92)90240-Y

Burkle-Vitzthum, V., Moulis, F., Zhang, J., Commenge, J. M., Schaer, E., & Marquaire, P. M. (2015). Annular flow microreactor: An efficient tool for kinetic studies in gas phase at very short residence times. *Chemical Engineering Research & Design*, *94*, 611–623. doi:10.1016/j.cherd.2014.10.003

Celik, I. B., Ghia, U., Roache, P. J., Freitas, C. J., Coleman, H., & Raad, P. E. (2008). Procedure for Estimation and Reporting of Uncertainty Due to Discretization in CFD Applications. *Journal of Fluids Engineering*, *130*, 078001-1–4.

Chen, P. C., Wu, M. H., & Wang, Y. N. (2014). Microchannel geometry design for rapid and uniform reagent distribution. *Microfluidics and Nanofluidics*, *17*(2), 275–285. doi:10.100710404-013-1312-8

Chen, S., & Doolen, G. D. (1998). Lattice Boltzmann Method For Fluid Flows. *Annual Review of Fluid Mechanics*, *30*(1), 329–364. doi:10.1146/annurev.fluid.30.1.329

Cheri, M. S., Latifi, H., Moghaddam, M. S., & Shahraki, H. (2013). Simulation and experimental investigation of planar micromixers with short-mixing-length. *Chemical Engineering Journal*, *234*, 247–255. doi:10.1016/j.cej.2013.08.067

Chukwudozie, C., & Tyagi, M. (2013). Pore scale inertial flow simulations in 3-D smooth and rough sphere packs using lattice Boltzmann method. *AIChE Journal. American Institute of Chemical Engineers*, *59*(12), 4858–4870. doi:10.1002/aic.14232

Chung, C. K., & Shih, T. R. (2007). A rhombic micromixer with asymmetrical flow for enhancing mixing. *Journal of Micromechanics and Microengineering*, *17*(12), 2495–2504. doi:10.1088/0960-1317/17/12/016

Daw, R., & Finkelstein, J. (2006). Lab on a chip. *Nature*, *442*(7101), 367. doi:10.1038/442367a

Delsman, E. R., De Croon, M. H. J. M., Pierik, A., Kramer, G. J., Cobden, P. D., Hofmann, Ch., ... Schouten, J. C. (2004). Design and operation of a preferential oxidation microdevice for a portable fuel processor. *Chemical Engineering Science*, *59*(22-23), 4795–4802. doi:10.1016/j.ces.2004.07.099

Derksen, J. J. (2015). Simulations of liquid-to-solid mass transfer in a fluidized microchannel. *Microfluidics and Nanofluidics*, *18*(5-6), 829–839. doi:10.100710404-014-1475-y

Ergun, S. (1952). Fluid flow through packed columns. *Chemical Engineering Progress*, *48*, 89–94.

Ferziger, J. H., & Peric, M. (2002). *Computational Methods for Fluid Dynamics* (3rd. ed.). Berlin, Germany: Springer-Verlag. doi:10.1007/978-3-642-56026-2

Galletti, C., Brunazzi, E., Siconolfi, L., Spaltro, D., & Mauri, R. (2017). Mixing Performance of Arrow-Shaped Micro-Devices. *Chemical Engineering Transactions*, *57*, 1309–1314.

Hong, C. C., Choi, J. W., & Ahn, C. H. (2004). A novel in-plane passive microfluidic mixer with modified Tesla structures. *Lab on a Chip*, *4*(2), 109–113. doi:10.1039/b305892a PMID:15052349

Hsiao, K. Y., Wu, C. Y., & Huang, Y. T. (2014). Fluid mixing in a microchannel with longitudinal vortex generators. *Chemical Engineering Journal*, *235*, 27–36. doi:10.1016/j.cej.2013.09.010

Inamuro, T., Yoshino, M., & Ogino, F. (1999). Lattice Boltzmann simulation of flows in a three-dimensional porous structure. *International Journal for Numerical Methods in Fluids*, *29*(7), 737–748. doi:10.1002/(SICI)1097-0363(19990415)29:7<737::AID-FLD813>3.0.CO;2-H

Iudicello, F. (1998). *Guide for the verification and validation of computational fluid dynamics simulations*. AIAA G-077-1998.

Jóźwik, P., Karcz, M., & Badur, J. (2011). Numerical Modelling of a Microreactor for Thermocatalytic Decompostion of Toxic Compounds. *Chemical & Process Engineering*, *32*(3), 215–227. doi:10.2478/v10176-011-0017-3

Kim, D. S., Lee, S. H., Kwon, T. H., & Ahn, C. H. (2005). A serpentine laminating micromixer combining splitting/recombination and advection. *Lab on a Chip*, *5*(7), 739–747. doi:10.1039/b418314b PMID:15970967

Kirby, B. (2010). *Micro- and Nanoscale Fluid Mechanics: Transport in Microfluidic Devices*. Cambridge, UK: Cambridge University Press. doi:10.1017/CBO9780511760723

Liu, H., Li, P., & Lew, J. V. (2010). CFD study on flow distribution uniformity in fuel distributors having multiple structural bifurcations of flow channels. *International Journal of Hydrogen Energy*, *35*(17), 9186–9198. doi:10.1016/j.ijhydene.2010.06.043

Maaref, S., Rokhforouz, M.-R., & Ayatollahi, S. (2017). Numerical Investigation of Two Phase Flow in Micromodel Porous Media: Effects of Wettability, Heterogeneity, and Viscosity. *Canadian Journal of Chemical Engineering*, *95*(6), 1213–1223. doi:10.1002/cjce.22762

Malecha, Z. M., & Malecha, K. (2014). Numerical Analysis of Mixing Under Low and High Frequency Pulsations at Serpentine Micromixers. *Chemical & Process Engineering*, *35*(3), 369–385. doi:10.2478/cpe-2014-0028

Mikaelian, D., Haut, B., & Scheid, B. (2015). Bubbly flow and gas–liquid mass transfer in square and circular microchannels for stress-free and rigid interfaces: CFD analysis. *Microfluidics and Nanofluidics*, *19*(3), 523–545. doi:10.100710404-015-1578-0

Mohamad, A. A. (2011). *Lattice Boltzmann Method: Fundamentals and Engineering Applications with Computer Codes*. London: Springer-Verlag London Limited. doi:10.1007/978-0-85729-455-5

Oberkampf, W. L., & Trucano, T. G. (2002). Verification and Validation in Computational Fluid Dynamics. *Progress in Aerospace Sciences*, *38*(3), 209–272. doi:10.1016/S0376-0421(02)00005-2

Ortega-Casanova, J. (2016). Enhancing mixing at a very low Reynolds number by a heaving square cylinder. *Journal of Fluids and Structures*, *65*, 1–20. doi:10.1016/j.jfluidstructs.2016.05.002

Pistoresi, C., Fan, Y., & Luo, L. (2015). Numerical study on the improvement of flow distribution uniformity among parallel mini-channels. *Chemical Engineering and Processing*, *95*, 63–71. doi:10.1016/j.cep.2015.05.014

Pohar, A., & Plazl, I. (2009). Process Intensification through Microreactor Application. *Chemical and Biochemical Engineering Quarterly*, *23*, 537–544.

Pradhan, T. K., & Panigrahi, P. K. (2016). Evaporation-induced natural convection of a liquid slug of binary mixture inside a microchannel: Effect of confinement. *Microfluidics and Nanofluidics*, *20*(8), 115. doi:10.100710404-016-1782-6

Rahmat, M., & Hubert, P. (2010). Two-phase simulations of micro heat pipes. *Computers & Fluids*, *39*(3), 451–460. doi:10.1016/j.compfluid.2009.09.014

Ranade, V. V. (2002). Multiphase Flow Processes. In Computational Flow Modeling for Chemical Reactor Engineering (pp. 85-122). Academic Press.

Roache, P. J. (1998). *Verification and Validation in Computational Science and Engineering*. Albuquerque, NM: Hermosa Publishers.

Santana, H. S., Silva, J. L. Jr, & Taranto, O. P. (2015a). Numerical simulation of mixing and reaction of *Jatropha curcas* oil and ethanol for synthesis of biodiesel in micromixers. *Chemical Engineering Science*, *132*, 159–168. doi:10.1016/j.ces.2015.04.014

Santana, H. S., Silva, J. L. Jr, & Taranto, O. P. (2015b). Numerical simulations of biodiesel synthesis in microchannels with circular obstructions. *Chemical Engineering and Processing*, *98*, 137–146. doi:10.1016/j.cep.2015.10.011

Santana, H. S., Tortola, D. S., Reis, E. M., Silva, J. L. Jr, & Taranto, O. P. (2016). Transesterification reaction of sunflower oil and ethanol for biodiesel synthesis in microchannel reactor: Experimental and simulation studies. *Chemical Engineering Journal*, *302*, 752–762. doi:10.1016/j.cej.2016.05.122

Santana, H. S., Tortola, D. S., Silva, J. L. Jr, & Taranto, O. P. (2017). Biodiesel synthesis in micromixer with static elements. *Energy Conversion and Management*, *141*, 28–39. doi:10.1016/j.enconman.2016.03.089

Shamloo, A., Madadelahi, M., & Akbari, A. (2016). Numerical simulation of centrifugal serpentine micromixers and analyzing mixing quality parameters. *Chemical Engineering and Processing*, *104*, 243–252. doi:10.1016/j.cep.2016.03.017

Shojaeian, M., & Kosar, A. (2016). Convective heat transfer of non-Newtonian power-law slip flows and plug flows with variable thermophysical properties in parallel-plate and circular microchannels. *International Journal of Thermal Sciences*, *100*, 155–168. doi:10.1016/j.ijthermalsci.2015.09.024

Silva, J. L. Jr, Santana, H. S., Sanchez, G. B., & Taranto, O. P. (2017). Numerical Simulation of Excess Ethanol Evaporation from Biodiesel in a Micro Heat-Exchanger. *Chemical Engineering Transactions*, *57*, 1123–1128.

Sobieszuk, P., Ilnicki, F., & Pohorecki, R. (2014). Contribution of Liquid- and Gas-Side Mass Transfer Coefficients to Overall Mass Transfer Coefficient in Tayloer Flow in a Microreactor. *Chemical & Process Engineering*, *35*(1), 35–45. doi:10.2478/cpe-2014-0003

Solehati, N., Bae, J., & Sasmito, A. P. (2014). Numerical investigation of mixing performance in microchannel T-junction with wavy structure. *Computers & Fluids*, *96*, 10–19. doi:10.1016/j.compfluid.2014.03.003

Soleymani, A., Kolehmainen, E., & Turunen, I. (2008). Numerical and experimental investigations of liquid mixing in T-type micromixers. *Chemical Engineering Journal*, *135S*, S219–S228. doi:10.1016/j.cej.2007.07.048

Stern, F., Wilson, R. V., Coleman, H. W., & Paterson, E. G. (1999). *Verification and Validation of CFD Simulations*. Iowa Institute of Hydraulic Research Report No. 407.

Stern, F., Wilson, R. V., Coleman, H. W., & Paterson, E. G. (2001a). Comprehensive Approach to Verification and Validation of CFD Simulations - Part 1: Methodology and Procedures. *Journal of Fluids Engineering*, *123*(4), 793–802. doi:10.1115/1.1412235

Stern, F., Wilson, R. V., Coleman, H. W., & Paterson, E. G. (2001b). Comprehensive Approach to Verification and Validation of CFD Simulations - Part 2: Application for Rans Simulation of a Cargo/Container Ship. *Journal of Fluids Engineering*, *123*, 803–810. doi:10.1115/1.1412235

Tabeling, P. (2010). *Introduction to Microfluidics*. Oxford University Press.

Tonkovich, A., Kuhlmann, D., Rogers, A., McDanie, J., Fitzgerald, S., Arora, R., Tu, J., Yeoh, G. H., & Liu, C. (2008). *Computational Fluid Dynamics: A Practical approach*. Oxford, UK: Butterwoth-Heinemann/Elsevier.

Versteeg, H. K., & Malalasekera, W. (2007). *An Introduction to Computational Fluid Dynamics: The Finite Volume Method* (2nd ed.). Edinburgh Gate, UK: Pearson Education Limited.

Wang, G. R., Yang, F., & Zhai, W. (2014). There can be turbulence in microfluidics at low Reynolds number. *Lab on a Chip*, *14*(8), 1452–1458. doi:10.1039/C3LC51403J PMID:24599543

Wang, H., Iovenitti, P., Harvey, E., & Masood, S. (2002). Optimizing layout of obstacles for enhanced mixing in microchannels. *Smart Materials and Structures*, *11*(5), 662–667. doi:10.1088/0964-1726/11/5/306

Wang, L., Ma, S., Wang, X., Bi, H., & Han, X. (2014). Mixing enhancement of a passive microfluidic mixer containing triangle baffles. *Asia-Pacific Journal of Chemical Engineering*, *9*(6), 877–885. doi:10.1002/apj.1837

Whitesides, G. M. (2006). The origins and the future of microfluidics. *Nature*, *442*(7101), 368–373. doi:10.1038/nature05058 PMID:16871203

Wilcox, D. C. (1998). *Turbulence Modeling for CFD*. La Cañada, CA: DCW Industries.

Woldemariam, M., Filimonov, R., Purtonen, T., Sorvari, J., Koiranen, T., & Eskelinen, H. (2016). Mixing performance evaluation of additive manufactured milli-scale reactors. *Chemical Engineering Science*, *152*, 26–34. doi:10.1016/j.ces.2016.05.030

Yang, D., Wang, Y., Ding, G., Jin, Z., Zhao, J., & Wang, G. (2017). Numerical and experimental analysis of cooling performance of single-phase array microchannel heat sinks with different pin-fin configurations. *Applied Thermal Engineering*, *112*, 1547–1556. doi:10.1016/j.applthermaleng.2016.08.211

Yoshino, M., & Inamuro, T. (2003). Lattice Boltzmann simulations for flow and heat/mass transfer problems in a three-dimensional porous structure. *International Journal for Numerical Methods in Fluids*, *43*(2), 183–198. doi:10.1002/fld.607

Yuan, F., & Isaac, K. M. (2017). A study of MHD-based chaotic advection to enhance mixing in microfluidics using transient three dimensional CFD simulations. *Sensors and Actuators. B, Chemical*, *238*, 226–238. doi:10.1016/j.snb.2016.07.063

Yuschak, T. (2005). Microchannel Technology Scale-Up to Commercial Capacity. *Chemical Engineering Research & Design*, *83*(6), 634–639. doi:10.1205/cherd.04354

Zhou, T., Liu, T., Deng, Y., Chen, L., Qian, S., & Liu, Z. (2017). Design of microfluidic channel networks with specified output flow rates using the CFD-based optimization method. *Microfluidics and Nanofluidics*, *21*(1), 11. doi:10.100710404-016-1842-y

ADDITIONAL READING

Alia, Z., Tucker, P. G., & Shahpar, S. (2017). Optimal mesh topology generation for CFD. *Computer Methods in Applied Mechanics and Engineering*, *317*, 431–457. doi:10.1016/j.cma.2016.12.001

Andreani, M., Haller, K., Heitsch, M., Hemström, B., Karppinen, I., Macek, J., ... Toth, I. (2008). A benchmark exercise on the use of CFD codes for containment issues using best practice guidelines: A computational challenge. *Nuclear Engineering and Design*, *238*(3), 502–513. doi:10.1016/j.nucengdes.2007.01.021

Bird, R. B., Stewart, W. E., & Lightfoot, E. N. (2002). *Transport Phenomena, 2nd*. New York, USA: John Wiley & Sons Inc.

Hirsch, C. (Ed.). (2007). Numerical Computation of Internal and External Flows.: Vol. 1. *The Fundamentals of Computational Fluid Dynamics* (2nd ed.). Burlington, MA, USA: Butterworth-Heinemann.

Leal, L. G. (2007). *Advanced Transport Phenomena: Fluid Mechanics and Convective Transport*. New York, USA: Cambridge University Press. doi:10.1017/CBO9780511800245

Panigrahi, P. K. (2016). *Transport Phenomena in Microfluidic Systems*. Singapore: John Wiley & Sons. doi:10.1002/9781118298428

Patankar, S. V. (1980). *Numerical Heat Transfer and Fluid Flow*. New York, USA: Taylor & Francis. doi:10.1201/9781482234213

Peiró, J., & Sherwin, Sp. Finite Difference, Finite Element and Finite Volume Methods for Partal Differential Equations. (2005). In Yip, S. (ed.), Handbook of Materials Modeling. Volume I: Methods and Models (pp. 1-32). Netherlands, Springer.

Pope, S. B. (2000). *Turbulent Flows*. Cambridge, UK: Cambridge University Press. doi:10.1017/CBO9780511840531

KEY TERMS AND DEFINITIONS

Eulerian Approach: Mathematical approach treating the system phases as continuous and interpenetrating fluids. The transport equations are weighted by the volumetric fraction of each phase.

Lagrangian Approach: Mathematical approach based on the force balance acting on the discrete phase.

Multicomponent Mixture: Fluid composed by three or more chemical species.

Multiphase Flow: Fluid flow composed by two or more immiscible phases.

Spatial Discretization: Subdivision of the computational domain in a finite number of control volumes or elements (i.e., the generation of the numerical grid).

Unstructured Grid: Spatial discretization method using nonuniform control volumes with an irregular connectivity of the neighbor elements.

Viscous Laminar Flow: Fluid flow regime where the viscous forces prevail over the inertial forces, characteristic of low Reynolds numbers.

Chapter 3
2D Numerical Study of a Micromixer Based on Blowing and Vortex Shedding Mechanisms

Maria Sanchez-Claros
Universidad de Málaga, Spain

Joaquin Ortega-Casanova
Universidad de Málaga, Spain

Francisco Jose Galindo-Rosales
Universidade do Porto, Portugal

ABSTRACT

In this chapter, a numerical study and assessment of the mixing efficiency of a novel microfluidic device for mixing two fluids are presented. The device under study consists of a two-dimensional straight microchannel with a square pillar centered across the channel. The main fluid flows through the microchannel from the main inlet to the outlet, while the second fluid is injected through the pillar as two small jets at its upstream corners. For different values of the Reynolds number, intensity ratio between the jets and the main channel stream and jets injection angle, the authors have conducted several numerical simulations to characterize both the mixing efficiency and the required input power to make the fluids flow. The optimum configuration has been revealed for high values of the Reynolds number, low intensity ratios, and high injection angles. Thanks to vortex shedding and the corresponding downstream oscillations, a mixing efficiency of around 90% can be reached. The worst mixing efficiency is obtained for a configuration without vortex shedding, having a mixing efficiency of only around 2%.

DOI: 10.4018/978-1-5225-7138-4.ch003

NOMENCLATURE

C_0: Molar concentration of solute at channel inlet [mol/m^3].
C_D: Drag coefficient [-]
C_L: Lift coefficient [-]
D: Mass diffusion coefficient [m^2/s].
F_x: Streamwise, horizontal, force [N/m]
F_y: Spanwise, vertical, force [N/m]
H: Channel width [m].
I: Jet intensity ratio [-].
L: Channel length [m].
n: Number of grid nodes at the channel outlet.
p: Pressure [-].
P: Input power energy [-].
Pe: Péclet number [-].
$\overline{P}_i$: Average static pressure at channel inlet [-].
$\overline{P}_j$: Average static pressure at jet injections [-].
q_i : Channel inlet mass flow rate [-].
q_j : Jet injection mass flow rate [-].
Re_i : Reynolds number at inlet of the channel [-].
Re_j : Jet Reynolds number at the injections [-].
Sc : Schmidt number [-].
St : Strouhal number [-].
t : Time [-].
U_i : Mean velocity at channel inlet [m/s].
U_j : Mean velocity at jet injections [m/s].
$\vec{v}$: Velocity vector [-].
W : Side of the inner square pillar [m].
W_j : Jet injection width [m].
Y : Mass fraction of one fluid [-].
Y_k : Mass fraction at outlet k-grid cell [-].
$\overline{Y}$: Average mass fraction at the outlet [-].
η : Mixing efficiency [%].
ρ: Density [kg/m3].
σ : Standard deviation of the mass fraction of one of the fluids [-].
$\sigma_{\max}$: Maximum standard deviation in the system [-].
υ : Kinematic viscosity [m^2/s].

INTRODUCTION

The present Chapter deals with the micromixing of two fluids in a particular geometry where the vortex shedding from a bluff body plays an important role in the mixing efficiency. The geometry consists of a straight microchannel, with the main fluid entering through the channel inlet, and having a square pillar in its middle, from which the second fluid enters by means of two small jets at the pillar's upstream face (this could be understood as a blowing technique). A sketch of the geometry can be seen in Figure 1. Therefore, three important aspects are addressed in this work: micromixing; blowing, or jets injection; and vortex shedding. Here below, each of these topics will be introduced and reviewed, and the main gaps of the state of the art will be discussed and referenced.

When two fluids have to be mixed at microscale where, normally, due to the device dimensions the Reynolds number is quite low and laminar diffusion governs the mixing process (see Capretto, Cheng, Hill, & Zhang, 2011, or Lee, Chang, Wang, & Fu, 2011, for a couple of reviews about micromixing), it is important to take any kind of action in order to enhance the mixing efficiency at the microdevice exit. Otherwise, both the efficiency and the microdevice performance will be very poor. Most of implemented actions to improve mixing in microdevices are classified as active or passive, depending on whether or not any kind of external energy is applied, respectively. On one hand, active micromixers are based on applying any kind of additional external energy to promote fluids interaction and, thus, to increase the mixing efficiency. In this regard, in the literature can be found many works as, among others, the one by Rahimi et al. (2013), who experimentally investigated how high frequency ultrasound waves can enhance mixing in a tubular sonoreactor. Other mixing mechanism was researched by Noroozi et al. (2009), who studied a centrifugal microfluidic-based device which, based on a Coriolis force mechanism, were able to achieve a good mixing, with almost complete mixing, with just three rotations of the centrifugal device. Another kind of active micromixer was studied by Niu et al. (2006), who designed and built a chaotic mixer based on electro-rheological valves to control some jets orthogonally injected into the main channel flow. They found that just 6 of these jets are enough to get fully chaotic mixing. Additionally, in the work by Wang et al. (2014) was investigated how an electromagnetically controlled microrobot, which can move and even rotate, can be used as micromixer in a microfluidic chip. Furthermore, Kuo and Li (2016), by means of both numerical and experimental techniques, were able to characterize the flow and mixing efficiency of a microfluidic mixer with four centrifuge-based compact disks. And to conclude this list of active micromixers, Ortega-Casanova (2017a, 2018) used external energy to heave and pitch a square pillar located in the middle of a straight channel to promote perturbation in the flow. In these works, it

Figure 1. Parameterized geometry of the system to analyze

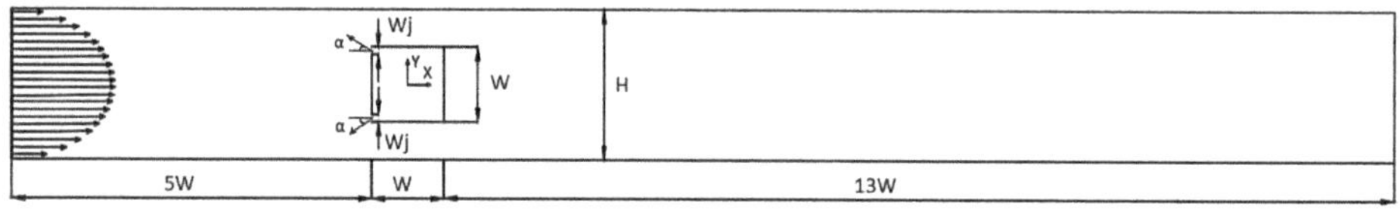

was shown how both motions could help to increase the mixing efficiency at the microdevice exit.

On the other hand, passive micromixers are based on a specific design of the geometry looking for increasing the surface contact between the fluids to promote the mixing. One can find in the literature many works focused in designing the geometry of the micromixer. In that sense, both Jain et al. (2013) and Hossain et al. (2011), used a grooved micromixer for mixing purposes which shape is optimized by numerical simulations. In the work by Kim et al. (2011), a ribbed microchannel was used and they looked for ribs angle, height, width and spacing, and channel length in order to have the highest mixing efficiency at the microdevice exit. Additionally, Tseng et al. (2011) studied how the mixing efficiency is affected by the shape, length, width and location of diamond-obstacles distributed along the micromixer, while Hsiao et al. (2014) used pairs of rectangular winglets mounted on the bottom of a T-type main channel and found winglet optimal geometric parameters in order to obtain the highest mixing efficiency as possible. Finally, to conclude these passive micromixer examples, Parsa and Hormozi (2014), by means of numerical and experimental techniques, were able to find the optimal phase shift between the sinusoidal side walls of a waving micromixer. Passive mixers are simple and operate with no moving parts and therefore preferred when working with standard softlithographic techniques, but their mixing efficiency at the micromixer exit is strongly coupled to flow rate and geometry (Ober, Foresti, & Lewis, 2015).

In this study, we focus on these latter two parameters: the flow rate (intensity ratio), promoted by blowing technique; and geometry (jets angle), which will promote vortex shedding downstream.

The injection of fluids from a static bluff object is a technique frequently called as *blowing* and has been previously used by different authors mainly to control and avoid both flow separation and vortex shedding from the object. These effects usually give to both high-pressure forces on the object and, eventually, to undesirable vibrations (known as Vortex Induced Vibrations and abbreviated as VIV) of the object. To know more about these topics, the reader is referred to, among others, the reviews by Willianson and Govardha (2004) or Sarpkaya (2004). To overcome those undesirable effects, both the high pressure forces and the body vibration, flow control techniques could have the capacity of creating flow field modifications without the necessity of changing the external shape of the object. In that sense, in the work by Dong et al. (2008) is presented a technique which is able to suppress the VIV of bluff bodies by small amounts of combined suction (upstream) and blowing (downstream). Additionally, in the work by Delaunay and Kasktsis (2001) was studied the flow around a circular cylinder and observed that, depending on the Reynolds number, just slight blowing or sufficiently high suction can stabilize the wake. On the other hand, in the work by Wu et al. (2007) was shown how the vortex shedding downstream a circular cylinder can be removed by generating a travelling wave by making the rear side of the cylinder of flexible material. Additionally, in the work by Chamoun et al. (2011) was showed, by means of the point vortex model, that the wake downstream a circular cylinder can be also stabilized with harmonic forcing and choosing appropriately the forcing parameters. Finally, Sohankar et al. (2015) investigated how a uniform suction and blowing through the porous surfaces of a square pillar can help to suppress vortex shedding, while Gonzalez-Mut and Ortega-Casanova (2016) shown how blowing and suction at different angles from the front and back sides of a square pillar can help to prevent vortex shedding and thus to reduce the drag coefficient. They also showed how just front blowing can increase the VIV and, consequently, the unsteadiness and oscillations of the aerodynamic coefficients.

Despite the results of some of the above mentioned works are based on numerical simulations of the corresponding micromixer, since CFD is the main tool used in this chapter to study the proposed microdevice, some additional numerical more works will be discussed next. In effect, one can found in literature many works where CFD is used to analyze and characterize micromixers looking for a high degree of mixing at their outlets. This is the case of the works by Wu and Chen (2018), Yang et al. (2015), Le et al. (2015), Solehati et al. (2014) and Santana et al. (2015), among others. In Wu and Chen (2018), a fractal micromixer based on Cantor fractal principle is numerically characterized for different Reynolds numbers and fractal complexity. They showed that a complete mixing can be achieved with 4 their fractal obstacles at $\mathrm{Re} = 100$. The work by Yang et al. (2015) deals with the design, simulation, fabrication and test, for Reynolds numbers ranging from 0.1 to 100, of a micromixer with Tesla structures inside resulting in an effective microdevice for bio-applications. In the work by Te et al. (2015) a micromixer with shifted trapezoidal blades along the main channel is analysed both experimental and numerically and found that a stable mixing efficiency of around 80% or above can be achieved for Reynolds number values in the range from 0.5 to 100. The work by Solehati et al. (2014) deals with a numerical investigation of a wavy microchannel with T-junction. The wavy shape helped to trigger secondary flows for enhancing chaotic mixing. And, finally, in the work by Santana et al. (2015) is shown that, thanks to the split and recombine technique done by means of circular obstructions in a T-channel, a high degree of mixing can be achieved for Reynolds number ranging from 1 to 160.

Finally, regarding mixing fluids by means of the vortex shedding mechanism and/or the blowing technique, few works, to the authors' knowledge, have been found in the literature. On the one hand, in the work by Ortega-Casanova (2017b) a numerical study is carried out to characterize the onset of vortex shedding from a confined square pillar for a given channel blockage ratio as function of the cylinder aspect ratio and Reynolds number. Moreover, it is also shown how the vortex shedding can help to increase mixing efficiency for a given Reynolds number and different cylinder aspect ratios. On the other hand, in the works by Rafferty and Kaminski (2014) and Nahum and Seifert (2006), twin parallel jets issuing into a confined domain are studied and flow transitions to different regimes characterized. They also assessed the mixing efficiency under different geometry configurations and Reynolds numbers. In particular, Rafferty and Kaminsky (2014) found different critical Reynolds numbers that separates different flow regimes. In the range of studied Reynolds numbers, as it increases, the flow evolves from a steady to an unsteady regime, then becomes again steady and finally again unsteady. This behavior is worth to mention since, as we shall see later, a similar behavior has been found here. Due to all this, the main aim of this work is to shed some light on the interaction of blowing and vortex shedding for mixing fluids purposes and help to increase the number of works on this topic.

To conclude this first Section, the structure of this Chapter is introduced: after this Introduction, the geometry under consideration, together with the governing equations and the introduction of some parameters, are discussed in second Section. Next, the third Section is devoted to present some numerical aspects of the simulations we have carried out, such as the numerical methods we have used, the implemented boundary conditions and a validation of the used grid and followed methodology. Later, all the results are presented and discussed in fourth Section, where different subsections are used to clearly present and compare all results. To conclude, in fifth Section are summarized all conclusions of the work.

GEOMETRY AND GOVERNING EQUATIONS

This numerical work focuses on the analysis of the efficiency of a two-dimensional passive micromixer device, designed to induce the highest interaction between the two fluids to be mixed. Briefly, the device consists (see Figure 1) of a straight channel, H meter width and L meters long, with an inner and fixed square pillar of side W located at constant distance from the inlet. One of the fluids to be mixed (named as fluid 1 in what follows) enters the device though the main inlet section (with mean velocity U_i) while the other (called as fluid 2 hereafter) is injecting by means of two small jets (of height W_j and mean velocity U_j) from both corners of the upstream face of the inner square pillar and at a certain known angle. It will be shown that, for certain values of the governing parameters, vortex shedding from the pillar appears and the oscillatory wake downstream is supposed to increase the interaction and contact between both fluids, and therefore will increase the efficiency of the mixing.

The problem to be tackled in this study is governed, on the one hand, by the continuity and Navier-Stokes equations, which contemplate the physical principles that describe the motion of a Newtonian incompressible fluid and, on the other hand, by the advection-diffusion mass conservation equation. The energy conservation equation is not taken into account since it is assumed that the entire domain is at the same temperature so there are not chemical reactions, heat contributions and friction does not dissipate energy. These three governing equations, in their dimensionless form, can be written as:

$$\nabla \cdot \vec{v} = 0 \tag{1}$$

$$\frac{\partial \vec{v}}{\partial t} + \vec{v} \cdot \nabla \vec{v} = -\nabla p + \frac{1}{\mathrm{Re}_i} \nabla^2 \vec{v} \tag{2}$$

$$\frac{\partial Y}{\partial t} + \vec{v} \cdot \nabla Y = \frac{1}{Pe} \nabla^2 Y \tag{3}$$

where v is the velocity vector, t is the time, p the pressure, Y the solute mass fraction, $Re_i = U_i W/\upsilon$ is the Reynolds number, $Pe = U_i W/D$ is the Péclet number, with υ and D being the kinematic viscosity and mass diffusion coefficient of the fluid, respectively (ρ would be the fluid density). The flow will take place at low Reynolds and high Péclet numbers, which implies that the device must be very long for good diffusion-based mixing. This is a practical limitation in microfluidic chips, which can be solved out by means of promoting vortex shedding. Additionally, the Péclet number can be written as $Pe = Rei.Sc$, where Sc is the Schmidt number defined as $Sc = \upsilon/D$. While Pe and Re_i are both fluid properties and geometry dimensions dependent, Sc is only fluid dependent being the ratio between two molecular diffusion mechanisms: the ones of momentum and mass. If the fluid properties are constant, Sc will be also constant even if the boundary conditions or geometry dimensions change, while Re_i and Pe and will change. For that reason, the governing parameters we have used are the Reynolds and Schmidt numbers, but not the Péclet number. To get the dimensionless form of the governing equations (1)-(3), we have used U_i, W, U_i/W, ρU_i^2 and C_0 as characteristic velocity, length, time, pressure and solute molar concentration.

A sketch of the geometry under study can be seen in Figure 1, while a detailed view of the upstream region is shown in Figure 2. In terms of the characteristic length W, it consists in a channel of length $L = 19W$ and height $H = 2W$, which represents a channel blockage ratio of $W/H = 0.5$, i.e. of 50%, with the square pillar located at $5W$ downstream the inlet. At the channel inlet, one of the fluid (fluid 1) enters with a fully developed parabolic velocity profile (with mean velocity U_i), while the fluid injected from the pillar (fluid 2) has a uniform velocity profile (with mean velocity U_j), being $W_j = W/10$ the width of the injectors. Both fluids will be mixed downstream the channel in one of these two ways: with a steady and symmetric wake downstream the pillar, so the mixing is supposed to be very poor; or with an oscillatory wake due to the vortex shedding from the cylinder, so the mixing is supposed to be better, as it will be shown later.

In this study, both fluids are the same liquid, specifically water liquid, and therefore have identical properties of density (ρ = 1000 kg/m^3), dynamic viscosity ($\mu = 10^{-3}$ kg/m-s), kinematic viscosity ($\upsilon = 10^{-6}$ m^2/s) and mass diffusion coefficient ($D = 10^{-10}$ m^2/s), but have different molar solute concentrations at their corresponding inlets. These values give a constant Schmidt number of $Sc = 10^4$, while the Reynolds number will depend on the used inlet average velocity U_i. In particular, three different values of the Reynolds number have been used: $\mathrm{Re}_i = 30, 70$ and 100. On the other hand, the mean velocity of the two jets U_j has been selected by giving values to the intensity ratio I defined as:

Figure 2. Zoom of geometry of the system to analyze

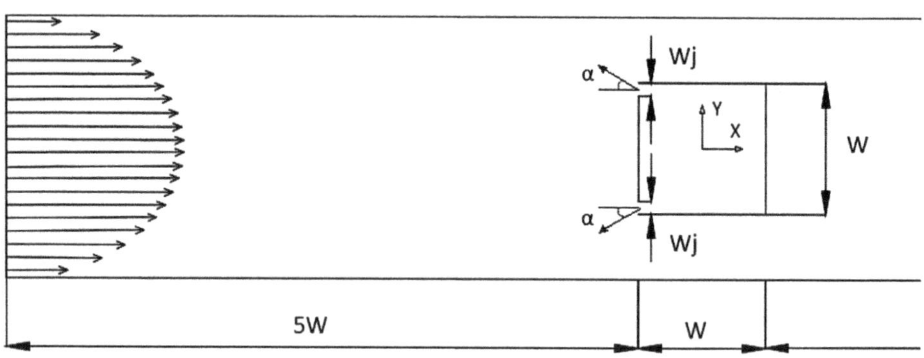

$$I = \frac{\mathrm{Re}_j}{\mathrm{Re}_i}$$

where Re_j is the jets Reynolds number: $\mathrm{Re}_j = U_j W_j / v$

Five different intensity ratios have been used: $I = 0,1/10,1/5,1/2.5$ and $1/1.5$. Another parameter related with the jets is the injecting angle α. It is the angle that establishes the direction in which the jets are injected. Its value is measured from the horizontal toward the walls of the channel. Two different values have been studied: $\alpha = 0°$ and $\alpha = 45°$.

A summary of the different parameter combinations can be seen in Table 1 where, as can be seen, a total of 27 different combinations have been considered.

There are different magnitudes that allow us to assess the quality and goodness of a mixing at the outlet of the microdevice. The main one is the standard deviation of the mass fraction of one of the fluids (σ) at the device exit defined as:

$$\sigma = \sqrt[2]{\frac{\sum_{k=1}^{n}\left(Y_k - \bar{Y}\right)^2}{n}} \quad (7)$$

Table 1. Analyzed cases in this project

Intensity Ratio	Injection Angle	$\mathrm{Re}_i = 30$	$\mathrm{Re}_i = 70$	$\mathrm{Re}_i = 100$
$I = 0$	-	✓	✓	✓
$I = 1/10$	$\alpha = 0°$	✓	✓	✓
	$\alpha = 45°$	✓	✓	✓
$I = 1/5$	$\alpha = 0°$	✓	✓	✓
	$\alpha = 45°$	✓	✓	✓
$I = 1/2.5$	$\alpha = 0°$	✓	✓	✓
	$\alpha = 45°$	✓	✓	✓
$I = 1/1.5$	$\alpha = 0°$	✓	✓	✓
	$\alpha = 45°$	✓	✓	✓

where k is the discretization cells counter along the width at the device exit, where there are n of them.

In this way, the standard deviation at the channel exit can take values between zero and one half, i.e. $0 \le \sigma \le 0.5$, with the perfect mixing corresponding to $\sigma = 0$ and no mixing at all taking place when $\sigma = 0.5$.

Once the standard deviation value at the outlet is measured, it can help to calculate the mixing efficiency according to (Sarkar, Singh, Shankar, & Shenoy, 2014):

$$\eta = \left(1 - \frac{\sigma}{\sigma_{\max}}\right) \cdot 100 \tag{8}$$

where $\sigma_{\max}$ is the maximum deviation of the mass fraction measured at the three geometry inlets. Additionally, the mixing efficiency defined in (8) can reach values between 0%, if mixing is non-existent, and 100%, if mixing is full and totally homogeneous, i.e. $0\ \% \le \eta \le 100\%$.

Apart from standard deviation and mixing efficiency at the device outlet, it is interesting to know the input power (P) consumed by the system to achieve the corresponding mixing. In that sense, devices giving high mixing efficiencies with low input power needs are desired. To evaluate the total input power, one has to take into account not only the power to generate the main stream, but also the input power to inject the two small jets from the inner pillar. This input power can be calculated according to the expression:

$$P = \bar{P}_i q_i + 2\bar{P}_j q_j \tag{9}$$

where $\bar{P}$ and q are the dimensionless time-averaged pressure and mass flow rate, respectively, at the channel inlet (sub-index i) and each small jet inlet (sub-index j).

Both η and P will help us to find the better device configuration in order to get the highest efficiency or lowest input power.

Additionally to these mixing parameters, two aerodynamic ones will be also calculated and shown when discussing the results. They are the drag and lift coefficient of the square pillar, C_D and C_L, respectively. They will also help us to validate the followed methodology by comparing the obtained results with those reported in literature. Both coefficients are defined as:

$$C_D = \frac{F_x}{\rho U_i^2 W} \text{ and } C_L = \frac{F_y}{\rho U_i^2 W} \tag{10}$$

where F_x and F_y are the horizontal and vertical forces, respectively.

NUMERICAL CONSIDERATIONS

This Section is devoted to present different numerical issues, such as the numerical methods and boundary conditions used in the simulations and the methodology validation, which are important to take into account in order to be confident with the results that will be shown in fifth Section.

Numerical Methods

The set of dimensionless equations (1)-(3) with the corresponding boundary conditions (see below) have been solved numerically by using the commercial software ANSYS-Fluent© by means of the pressure-based formulation with the pressure-velocity coupling handled by the 'Coupled' method, which solves the momentum and pressure-based continuity equations simultaneously. With respect to the spatial discretization, the pressure equation was solved with the 'Second Order' method, while the momentum and mass fraction equations with the 'Second Order Upwind' scheme.

Boundary Conditions

The following boundary conditions have been imposed:

- The walls of the channel and the outer wall of the square pillar have been considered as stationary walls with no slip;
- At the entrance of the channel, a velocity inlet boundary condition has been considered, with a horizontal and fully developed parabolic velocity profile with average velocity U_i;
- The injections from the inner pillar have been considered also as velocity inlet with constant velocity U_j but with adjustable magnitude and direction, depending on the intensity ratio and injection angle; and
- The outlet of the channel has been considered as a pressure outlet boundary condition, with null pressure difference with respect to the atmospheric one.

Validation

The followed methodology to get the present results has been previously validated by reproducing previous results from different authors. Two kind of validations have been carried out: the first one, from an aerodynamic point of view, to validate the drag and lift coefficients for a blockage ratio of 1/2 and different Reynolds numbers with those by Sharma and Eswaran (2004) and Suzuki et al. (1993); and the second one, from a mixing point of view, by reproducing the mixing results by Ortega-Casanova (2017b) for a blockage ratio of 1/5, inner cylinder aspect ratio of 1/2, Reynolds number of 40 and the same Schmidt number of 10^4.

On the one hand, and regarding the aerodynamic validation, Figure 3 shows the comparison between the drag and lift coefficients, and Strouhal number from Sharma and Eswaran (2004), Suzuki et al. (1993) and present ones for the same channel blockage ratio and three different Reynolds numbers: one corresponding to a steady flow and the other two to unsteady ones because of the vortex shedding from square pillar. As can be seen, the agreement between the present and previous results is quite good.

Figure 3. Comparison of aerodynamic parameters from this study and previous ones: (a) Drag coefficient; (b) Root mean square of the lift coefficient, and; (c) Strouhal number

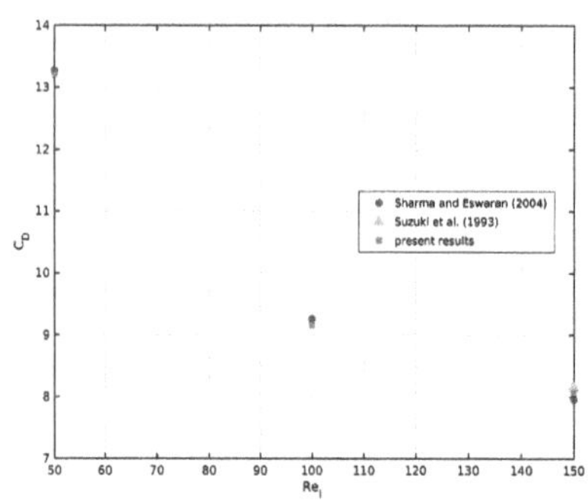

(a)

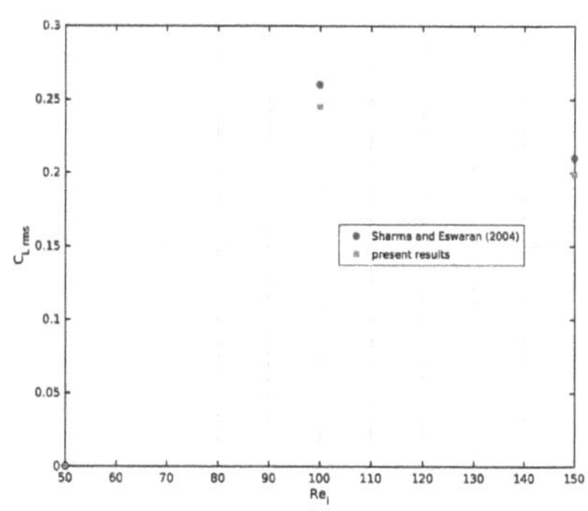

(b)

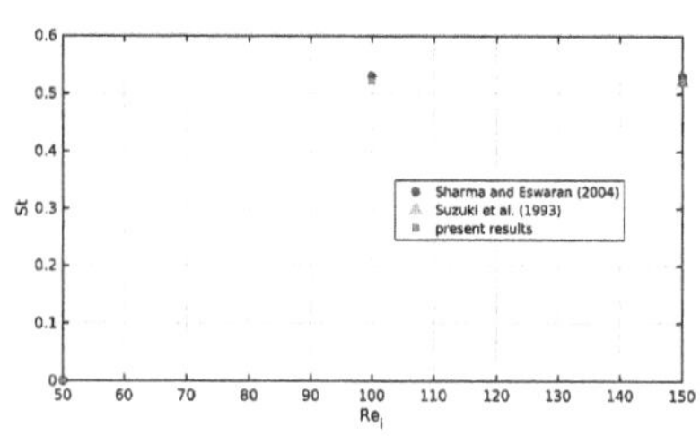

(c)

On the other hand, Table 2 shows, in terms of the mixing efficiency, the comparison between present and previous results. As can be seen, the agreement is very good since a small deviation of less than 1.5% is obtained. This has also allowed us to determine the size of the grid to be used in the simulations.

RESULTS

This section is devoted to present and discuss all obtained results. It will be organized as follows. Different subsections will show the results for each constant Reynolds number, where figures showing isosurfaces of mass fraction and streamlines will help us to understand and discuss the results. Later, in a final subsection, all results, in terms of mixing efficiency and input power, will be compared.

Results for $\mathrm{Re}_i = 30$

In Table 3 are summarized both the different input parameter combinations and the values reached by the main output magnitudes for all cases in which the Reynolds number of the flow at the inlet of the channel is $\mathrm{Re}_i = 30$.

As can be seen, at this lowest Reynolds number, all analyzed configurations reach a steady state regime. In particular, when $I = 0$ and fluid 2 is not being injected, results are in agreement with Sharma and Eswaran (2004), where also a steady state was found for Reynolds number below ~ 60 with a similar geometry. Consequently, always that $\mathrm{Re}_i = 30$, the flow response at the outlet of the channel is constant in time, regardless the intensity ratio, being symmetrical the flow pattern in the geometry, as can be seen in Figure 4, where the streamlines for an injection angle of 0° and all intensity ratios are depicted (quite similar results are obtained for an injection angle of 45° and are not shown). Because of the flow symmetry, the lift coefficient is null and vortex shedding from the square pillar is not observed: at a short distance downstream from the pillar, the streamlines are parallel and horizontal. From Table 2, one can also observed that for a constant injection angle, the drag coefficient increases when I increases because the mass flow rate above and below

Table 2. Mixing efficiency validation with previous results

	Ortega-Casanova (2017b)	**Present Results**
Mixing efficiency η [%]	12.8	12.95

Table 3. Summary results for $\mathrm{Re}_i = 30$

Configuration	Temporal Response	C_D	C_L	η [%]	P [-]
$I = 0$	Stationary	15.537	0	-	15.8160
$I = 1/10, \alpha = 0°$	Stationary	17.628	0	51.7142	20.1528
$I = 1/10, \alpha = 45°$	Stationary	16.906	0	61.4386	19.2913
$I = 1/5, \alpha = 0°$	Stationary	19.929	0	31.0003	25.4042
$I = 1/5, \alpha = 45°$	Stationary	18.313	0	42.0019	23.4950
$I = 1/2.5, \alpha = 0°$	Stationary	25.492	0	10.1805	38.4509
$I = 1/2.5, \alpha = 45°$	Stationary	21.300	0	21.2647	33.8834
$I = 1/1.5, \alpha = 0°$	Stationary	35.985	0	1.9983	62.0134
$I = 1/1.5, \alpha = 45°$	Stationary	26.393	0	7.7809	53.4216

the square pillar also increases, which creates larger separation regions that reduces the area through which fluid 1 flows and this results in an increase of the pressure at the front face of the pillar and, subsequently, C_D also increases. Additionally, for a constant intensity ratio, the drag coefficient decreases when the injection angle increases from 0 to 45° because at the lowest injection angle the fluid 2 is forced to flow far from the square pillar side surface than for a 45° angle; this results in a higher recirculation area above and below the pillar, and again the area through which Fluid 1 flows is reduced and resulting in an increase of the pressure at the front face of the pillar and, subsequently, also of C_D. From the streamlines showed in Figure 4 another interesting feature of the flow at this Reynolds number can be observed: regardless the injection angle, the length of the wake just downstream the pillar increases when the intensity ratio does. This is due to the fact that local upper and lower velocities at the end of the square increase when I increases. As consequence, the inertia of the fluid leaving the square increases and so does the length of the wake. Similar behavior, with the size of the recirculation bubble increasing when the Reynolds number increases from 100 to 150, has been observed by Sharma and Eswaran (2004) for a similar blockage ratio.

Regarding the mixing process at this lowest Reynolds number, Figure 5 depicts the mass fraction contours for an injection angle of 0º (quite similar results are obtained for an injection angle of 45º and are not shown). As can be again seen, the mass fraction pattern downstream is symmetrical and steady with low interaction between fluid 1 and fluid 2 as they flow along the device. This is consequence of the lack of vortex shedding from the inner pillar, which clearly shows that the mixing efficiency is relatively low at this lowest Reynolds number. Additionally, Figure 6 depicts the evolution of the mixing efficiency as the intensity ratio changes. When the latter decreases, the efficiency increases, regardless the injection angle because of two complementary effects: the fluid 1 passes through a larger area that reduces its local velocities and therefore its local Péclet numbers; and additionally, there is less amount of fluid 2 being injected, which also results in lower local velocities and therefore lower local Péclet numbers. It can be also seen in Figure 6 that the efficiency also reaches better results for configurations with injection angles of 45º. As stated above, a larger injection angle induces a smaller separation region above and below the pillar and therefore, the fluid 1 moves through a larger effective area, resulting in lower local velocities and lower local Péclet numbers. As a consequence of this, the efficiency increases when the injection angle is larger. In particular, the highest mixing efficiency is around 62%, which is relatively low and is obtained when $I = 1/10$ and $\alpha = 45\,^{\circ}$, the configuration for the largest local Péclet number.

To conclude, Figure 7 shows how P evolves as function of I for both injection angles. As can be seen, when I increases, the input power also increases because of the combination of two effects: on the one hand, more fluid is being injected through the jets, which needs more power; and, on the other hand, when I increases, the effective area for the fluid 1 to pass through decreases, resulting in a higher pressure drop and therefore more power is needed to keep constant the mass flow of fluid 1. Regarding the injection angle, the input power when $\alpha = 0\,^{\circ}$ is higher also due to two contributions: on one side, fluid 2 is being injected completely against fluid-1 direction, which requires higher power to drag it downstream; and on the other side, smaller angle results in a larger separation region, reducing the area through which fluid 1 passes through, what increases the pressure drop and therefore the input power.

Results for $\mathrm{Re}_i = 70$

In Table 4 are summarized the main output magnitudes reached for the cases with $Re_i = 70$. According with Sharma and Eswaran (2004), for the same blockage ratio and Reynolds number, the flow when $I = 0$, i.e. without injecting fluid 2, must be oscillatory with vortexes being shed from the inner pillar and with fluid properties

Figure 4. Streamlines for $\mathrm{Re}_i = 30$ *and* $\alpha = 0°$*: (a)* $I = 1/10$*; (b)* $I = 1/5$*; (c)* $I = 1/2.5$*; and (d)* $I = 1/1.5$

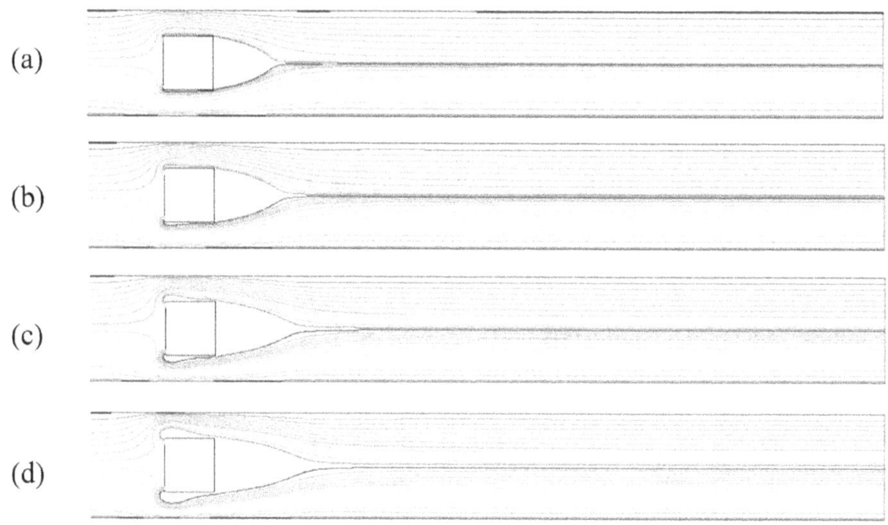

Figure 5. Mass fraction contours for $\mathrm{Re}_i = 30$ *and* $\alpha = 0°$*: (a)* $I = 1/10$*; (b)* $I = 1/5$*; (c)* $I = 1/2.5$*; and (d)* $I = 1/1.5$*. In (e) is shown the mass fraction colorbar.*

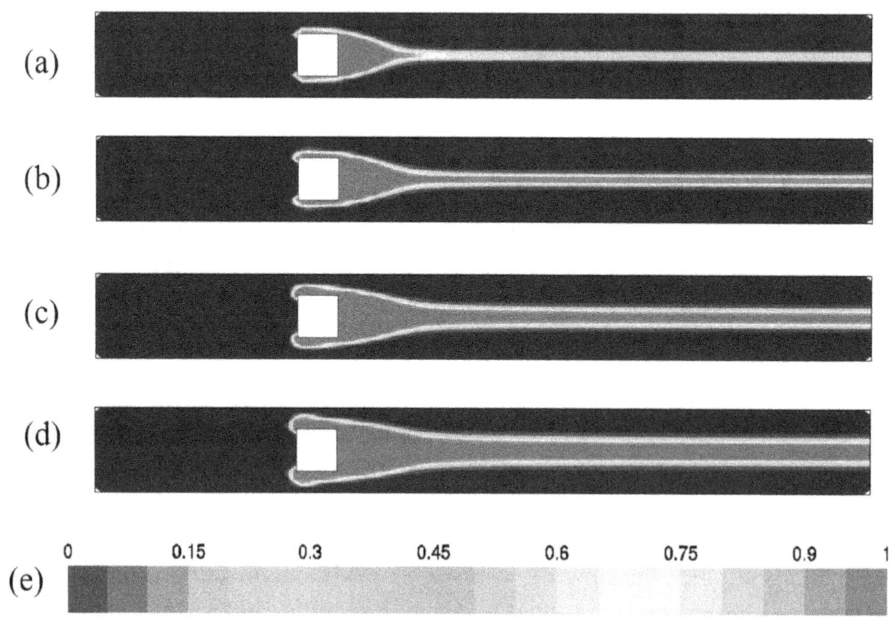

**For a more accurate representation see the electronic version.*

Figure 6. Mixing efficiency vs. the intensity ratio at the channel exit for $\mathrm{Re}_i = 30$ *and the indicated injection angles*

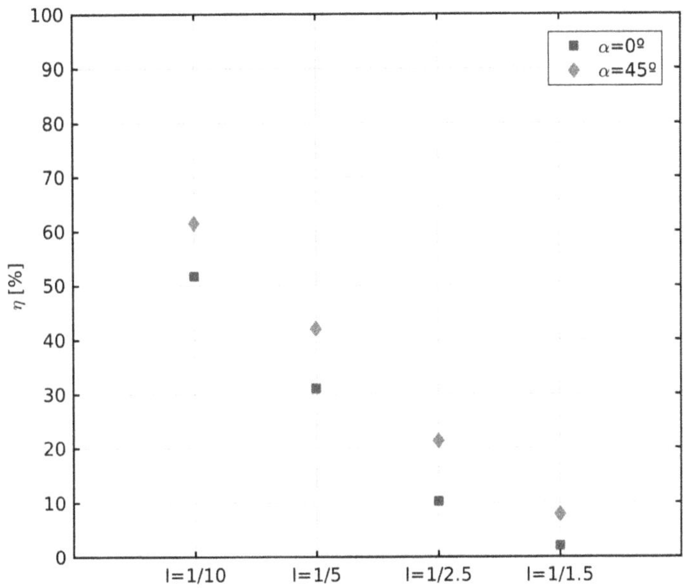

Figure 7. Input power vs. the intensity ratio at the channel exit for $\mathrm{Re}_i = 30$ *and the indicated injection angles*

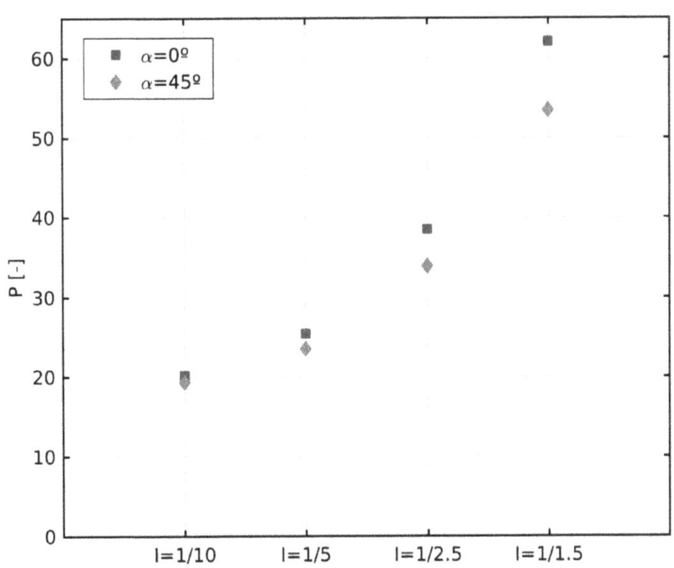

changing with time. In effect, as can be seen in Table 4, when $I = 0$ all properties are oscillatory, with maximum and minimum parameter values indicated between brackets and in bold its time-averaged value. Additionally, the numerically obtained drag coefficient when $I = 0$ and $\mathrm{Re}_i = 70$, $C_D \cong 9.5$, is between the values given by Sharma and Eswaran (2004) for $\mathrm{Re}_i = 50$, with $C_D \cong 13$, and $\mathrm{Re}_i = 100$, with $C_D \cong 9$.

At this Reynolds number, all configurations have achieved oscillatory response in time, except the case with $I = 1/1.5$ and $\alpha = 0°$ that will be discussed later. This oscillatory behavior is due to the vortex shedding which promotes an oscillatory wake downstream the inner pillar, as the streamlines depicted in Figure 8 and Figure

Table 4. Summary of the results when $\mathrm{Re}_i = 70$. *Between square brackets the minimum and maximum values of the temporal evolution of the corresponding property and in bold its time-averaged value.*

Configuration	Response	C_D	C_L	η [%]	P [-]
$I = 0$	Oscillatory	[9.4846, 9.4848] **9.4847**	[-0.0426, 0.0418] **0.0**	-	8.6044
$I = 1/10, \alpha = 0°$	Oscillatory	[11.1922, 11.1944] **11.1933**	[-0.2317, 0.2317] **0.0**	78.5130	11.3608
$I = 1/10, \alpha = 45°$	Oscillatory	[10.5982, 10.6000] **10.5991**	[-0.2021, 0.2021] **0.0**	83.4833	10.7600
$I = 1/5, \alpha = 0°$	Oscillatory	[13.3997, 13.4034] **13.4015**	[-0.3352, 0.3352] **0.0**	60.7028	14.9376
$I = 1/5, \alpha = 45°$	Oscillatory	[11.8751, 11.8780] **11.8766**	[-0.2836, 0.2836] **0.0**	70.5645	13.4281
$I = 1/2.5, \alpha = 0°$	Oscillatory	[19.7453, 19.7504] **19.7478**	[-0.4863, 0.4863] **0.0122**	47.8896	25.2599
$I = 1/2.5, \alpha = 45°$	Oscillatory	[15.1965, 15.1988] **19.1942**	[-0.4058, 0.4058] **0.0**	52.2680	21.0351
$I = 1/1.5, \alpha = 0°$	Stationary	**37.2807**	**0.1629**	3.2660	50.8525
$I = 1/1.5, \alpha = 45°$	Oscillatory	[24.2617, 24.2628] **24.2622**	[-0.2348, 0.2348] **0.0**	11.0126	39.5701

9 for $\alpha = 0°$ and 45°, respectively, show (when the flow is oscillatory the contours shown hereinafter correspond to a time instant once the wake reached a periodic and repetitive behavior). However, when $\alpha = 0°$ and the intensity ratio increases, there is a critical value of I above which the flow eventually becomes steady, as the constant values of the coefficient and the streamlines reveal, see Figure 8(d), and with an asymmetric wake downstream the pillar. This transition, from oscillatory to steady solutions, has been also previously observed by Rafferty and Kaminsky (2014), in a configuration quite similar to the one here studied, when the Reynolds number increases. On the contrary, the transition to a steady solution is not observed when $\alpha = 45°$ (see Figure 9), which means that the critical intensity ratio is higher than when $\alpha = 0°$ and thus depends on the injection angle. Regarding the aerodynamic coefficients C_D and C_L, both show an oscillatory behavior, except for the steady case, with C_D increasing with the increase of the intensity ratio, as happened when $\mathrm{Re}_i = 30$, and being the drag coefficient smaller when $\alpha = 45°$ than when $\alpha = 0$ °, as also happened when $\mathrm{Re}_i = 30$. The time-averaged value of C_L is always close to zero, except for the steady case, for which it is positive, as one could expect from the asymmetric streamline pattern shown in Figure 8(d).

When vortex shedding takes place, the mixing efficiency can be increased thanks to the waving of the wake downstream the pillar. As a consequence of this waving, the surface contact between fluid 1 and fluid 2 increases and, consequently, the mixing efficiency is supposed to increase, too. The pattern of the fluid mass fractions can be seen for all intensity ratios in Figure 10, for $\alpha = 0°$, and Figure 11, for $\alpha = 45°$. Except for the steady case, Figure 10(d), all other contours clearly show how the vortex shedding helps to increase the interaction between the fluids to mix.

The effect of the vortex shedding on the mixing efficiency can be seen in Figure 12, where it is plotted as function of the intensity ratio for the different injection angles. As one can see, the efficiency is a little higher when $\alpha = 45°$ and decreases when the intensity ratio increases, corresponding the lowest efficiency to the stationary configuration with $\alpha = 0°$ and $I = 1/1.5$. As previously commented, the increase of the intensity ratio tends to suppress the downstream flow oscillations, becoming time independent above a critical intensity ratio. As a consequence, the increase of the intensity ratio gives as result a decrease of the mixing efficiency since the vortex shedding also decreases until it vanishes for the steady configuration. Due to this behavior, the highest efficiency corresponds to the lowest intensity ratio since it hardly affects the vortex shedding. This, together with the small amount of fluid 2 being injected, makes this configuration with $I = 1/10$, regardless the injection angle, the best one from a mixing point of view, with efficiencies of around 80%, see Figure 10(a) and Figure 11(a).

Figure 8. Streamlines for $\mathrm{Re}_i = 70$ *and* $\alpha = 0°$*: (a)* $I = 1/10$*; (b)* $I = 1/5$*; (c)* $I = 1/2.5$*; and (d)* $I = 1/1.5$

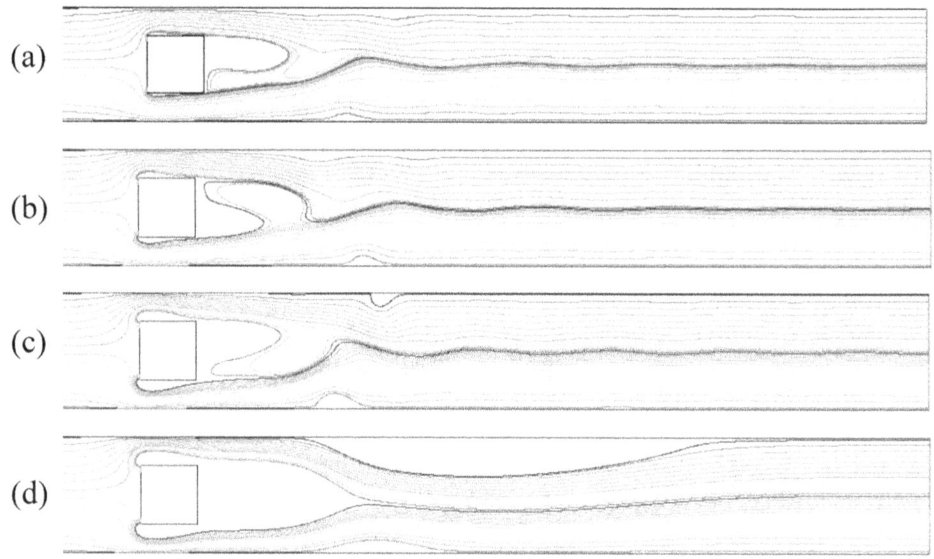

*For a more accurate representation see the electronic version.

Figure 9. Streamlines for $\mathrm{Re}_i = 70$ *and* $\alpha = 45°$*: (a)* $I = 1/10$*; (b)* $I = 1/5$*; (c)* $I = 1/2.5$*; and (d)* $I = 1/1.5$

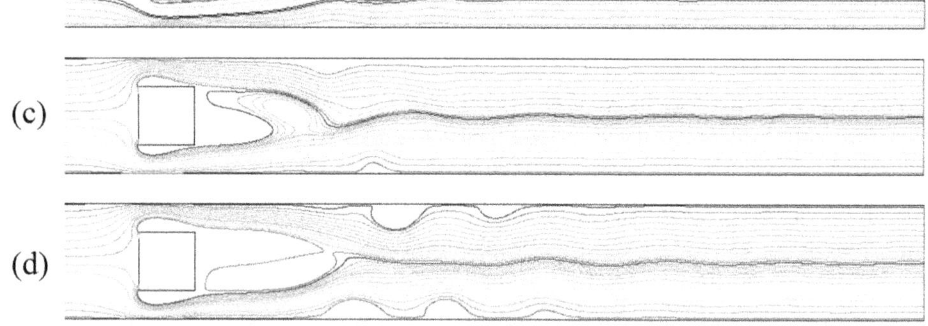

*For a more accurate representation see the electronic version.

Figure 10. Mass fraction contours for $\mathrm{Re}_i = 70$ *and* $\alpha = 0°$*: (a)* $I = 1/10$*; (b)* $I = 1/5$*; (c)* $I = 1/2.5$*; and (d)* $I = 1/1.5$*. For the color scale see Figure 5(e)*

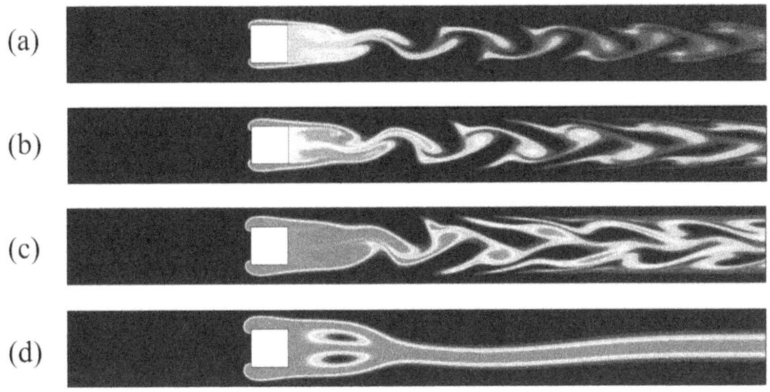

**For a more accurate representation see the electronic version.*

Figure 11. Mass fraction contours for $\mathrm{Re}_i = 70$ *and* $\alpha = 45°$*: (a)* $I = 1/10$*; (b)* $I = 1/5$*; (c)* $I = 1/2.5$*; and (d)* $I = 1/1.5$*. For the color scale see Figure 5(e)*

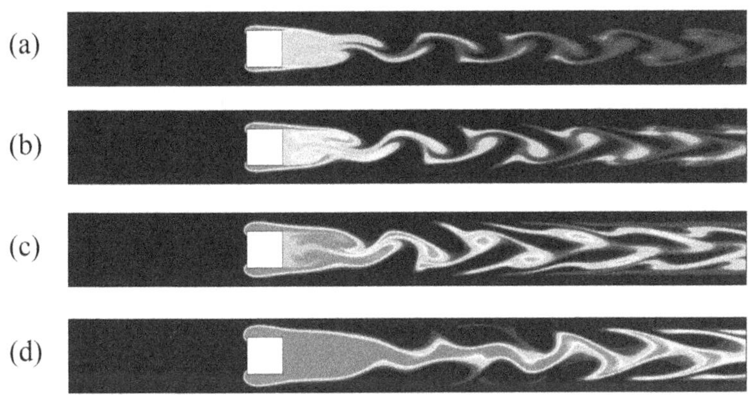

**For a more accurate representation see the electronic version.*

Regarding the input power, Figure 13 shows the amount of input power needed by each configuration. As can be seen, it increases as the intensity ratio increases, regardless the injection angle, being always, and for the same reasons than when $\mathrm{Re}_i = 30$, higher the power needs when $\alpha = 0°$.

Results for $\mathrm{Re}_i = 100$

At this higher Reynolds number, a similar behavior than the previous lower one has been observed. A summary of all output parameters can be seen in Table 5, where

Figure 12. Mixing efficiency vs. the intensity ratio for $\mathrm{Re}_i = 70$ *and the indicated injection angles*

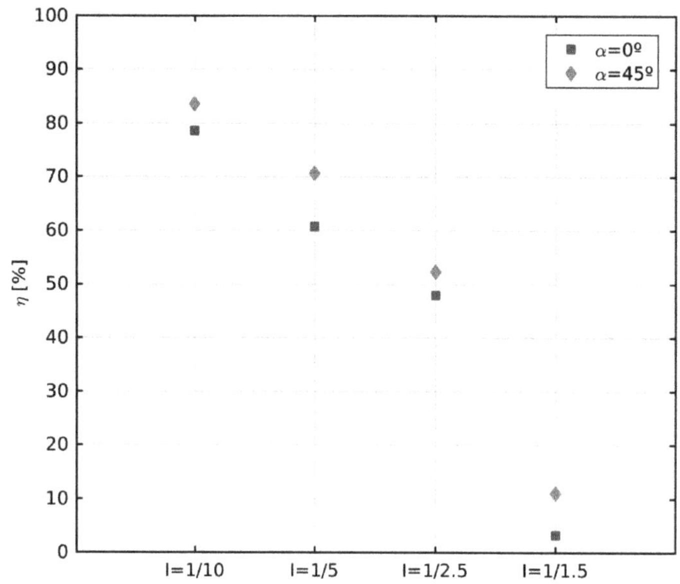

Figure 13. Input power versus the intensity ratio for $\mathrm{Re}_i = 70$ *and the indicated injection angles*

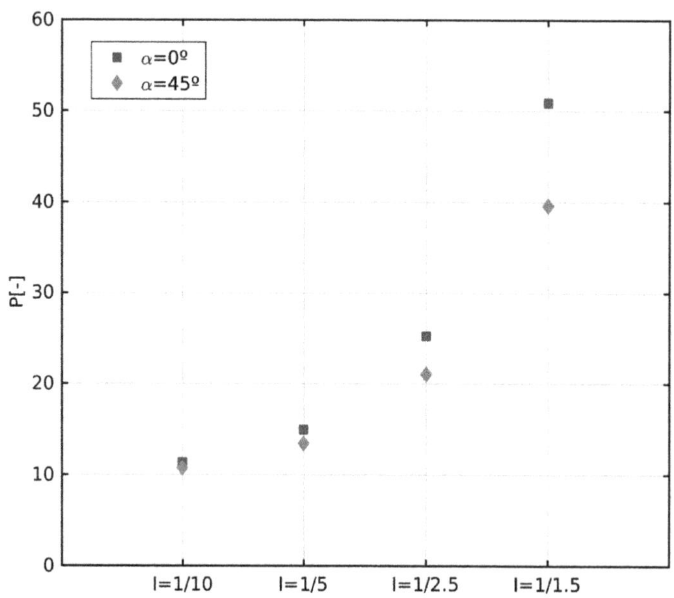

one can see that the highest intensity ratios are above a critical value and stationary solutions are reached, i.e. without vortex shedding, not only for $\alpha = 0°$, as when $\mathrm{Re}_i = 70$, but also for $\alpha = 45°$. Briefly, regarding the aerodynamic coefficients the drag coefficient increases with the increase of the intensity ratio and is always higher at $\alpha = 0°$ than at 45° due to the reasons previously explained, while the lift coefficient has always null average except for those configurations where a final steady state is reached, for which the average value can be either null, if the wake is symmetric, or not, if the wake is asymmetric.

As previously mentioned, for intensity ratios above a critical value, which is different depending on the injection angle, the flow is steady as can be seen from the aerodynamic values from Table 5 and also from Figure 14, for $\alpha = 0°$, and Figure 15, for $\alpha = 45°$, where the streamlines for all configurations are depicted. In effect, for $\alpha = 0°$ and $I < 1/2.5$, see Figure 14(a)-(b), the flow is oscillatory with vortex shedding from the inner pillar and being thus the wake oscillatory, too. However, for $I = 1/2.5$, see Figure 14(c), the wake is steady and symmetrical, while for $I = 1/1.5$, see Figure 14(d), it is also steady but asymmetrical. This means there is a new transition from symmetrical to asymmetrical wake downstream depending on the value of the intensity ratio when $I \geq 1/2.5$. On the other hand, for $\alpha = 45°$ and $I < 1/1.5$, see Figure 15(a)-(c), the flow is unsteady and oscillatory due to the vortex shedding, while for $I = 1/1.5$, see Figure 15(d), the flow is steady and asymmetric. In this case, the symmetric solution has not been observed and it is supposed to occur for intensity ratios between 1/2.5 and 1/1.5. Rafferty and Kaminsky (2014) also detected the transition from symmetric to asymmetric solutions depending on the Reynolds number of the flow.

Regarding the mixing efficiency, Figure 16 and Figure 17 show, for $\alpha = 0°$ and 45°, respectively, the mass fraction contours for all intensity ratios. As can be seen, in those configurations with vortex shedding and oscillatory wake, the high fluids interaction gives high values of mixing efficiency, see Figure 18, while those steady configurations give poor values of efficiency of around 20%. It is remarkable that the worst mixing efficiency, of around 12%, corresponds to the case with symmetric wake, see Figure 16(c), which also corresponds to the case with lowest fluid residence time, being the efficiency higher in those configurations with asymmetric wake, in which, thanks to the wake curvature, the fluids residence time is a little higher.

Regarding the input power to mix the fluids, Figure 19 depicts its evolution as function of the intensity ratio for both injection angles. The same behavior, and due to the same reasons than when $\mathrm{Re}_i = 30$, applies now: the input power increases when the intensity ratio increases and it is higher when fluid-2 jets are injected completely against the incoming fluid 1, i.e. when $\alpha = 0°$, than when they are at 45°.

Table 5. Summary of the results when $\mathrm{Re}_i = 100$. *Between square brackets the minimum and maximum values of the temporal evolution of the corresponding property and in bold its time-averaged value.*

Configuration	Response	C_D [-]	C_L [-]	η [%]	P [-]
$I = 0$	Oscillatory	[9.149, -9.155] **9.152**	[+0.3464, -0.3464] **0.002**	-	6.8643
$I = 1/10, \alpha = 0°$	Oscillatory	[9.9237, 9.9181] **9.9210**	[+0.3299, -0.3299] **0.0021**	84.1055	9.3782
$I = 1/10, \alpha = 45°$	Oscillatory	[9.2894, 9.2843] **9.2868**	[+0.3287, -0.3287] **-0.0079**	88.0739	8.7518
$I = 1/5, \alpha = 0°$	Oscillatory	[12.3386, 12.3333] **12.3359**	[+0.3295, -0.3295] **-0.0001**	71.2939	12.9950
$I = 1/5, \alpha = 45°$	Oscillatory	[10.6819, 10.6776] **10.6797**	[+0.3267, -0.3267] **-0.0021**	78.6613	11.3827
$I = 1/2.5, \alpha = 0°$	Stationary	**20.6842**	**0.0009**	14.3847	24.3096
$I = 1/2.5, \alpha = 45°$	Oscillatory	[15.1016, 15.1003] **15.1010**	[+0.1929, -0.1929] **-0.0011**	57.8043	19.5093
$I = 1/1.5, \alpha = 0°$	Stationary	**44.9811**	**0.3432**	17.1689	63.3281
$I = 1/1.5, \alpha = 45°$	Stationary	**23.4357**	**-2.7848**	25.7768	37.8780

Result Comparisons

Once all cases have been analyzed, it is possible to compare both the mixing efficiency, in Figure 20, and input power, in Figure 21, as function of the intensity ratio. The tendency of both parameters is, in general, repetitive, regardless the Reynolds number and the injection angle: the mixing efficiency decreases and the input power increases when the intensity ratio increases. In addition, as a general rule, and regardless the intensity ratio, the best results are obtained for $\mathrm{Re}_i = 100$ and injection angles of $\alpha = 45$ °. In effect, for any value of I, the efficiency is always the highest and the input power the lowest when other two governing parameters are $\mathrm{Re}_i = 100$, $\alpha = 45$ °. Additionally, it is worth to say that for given

Figure 14. Streamlines for $\mathrm{Re}_i = 100$ *and* $\alpha = 0\,^o$*: (a)* $I = 1/10$*; (b)* $I = 1/5$*; (c)* $I = 1/2.5$*; and (d)* $I = 1/1.5$

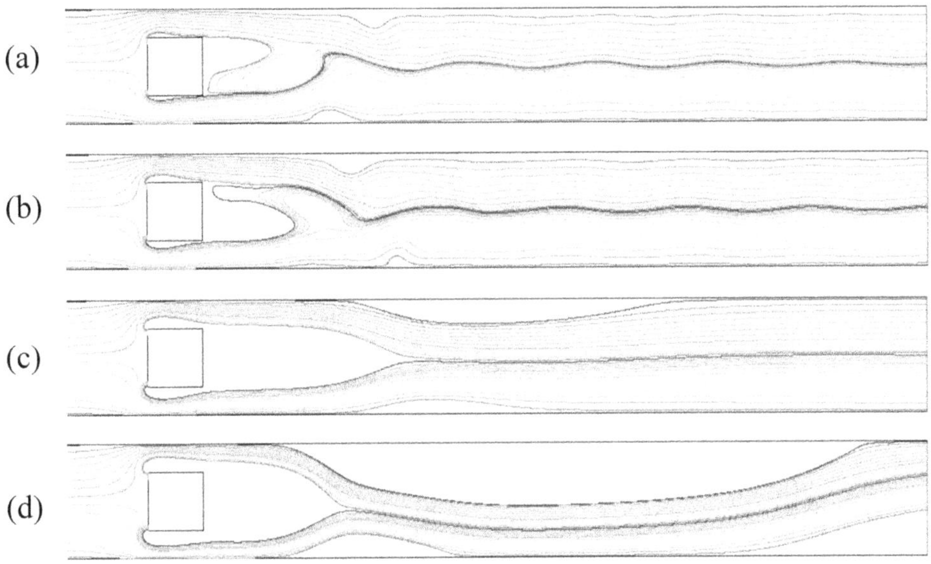

**For a more accurate representation see the electronic version.*

Figure 15. Streamlines for $\mathrm{Re}_i = 100$ *and* $\alpha = 45\,^o$*: (a)* $I = 1/10$*; (b)* $I = 1/5$*; (c)* $I = 1/2.5$*; and (d)* $I = 1/1.5$

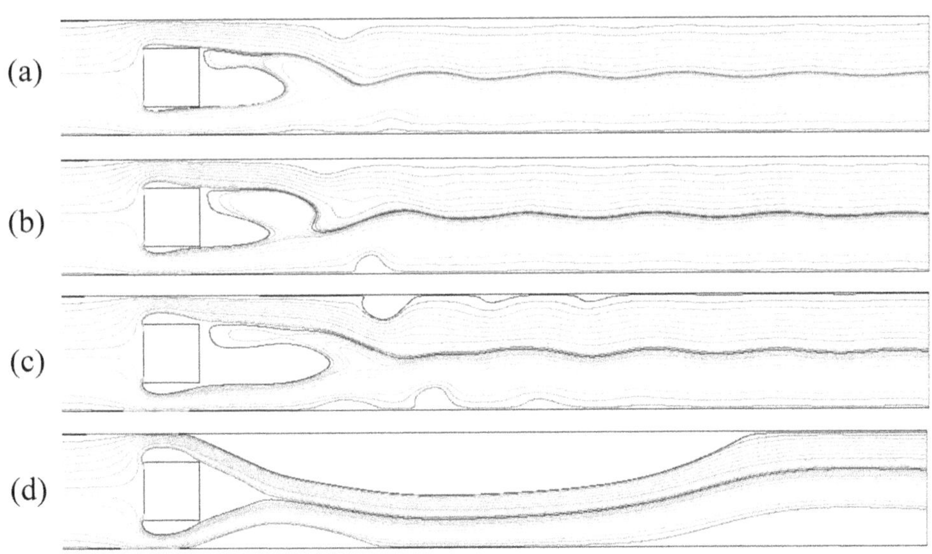

**For a more accurate representation see the electronic version.*

Figure 16. Mass fraction contours for $\mathrm{Re}_i = 100$ *and* $\alpha = 0^o$*: (a)* $I = 1/10$*; (b)* $I = 1/5$*; (c)* $I = 1/2.5$*; and (d)* $I = 1/1.5$*. For the color scale see Figure 5(e).*

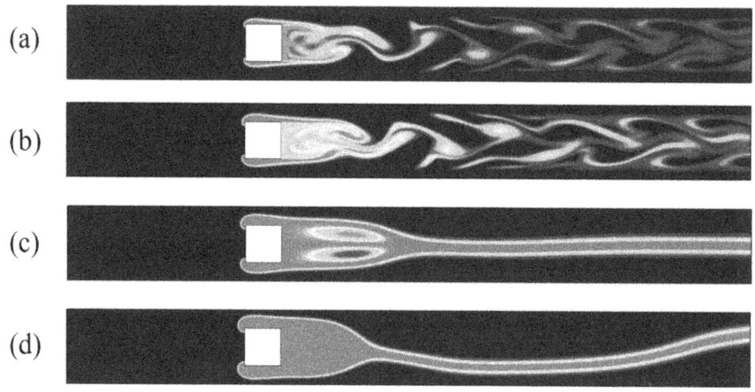

**For a more accurate representation see the electronic version.*

Figure 17. Mass fraction contours for $\mathrm{Re}_i = 100$ *and* $\alpha = 45^o$*: (a)* $I = 1/10$*; (b)* $I = 1/5$*; (c)* $I = 1/2.5$*; and (d)* $I = 1/1.5$*. For the color scale see Figure 5(e).*

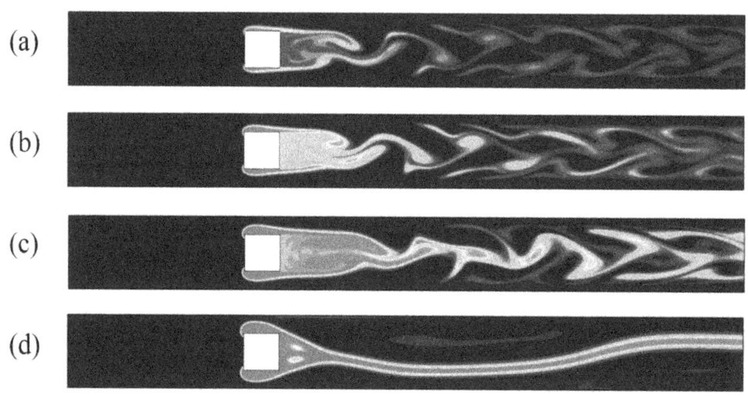

**For a more accurate representation see the electronic version.*

values of Re_i and I, the configuration with $\alpha = 45^\circ$ is always slightly better than when $\alpha = 0^\circ$, with the former having higher mixing efficiencies and lower input power needs than the latter. On the contrary, the configuration with $Re_i = 30$ and $\alpha = 0^\circ$ is, in general, the most unfavorable one in terms of both mixing efficiency and input power.

Figure 18. Mixing efficiency vs. the intensity ratio for $\mathrm{Re}_i = 100$ *and the indicated injection angles*

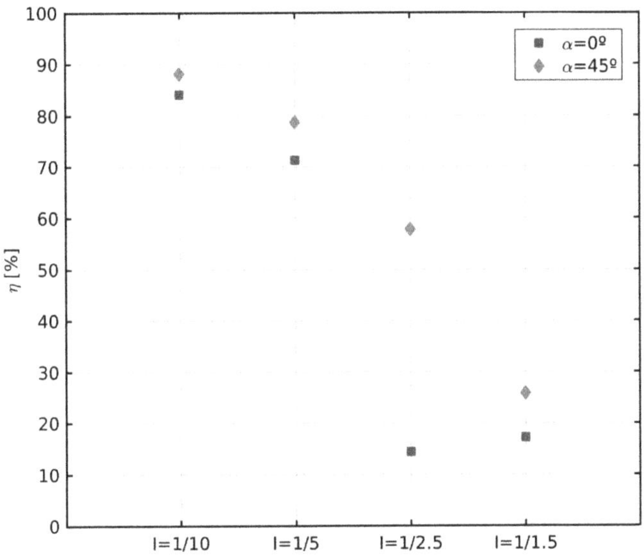

Figure 19. Input power vs. the intensity ratio for $\mathrm{Re}_i = 100$ *and the indicated injection angles*

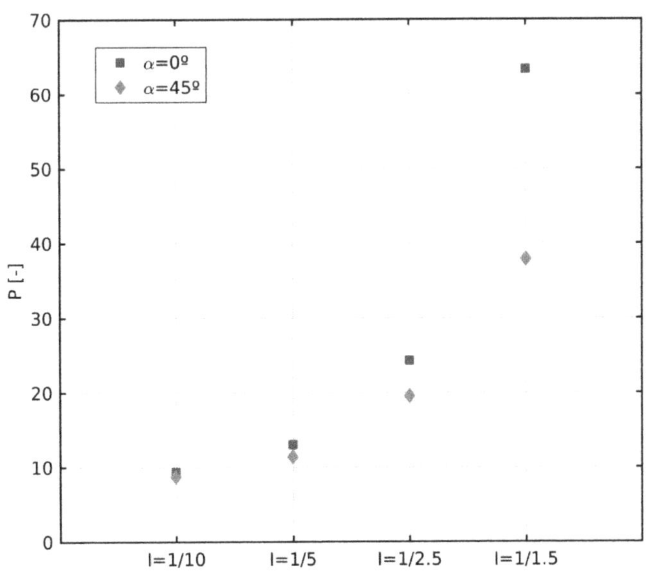

Figure 20. Comparison of mixing efficiency versus the intensity ratio for all studied configurations

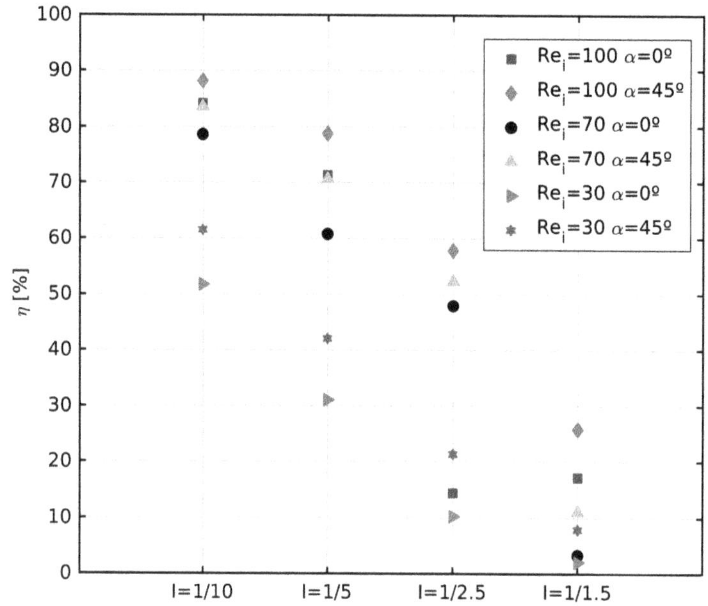

Figure 21. Comparison of input power versus the intensity ratio for all studied configurations

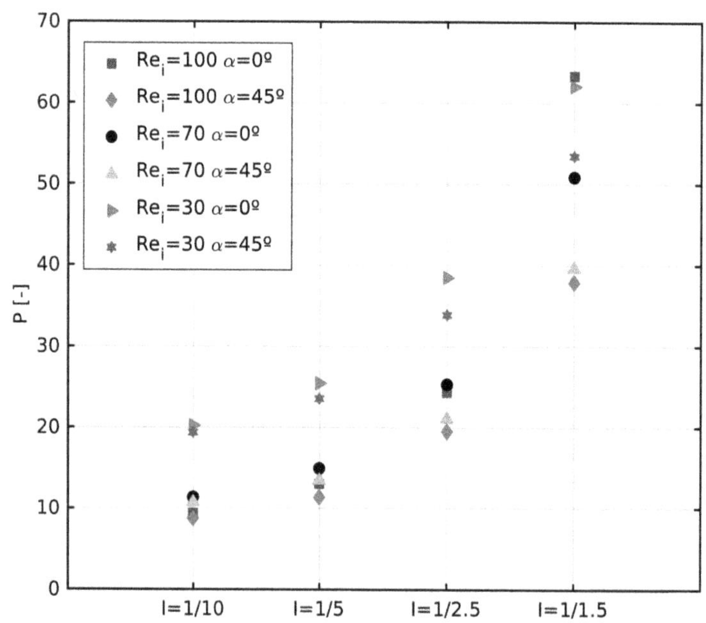

CONCLUSION

In this chapter, a passive micromixer, consisting of a straight channel with a pillar inside it, with mixing based on blowing and vortex shedding mechanisms, used to promote an irregular flow downstream, has been numerically modeled and characterized. The flow along the microdevice has been considered unsteady, 2D, laminar and with the mixing taking place at high Péclet and Schmidt numbers. For this purpose, ANSYS-Fluent software has been used to model different combinations of the governing parameters. Specifically, it has been studied three different Reynolds numbers ($\mathrm{Re}_i = 30{,}70$ and 100), five different jet intensity ratios ($I = 0{,}1/10{,}1/5{,}1/2.5$ and $1/1.5$) and two different injection angles ($\alpha = 0°$ and $45°$).

The main objective was to find out the values of the governing parameters that provide the highest mixing efficiency at the exit of the channel and the required input power to make both fluids to flow and mix. Regarding the mixing efficiency, the highest value is obtained for the highest studied Reynolds number, $\mathrm{Re}_i = 100$, the lowest intensity ratio, $I = 1/10$, and an injection angle of $\alpha = 45°$, for which a mixing efficiency of ~90% is reached, with a needed input power of ~9 units, while the worst configuration ($\mathrm{Re}_i = 30$, $\alpha = 0°$ and $I = 1/1.5$) gives an efficiency of only ~2%, with ~60 units of needed input power. Apart from this, the results have revealed additional interesting information related with the mixing, which will be discussed below.

On the one hand, it has been verified that the mixing efficiency increases when the Reynolds number increases and decreases when the intensity ratio between the fluids to mix increases. The latter effect tends to make the flow stationary and to eliminate the vortex shedding, which reduces the mixing efficiency. In general, we could say that the higher the Reynolds number, the higher the mixing efficiency, since high values of the Reynolds number promote vortex shedding, which enhances the mixing of both fluids. It is worth to note that for similar Reynolds numbers but in a completely horizontal channel without the square pillar, the opposite would have been obtained: very poor mixing at high Reynolds numbers because the residence time of the fluids in the channel would have been smaller and, therefore, the time they would have been for mixing.

On the other hand, it has also shown that higher mixing efficiencies are achieved when the path followed by the fluids downstream the pillar is longer and, therefore, there is more time for taking place fluids diffusion and interaction. This is the reason why, when obtaining stationary solutions, it is worth to get asymmetric flow patterns than symmetrical ones. In the future, new numerical simulations will be conducted in order to determine the critical intensity ratios that provide steady or oscillatory responses.

Regarding the input power needed to achieve the corresponding mixing, it has been shown that the power increases with the intensity ratio, consequence of the higher pressure needed both to inject a higher mass flow rate of fluid 2 and to be overcome by fluid 1. In addition, the input power decreases in those cases where the injection angle is $\alpha = 45\,°$. This is due to a double contribution: the cost of the main stream to redirect the jets from the square pillar is smaller than with $\alpha = 0\,°$; and the effective area above and below the pillar is larger than with $\alpha = 0$, resulting in a lower pressure drop for fluid 1.

With the work described in this Chapter it has shown that micromixers based on the blowing of one of the fluid from an inner pillar, together with the vortex shedding from it, are, in fact, a good option to take into account for developing geometrically simple micromixing devices, with relatively high mixing efficiencies.

ACKNOWLEDGMENT

The second author thanks the Ministerio de Ciencia, Educación y Universidades and the program "Salvador de Madariaga" 2018, reference PRX18/00334, for supporting his stay at Universidade do Porto, Faculdade de Engenharia, during which part of this work was carried out. FJGR acknowledges the financial support provided by Fundação para a Ciência e a Tecnologia (FCT), COMPETE and FEDER through IF/00190/2013.

REFERENCES

Capretto, L., Cheng, W., Hill, M., & Zhang, X. (2011). Micromixing within Microfluidic Devices. *Microfluidics*, *304*, 27–68. doi:10.1007/128_2011_150 PMID:21526435

Chamoun, G. C., Schilder, F., & Brns, M. (2011). Stabilization of vortices in the wake of a circular cylinder using harmonic forcing. *Physical Review. E*, *83*(6), 066308. doi:10.1103/PhysRevE.83.066308 PMID:21797478

Delaunay, Y., & Kaiktsis, L. (2001). Control of circular cylinder wakes using base mass transpiration. *Physics of Fluids*, *13*(11), 3285–3302. doi:10.1063/1.1409968

Dong, S., Triantafyllou, G. S., & Karniadakis, G. E. (2008). Elimination of vortex streets in bluff-body flows. *Physical Review Letters*, *100*(20), 204501. doi:10.1103/PhysRevLett.100.204501 PMID:18518541

Gonzalez-Mut, P., & Ortega-Casanova, J. (2016). CFD Study on the Drag Reduction of a Bluff Body by a Combined Blowing/Suction Technique. In G. Powell (Ed.), *Computational Fluid Dynamics (CFD): Characteristics, Applications and Analysis* (pp. 117–142). New York: Nova Science Publishers.

Hossain, S., Husain, A., & Kim, K. Y. (2011). Optimization of micromixer with staggered herringbone grooves on top and bottom walls. *Engineering Applications of Computational Fluid Mechanics*, *5*(4), 506–516. doi:10.1080/19942060.2011.11015390

Hsiao, K. Y., Wu, C. Y., & Huang, Y. T. (2014). Fluid mixing in a microchannel with longitudinal vortex generators. *Chemical Engineering Journal*, *235*, 27–36. doi:10.1016/j.cej.2013.09.010

Jain, M., Rao, A., & Nandakumar, K. (2013). Numerical study on shape optimization of groove micromixers. *Microfluidics and Nanofluidics*, *15*(5), 689–699. doi:10.100710404-013-1169-x

Kim, B. S., Kwak, B. S., Shin, S., Lee, S., Kim, K. M., Jung, H. I., & Cho, H. H. (2011). Optimization of microscale vortex generators in a microchannel using advanced response surface method. *International Journal of Heat and Mass Transfer*, *54*(1), 118–125. doi:10.1016/j.ijheatmasstransfer.2010.09.061

Kuo, J. N., & Li, Y. S. (2017). Centrifuge-based micromixer with three-dimensional square-wave microchannel for blood plasma mixing. *Microsystem Technologies*, *23*(7), 2343–2354. doi:10.100700542-016-2950-z

Le, T. H., Le, T. H., Dong, T., Ta, Q. B., Tran, M. N., & Karlsen, F. (2015). An effective passive micromixer with shifted trapezoidal blades using wide Reynolds number range. *Chemical Engineering Research & Design*, *93*, 1–11. doi:10.1016/j.cherd.2014.12.003

Lee, C. Y., Chang, C. L., Wang, Y. N., & Fu, L. M. (2011). Microfluidic mixing: A review. *International Journal of Molecular Sciences*, *12*(5), 3263–3287. doi:10.3390/ijms12053263 PMID:21686184

Nahum, A., & Seifert, A. (2006). On the application of confined twin-jet instability to micro-mixing enhancement. *Physics of Fluids*, *18*(6), 064107. doi:10.1063/1.2214882

Niu, X., Liu, L., Wen, W., & Sheng, P. (2006). Active microfluidic mixer chip. *Applied Physics Letters*, *88*(15), 153508. doi:10.1063/1.2195567

Noroozi, Z., Kido, H., Micic, M., Pan, H., Bartolome, C., Princevac, M., ... Madou, M. (2009). Reciprocating flow-based centrifugal microfluidics mixer. *The Review of Scientific Instruments*, *80*(7), 075102. doi:10.1063/1.3169508 PMID:19655976

Ober, T. J., Foresti, D., & Lewis, J. A. (2015). Active mixing of complex fluids at the microscale. *Proceedings of the National Academy of Sciences of the United States of America*, *112*(40), 12293–12298. doi:10.1073/pnas.1509224112 PMID:26396254

Ortega-Casanova, J. (2016). Enhancing mixing at a very low Reynolds number by a heaving square cylinder. *Journal of Fluids and Structures*, *65*, 1–20. doi:10.1016/j.jfluidstructs.2016.05.002

Ortega-Casanova, J. (2017a). CFD study on mixing enhancement in a channel at a low Reynolds number by pitching a square cylinder. *Computers & Fluids*, *145*, 141–152. doi:10.1016/j.compfluid.2016.12.022

Ortega-Casanova, J. (2017b). On the onset of vortex shedding from 2D confined rectangular cylinders having different aspect ratios: Application to promote mixing fluids. *Chemical Engineering and Processing: Process Intensification*, *120*, 81–92. doi:10.1016/j.cep.2017.06.014

Parsa, M. K., & Hormozi, F. (2014). Experimental and CFD modeling of fluid mixing in sinusoidal microchannels with different phase shift between side walls. *Journal of Micromechanics and Microengineering*, *24*(6), 065018. doi:10.1088/0960-1317/24/6/065018

Rafferty, I., & Kaminski, D. (2014). Non-dimensionalization and mixing quantification of laminar twin semi-confined jets. *International Journal of Heat and Fluid Flow*, *48*, 15–23. doi:10.1016/j.ijheatfluidflow.2014.03.007

Rahimi, M., Azimi, N., & Parvizian, F. (2013). Using microparticles to enhance micromixing in a high frequency continuous flow sonoreactor. *Chemical Engineering and Processing: Process Intensification*, *70*, 250–258. doi:10.1016/j.cep.2013.03.013

Santana, H. S., Júnior, J. L. S., & Taranto, O. P. (2015). Numerical simulations of biodiesel synthesis in microchannels with circular obstructions. *Chemical Engineering and Processing: Process Intensification*, *98*, 137–146. doi:10.1016/j.cep.2015.10.011

Sarkar, S., Singh, K. K., Shankar, V., & Shenoy, K. T. (2014). Numerical simulation of mixing at 1–1 and 1–2 microfluidic junctions. *Chemical Engineering and Processing: Process Intensification*, *85*, 227–240. doi:10.1016/j.cep.2014.08.010

Sarpkaya, T. (2004). A critical review of the intrinsic nature of vortex-induced vibrations. *Journal of Fluids and Structures*, *19*(4), 389–447. doi:10.1016/j.jfluidstructs.2004.02.005

Sharma, A., & Eswaran, V. (2004). Heat and fluid flow across a square cylinder in the two-dimensional laminar flow regime. *Numerical Heat Transfer Part A*, *47*, 79–107. doi:10.1080/10407780490520760

Sohankar, A., Khodadadi, M., & Rangraz, E. (2015). Control of fluid flow and heat transfer around a square cylinder by uniform suction and blowing at low Reynolds numbers. *Computers & Fluids*, *109*, 155–167. doi:10.1016/j.compfluid.2014.12.020

Solehati, N., Bae, J., & Sasmito, A. P. (2014). Numerical investigation of mixing performance in microchannel T-junction with wavy structure. *Computers & Fluids*, *96*, 10–19. doi:10.1016/j.compfluid.2014.03.003

Suzuki, H., Inoue, Y., Nishimura, T., Fukutani, K., & Suzuki, K. (1993). Unsteady flow in a channel obstructed by a square rod (crisscross motion of vortex). *International Journal of Heat and Fluid Flow*, *14*(1), 2–9. doi:10.1016/0142-727X(93)90034-K

Tseng, L. Y., Yang, A. S., Lee, C. Y., & Hsieh, C. Y. (2011). CFD-based optimization of a diamond-obstacles inserted micromixer with boundary protrusions. *Engineering Applications of Computational Fluid Mechanics*, *5*(2), 210–222. doi:10.1080/19942060.2011.11015365

Wang, J., Jiao, N., Tung, S., & Liu, L. (2014). Magnetic microrobot and its application in a microfluidic system. *Robotics and Biomimetics*, *1*(1), 1–8. doi:10.118640638-014-0018-z PMID:26613074

Williamson, C. H. K., & Govardhan, R. (2004). Vortex-induced vibrations. *Annual Review of Fluid Mechanics*, *36*(1), 413–455. doi:10.1146/annurev.fluid.36.050802.122128

Wu, C. J., Wang, L., & Wu, J. Z. (2007). Suppression of the von Karman vortex street behind a circular cylinder by a travelling wave generated by a flexible surface. *Journal of Fluid Mechanics*, *574*, 365–391. doi:10.1017/S0022112006004150

Wu, Z., & Chen, X. (2018). A novel design for 3D passive micromixer based on Cantor fractal structure. *Microsystem Technologies*, 1–12.

Yang, A. S., Chuang, F. C., Chen, C. K., Lee, M. H., Chen, S. W., Su, T. L., & Yang, Y. C. (2015). A high-performance micromixer using three-dimensional Tesla structures for bio-applications. *Chemical Engineering Journal*, *263*, 444–451. doi:10.1016/j.cej.2014.11.034

ADDITIONAL READING

Cai, G., Xue, L., Zhang, H., & Lin, J. (2017). A review on micromixers. *Micromachines*, *8*(9), 274. doi:10.3390/mi8090274

Gambhire, S., Patel, N., Gambhire, G., & Kale, S. (2016). A Review on Different Micromixers and its Micromixing within Microchannel. *International Journal of Current Engineering and Technology*, *4*, 409–413.

Gorjikhah, F., Davaran, S., Salehi, R., Bakhtiari, M., Hasanzadeh, A., Panahi, Y., ... Akbarzadeh, A. (2016). Improving "lab-on-a-chip" techniques using biomedical nanotechnology: A review. *Artificial Cells, Nanomedicine, and Biotechnology*, *44*(7), 1609–1614. doi:10.3109/21691401.2015.1129619 PMID:26758969

Gupta, S., Ramesh, K., Ahmed, S., & Kakkar, V. (2016). Lab-on-Chip Technology: A Review on Design Trends and Future Scope in Biomedical Applications. *International Journal of Bio-Science and Bio-Technology*, *8*(5), 311–322. doi:10.14257/ijbsbt.2016.8.5.28

Kim, K. Y., Ansari, M. A., & Afzal, A. (Eds.). (2018). *Passive Micromixers*. MDPI.

Lee, C. Y., Chang, C. L., Wang, Y. N., & Fu, L. M. (2011). Microfluidic mixing: A review. *International Journal of Molecular Sciences*, *12*(5), 3263–3287. doi:10.3390/ijms12053263 PMID:21686184

Mancera-Andrade, E. I., Parsaeimehr, A., Arevalo-Gallegos, A., Ascencio-Favela, G., & Parra, R. S. (2018). Microfluidics technology for drug delivery: A review. *Frontiers in Bioscience (Elite Edition)*, *10*(1), 74–91. doi:10.2741/e809 PMID:28930605

Pranzo, D., Larizza, P., Filippini, D., & Percoco, G. (2018). Extrusion-Based 3D Printing of Microfluidic Devices for Chemical and Biomedical Applications: A Topical Review. *Micromachines*, *9*(8), 374. doi:10.3390/mi9080374

Samiei, E., Tabrizian, M., & Hoorfar, M. (2016). A review of digital microfluidics as portable platforms for lab-on a-chip applications. *Lab on a Chip*, *16*(13), 2376–2396. doi:10.1039/C6LC00387G PMID:27272540

KEY TERMS AND DEFINITIONS

Computational Fluid Dynamics (CFD): It is the part of fluid mechanics that computationally (i.e., by means of computers) studies the flow and motion of fluids.

Lab on a Chip (LOC): This term refers to a device which integrates functions of a laboratory in a microchip. Due to its small dimensions, small amount of fluids and with a very short time of response can be analyzed.

Laminar Regime: The flow of a fluid is said to take place in laminar regime when the Reynolds number of the flow is low and below a critical value, which is usually around of few thousand.

Microfluidics: This term is used to refer the study of fluids in flow paths with at least one dimension of the order of microns.

Micromixing: Process on which the mixing of two or more fluids takes place at microscale.

Reynolds Number: One of the most important dimensionless numbers in fluid mechanics. It compares the viscous forces versus the inertial ones. When the former are dominant, the Reynolds number will be low (laminar regime), and if the latter are more important, the Reynolds number will be high (turbulent regime).

Vortex Shedding: When a stream of a fluid past a bluff object, above a certain fluid velocity, vortices, which are shed from the object, can be observed downstream and make the wake oscillatory and unsteady.

Chapter 4
Analytical and Numerical Modelling of Liquid Penetration in a Closed Capillary

Goutam Kumar Bose
Haldia Institute of Technology, India

Pritam Ghosh
Haldia Institute of Technology, India

Debashis Pal
Indian Institute of Engineering Science and Technology, India

ABSTRACT

The chapter explores the dynamics of liquid penetration in a closed end vertical capillary. This model is very important for impedance spectroscopy methodology where oxidized porous silicon provides an in vitro medium, and one important criteria of this methodology is the liquid penetration depth inside the silicon pores as the impedance is greatly affected by this penetration depth. This problem is also important in order to understand how the presence of entrapped air inside a micro pore can influence the dynamics of capillary flow. For this purpose, the model is studied both analytically and numerically. In this study, different pore size (500 nm and 2 μm diameter) with equal pore depth (~10 μm) have been used. Finally, the analytical solution is compared with the numerical results. In addition, the linearization of the system is also investigated and found the critical viscosity of which demarcates the over-damped and under-damped regimes. Further, this study is extended by incorporating the dynamic contact angle effects on the meniscus dynamics.

DOI: 10.4018/978-1-5225-7138-4.ch004

INTRODUCTION

The applications of capillary driven flow phenomena is spread over our everyday life as well as in technological applications (Fries & Dreyer, 2008; Fries & Dreyer, 2009; Guy Ramon & Alexander Oron, 2008; Deutsch, 1979; Popescu, Ralston, & Sedev, 2008; Hoffman & R. L., 1975) where surface tension mainly acts as an effective driving mechanism for this capillary flow. The rise of underground water in soil, water transport from the roots to the leaves of a tree, soaking of water in towel, flow of oil in wicks etc. are some of the daily life examples of capillary-driven flow. Among the technological applications oil extraction through porous rocks, ink-jet printing etc. are some major applications of capillary-driven flow. Instead of the above stated examples there is also an important implementation of capillary driven flow in Impedance Spectroscopy (Mondal, Pal, & Chaudhuri, 2015) methodology for continuous monitoring of cell-cell and cell-substrate interactions of HaCaT cells (immortal human skin keratinocytes) on oxidized silicon pores that provides a distinct in vitro environment for better understanding of in vivo behavior. This impedance spectroscopy not only monitors proliferation, cellular adhesion and spreading but also estimates the average values of cell membrane substrate distance, cell-cell junction resistance and also their fluctuations due to micro-motion. The oxidized silicon which is used for this study, consists of large number of fine pores and choosing of proper pore depth is one of the vital criterion for Impedance Spectroscopy process. Because when the culture medium, used for culturing the cells, is placed on it the medium starts penetrating inside the pore. For this Impedance Spectroscopy methodology the penetration depth inside the silicon pore plays a vital role because based on this penetration depth, pore depth is chosen for the experimental study. Another important thing for continuous monitoring of HaCaT cells by Impedance Spectroscopy is the restriction of electric field lines through the pores. Because when the whole setup is connected to electrical circuit for measuring and monitoring the impedance, cells behaviors etc. and at the same time if the medium penetrates through the bottom of the pore, most of the electric field lines will pass through the medium shunting the cellular layer on the top as electrical current always flows through the least resistance path. For that reason first estimation of the penetration depth of the medium inside the pore is necessary, based on which the pore depth are fabricated so that the medium does not fill the pore totally. Thus this will force the electric field lines to pass through the silicon oxide column at the bottom of the pore, which will increase the impedance of this path compared to the top. A schematic representation is shown in Figure 1.

In this study two pores having diameter 500 nm and 2 μm is used. Since the pore diameter is very small and one end of this pore is closed, these pores will act like a closed end capillary. While the culture medium starts penetrating inside the pore it will obviously entrap some gas inside it and a liquid-gas interphase will be formed and it will start penetration inside pores. The parameters which will affect the dynamics of liquid-air interface in a vertical closed-end capillary with entrapped gas are inertial effects, viscous effects, surface tension effects and entrapped gas effects. Here the gravitational force is negligible due to very small geometry. A force balance equation is used to derive the governing equation for liquid flow in a closed-end capillary. Then the governing equation is converted into a dimensionless equation and solved that equation analytically for finding out the equilibrium penetration depth and it has also studied whether the negligence of gravitational force makes any affect to the liquid dynamics or not. Besides, this problem is solved numerically to compare with the analytical results. Along with this dynamic contact angle is also studied because when the liquid-gas meniscus is in motion the contact angle is no longer static rather it varies with the meniscus velocity. The references (Nichita, Zun, & Thome, 2010; Shikhmurzaev & Sprittles, 2013; Friedman, 1999; Sung, Kim, & Lee, 2012; Mourik, Veldman, & Dryer, 2005) provide more information about the modelling of dynamic contact angle in capillary flow.

MATHEMATICAL MODELLING

In this a vertical capillary is considered whose bottom end is closed and the top end is kept in touch with a liquid reservoir, as depicted in Figure 2. At the very beginning when there is no contact between the capillary and fluid reservoir at that time air will be there inside capillary. And when this capillary gets in touch with fluid reservoir due to surface tension fluid will starts entering inside the pore compressing the air.

Figure 1. Impedance Spectroscopy. (a) Schematic of the sensor. (b) 2D cross sectional representation of the porous structure with confluent cell growth in the region of working electrode.
(Taken from Mondal, Pal & Chaudhuri,. 2015)

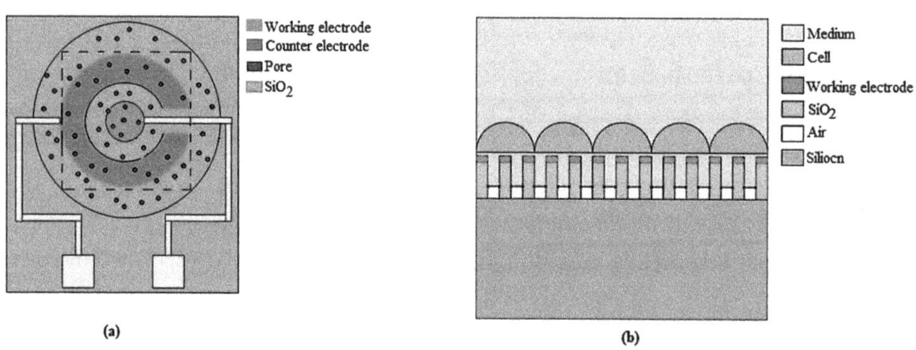

Due to this compression air will give an opposite pressure on the liquid-air meniscus against the entering fluid. So when the liquid-air interphase is penetrating into the capillary pore the forces acting on the meniscus are: inertial force, surface tension force, viscous force, gravity force and force due to entrapped air inside the capillary (Radiom, Chan, & Yang, 2010). But, as in this study very small pore diameter and pore depth is used, the effect of gravity becomes negligible.

Following the above stated assumptions, the force balance equation becomes

$$F_i = F_s - F_v + F_e \tag{1}$$

where,

F_i = inertia force
F_s = surface tension force
F_v = viscous force
F_e = entrapped gas force

The inertia force, acts on the liquid column, depends on mass and velocity of liquid air interface. Thus

Figure 2. Liquid penetration in a closed-end capillary

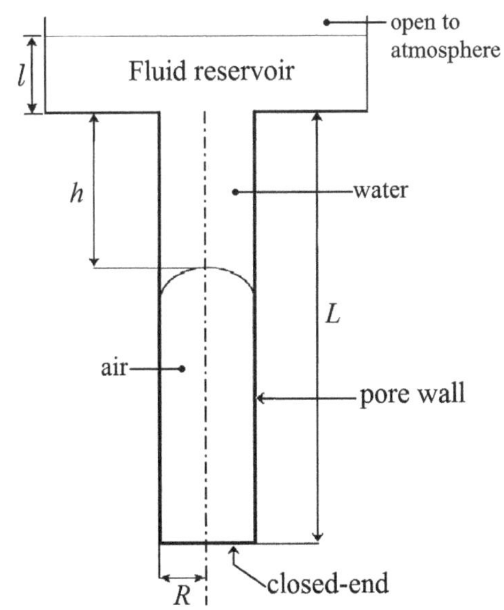

Figure 3. Forces acting on the liquid-air interface

Where, L = length of the capillary; l = water height in the reservoir; h = h (t) = meniscus height at time t R = radius of capillary; θ = contact angle.

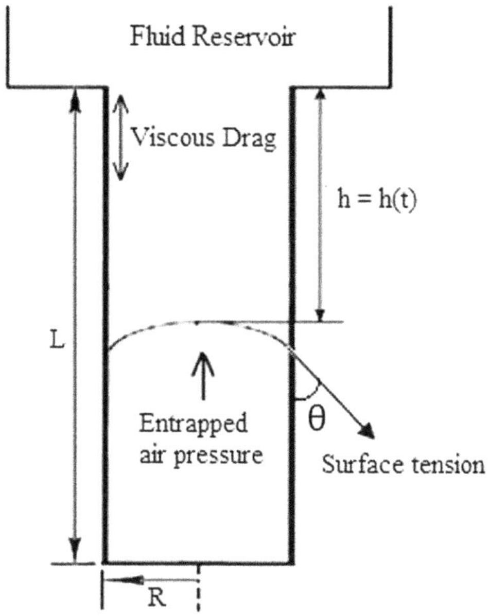

$$F_i = \frac{d}{dt}\left(mu\right) = \rho A \frac{d}{dt}\left(hh'\right) \tag{2}$$

The surface tension force is given as:

$$F_s = 2\frac{\gamma \cos\theta_s}{R} A \tag{3}$$

The viscous force is given as:

$$F_v = 8\frac{\mu}{R^2} hh' A \tag{4}$$

and, let entrapped gas force,

$$F_e = \varphi\left(h, L\right) A \tag{5}$$

where A is cross-sectional area of capillary tube, ρ and μ are the density and viscosity of liquid respectively, γ is the liquid-air surface tension, R is the radius of capillary, θ_s is the static contact angle of the liquid, h is the height of meniscus, L is the height of the capillary. As a closed-end capillary is considered, the entrapped gas action has to be taken into account by adding a term $\varphi(h, L)$ in the governing equation. Considering adiabatic compression of gas i.e. $p_a V^n$ = constant, (Fazio & Iacono, 2009) modeled the equation for the pressure that is applied to the liquid-air interface due to compression of air inside the capillary as:

$$\varphi(h, L) = p_a - p_a \left(\frac{L}{L-h}\right)^n \tag{6}$$

where, p_a is the atmospheric pressure and n is polytropic index.

Thus substituting all the parameters in Equation (1) (from Equation (2) to Equation (6), and dividing both sides by A, Equation (1) yields:

$$\rho \frac{d}{dt}\left(hh'\right) = 2\frac{\gamma \cos\theta_s}{R} - 8\frac{\mu}{R^2} hh' + p_a - p_a \left(\frac{L}{L-h}\right)^n \tag{7}$$

With boundary conditions:

$$h(0) = 0 \tag{8a}$$

$$h'(0) = 0 \tag{8b}$$

Equation (7) is the main governing equation of this problem which describes the meniscus dynamics inside a closed end capillary. Since the geometry is too small the gravity term becomes insignificant and hence it is neglected and due to entrapment of air inside the pore entrapped air pressure term is added in the governing equation, these make Equation (11) differ from the Washburn equation (Washburn, 1921).

DIMENSIONLESS SCALING

For simplification of the governing equation a dimensionless model is developed. Here the meniscus height (h) is normalized using the length of the pore (L). For this case the following dimensionless parameters are taken for normalizing the governing equation (Equation 7).

$$H = \frac{h}{L} \tag{9a}$$

$$T = \frac{t}{\tau_L} \tag{9b}$$

Here L and τ_L are used to normalize meniscus height and time respectively. Using H and T in Equation (7) and multiplying all terms by $\frac{R}{\gamma}$ lead to the following expression:

$$\frac{\rho L^2 R}{\gamma \tau_L^{\ 2}} \frac{d}{dT}\left(HH'\right) = 2\cos\theta_s - \frac{8\mu L^2}{R\gamma\tau_L} HH' + \frac{RP_a}{\gamma}\left[1 - \left(\frac{1}{1-H}\right)^n\right] \tag{10}$$

The coefficient on the left hand side of Equation (10) is assumed to be unity to find τ_R as follows:

$$\tau_L = \sqrt{\frac{\rho L^2 R}{\gamma}} \tag{11}$$

Now, substituting the value of τ_R from Equation (11) in Equation (10) and assuming the second coefficient on the right side of Equation (10) to be ω, Equation (10) transforms into the following equation:

$$\frac{d}{dT}\left(HH'\right) = 2\cos\theta_s - 8\omega HH' + \frac{RP_a}{\gamma}\left[1 - \left(\frac{1}{1-H}\right)^n\right] \tag{12}$$

with the boundary conditions:

$$H(0) = 0 \tag{13a}$$

$$H'(0) = 0 \tag{13b}$$

In Equation (16), ω is a positive dimensionless number, obtained by:

$$\omega = \frac{\mu L}{R\sqrt{\rho \gamma R}} \tag{14}$$

and in Equation (12) the value of ω indicates the value of viscosity in dimensionless form.

The Equation (12) is the dimensionless form of the governing Equation (7). But the problem with this equation is that, this equation gives infinite velocity at time $T = 0$, which is impossible. This impossible prediction was removed by (Szekely, Neumann, & Chuang, 1971) by applying correct momentum balance for the entry flow. To eliminate this problem an effective initial height λ is considered which is proportional to R as described by (Zhmud, Tiberg, & Hallstensson, 2000). According to Szekely et al. (1971), $\lambda = cR$, where c the coefficient of apparent mass and R stands for radius of capillary. In this study the value of c is taken as 1. Thus incorporating this apparent mass effect Equation (12) becomes:

$$\left[\left(\lambda + H\right)H'' + H'^2\right] = 2\cos\theta_s - 8\omega HH' + \frac{RP_a}{\gamma}\left[1 - \left(\frac{1}{1-H}\right)^n\right] \tag{15}$$

Equation (15) is the modified Washburn equation considering static contact angle (θ_s).

LINEARIZATION OF GOVERNING EQUATION

Equation (7) differs from Washburn equation through the addition of entrapped air pressure inside pore and neglecting the gravity term. Firstly the equilibrium of the system is taken care. Under steady state conditions, the governing Equation (7) can be reduced to:

$$\frac{2\gamma\cos\theta_s}{R} + p_a - p_a\left(\frac{L}{L - H_{eq}}\right)^n = 0 \tag{16}$$

where H_{eq} is the equilibrium position of the interface, which can be easily obtained as:

$$H_{eq} = L\left[1 - \left(1 + \frac{F_{ca}}{p_a}\right)^{-\frac{1}{n}}\right] \tag{17}$$

Here $F_{ca} = \frac{2\gamma\cos\theta_s}{R}$ is used representing the contribution of capillarity.

Now the stability of the steady meniscus position H_{eq} has considered by introducing a small disturbance to the equilibrium in the form:

$$h = H_{eq} + \varepsilon(t) \tag{18}$$

where $\varepsilon(t)$ represents a small departure of the interface elevation from its equilibrium.

Substituting Equation (18) into Equation (7) yields:

$$\rho\frac{d}{dt}\left[\left(H_{eq} + \varepsilon\right) + \varepsilon'\right] = F_{ca} - \frac{8\mu}{R^2}\left(H_{eq} + \varepsilon\right)\varepsilon' + p_a - p_a\left(\frac{L}{L - H_{eq} - \varepsilon}\right)^n \tag{19}$$

Linearizing Equation (19) with respect to $\varepsilon(t)$, the following equation is obtained:

$$\varepsilon''(t) + \alpha\varepsilon'(t) + \beta\varepsilon(t) = 0 \tag{20}$$

where the primes denote differentiation with respect to time and:

$$\alpha = \frac{8\mu}{\rho R^2} \tag{21a}$$

$$\beta = \frac{np_a}{\rho L H_{eq}} \left(\frac{L}{L - H_{eq}} \right)^{n+1} \tag{21b}$$

When a stable equilibrium exists, equation (20) predicts the following: when the discriminant $\delta \equiv \alpha^2 - 4\beta > 0$, the system is over damped and a monotonic behavior is expected, where for $\delta < 0$ damped oscillation occur.

The transition from the oscillatory to monotonic evolution of the interface can be found at a point where the viscous drag is sufficient to suppress any "overshoot" due to liquid inertia. The transition threshold is determined by $\delta = 0$, namely:

$$\delta_0 = \alpha^2 - 4\beta \tag{22}$$

or,

$$\alpha = 2\sqrt{\beta} \tag{23}$$

$$\therefore \mu_{critical} = \frac{\rho R^2}{4} \sqrt{\frac{np_a}{\rho L H_{eq}} \left(\frac{L}{L - H_{eq}} \right)^{n+1}} \tag{24}$$

Equation (24) gives the critical value of viscosity, above which the system will be over-damped and below which damped oscillation will occur.

NUMERICAL SOLUTION METHODOLOGY

For numerical analysis commercial CFD software has been used to model two-dimensional, axisymmetric, transient, two-phase flow in the micro-pore. The dimensions of the geometry are shown in Figure 4 and other variables that are used for solving this problem are given in Table 1 and Table 2. Appropriate grid generation of the working geometry is very important for numerical study. In this present problem for numerical solution Mapped Face Meshing (Quadrilaterals method) has used. As the geometry is very small, a portion of meshing has shown in figure 5. Since in this case two immiscible liquids (air and water) are used, VOF model is appropriate for the numerical simulation. VOF method is used to identify the liquid-gas interface by solving a volume fraction equation for one of the phases. An explicit geometric-reconstruction scheme is used to represent the interface by using a piecewise-linear approach. A surface tension effect is modelled by using the continuum surface force model proposed by (Brackbill, Kothe, & Zemach, 1992). No-slip boundary is used for pore wall. Besides heat flux through the pore wall is takes as zero which means it's an adiabatic wall. The main reason behind consideration of an adiabatic wall is that above the pore huge amount of liquid is there, so whatever amount of heat will be generated within the pore during movement of meniscus that will be immediately absorbed by that liquid which confirms there will be no such chemical change in the liquid. Surface tension induced stress is approximated by the CSF model by a body force which acts throughout a small but finite fluid region surrounding the interface. In this present computation, a PRESTO interpolation scheme is used to compute face pressure. The QUICK scheme is used for the discretization of the momentum equations. A time-step-size of 1×10-9 is used for this present simulation.

While solving those governing equation a problem has arrived on the value of polytropic index and that's why some arbitrary values of 'n' has taken just for getting a little bit idea about the meniscus dynamics. But after performing numerical simulations using commercial CFD software a new thing has been identified which is using the results from numerical solution, indirectly the actual value of polytropic index can be obtained for a given pore diameter. Because after numerical simulation equilibrium meniscus height can be obtained and using Equation (17) the proper value of 'n' can be easily calculated. This is the major contribution of CFD analysis in this study. The following sections will describe the details of various numerical schemes used during solving this problem numerically.

RESULTS AND DISCUSSIONS

Analytical Results

Equation (15) has been solved both for 500 nm and 2μm diameter pores using Runge Kutta method. While solving Equation (15) the one main problem has arrived which is about the value of polytropic index 'n'. Because at this point proper value of 'n'

Table 1. Properties of water and air used in simulation

Fluid	Density	Viscosity	Surface Tension
	(kg/m^3)	(kg/m-s)	(N/m)
Water	998.2	0.001003	0.0728
Air	Ideal-gas	1.7894×10^{-05}	

Figure 4. Schematic representation of the geometry and boundary conditions used in CFD simulations. (a) pore diameter 500 nm; (b) pore diameter 2 μm.

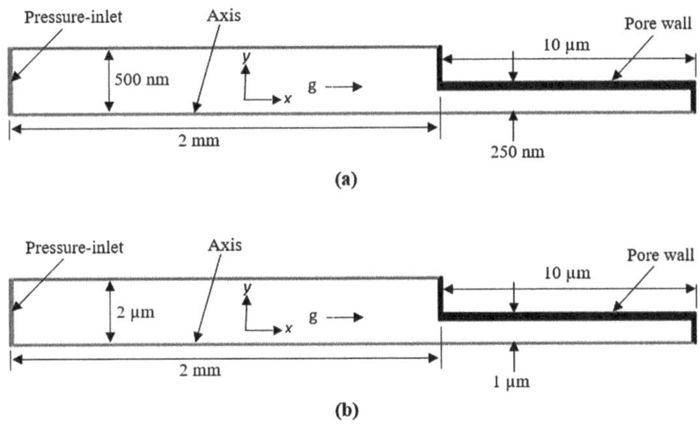

Figure 5. Meshing of the geometry

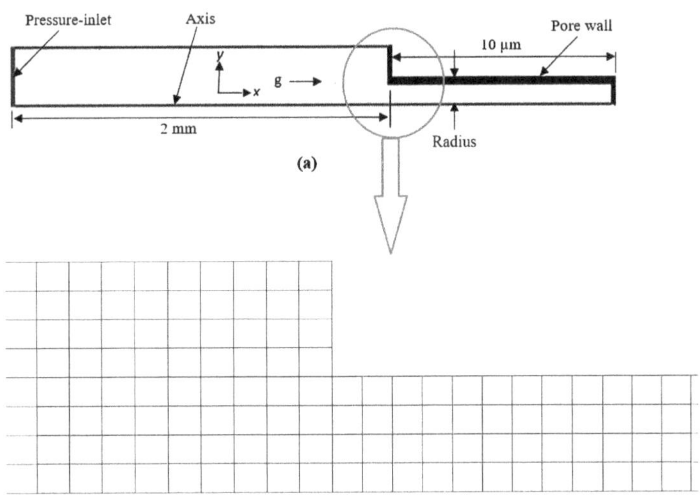

for this particular process is unknown. And also the value of 'n' may be different for different diameter pores. So for solving those equations different values of 'n' has been taken and Equation (15) is solved analytically for observation of meniscus dynamics inside pores. Figure 6 shows variation of meniscus height with time at values of n and it is seen that the equilibrium height decrease with the increase in 'n'. This is because as the value of polytropic index 'n' increases the entrapped gas pressure also increases. This opposes the liquid to penetrate into the pore, thus the overall equilibrium height decreases.

Table 2. Parameter used for calculation

Parameter	Value	Unit
L	10	μm
p_a	101325	Pa
γ	0.0728	N/m
θ	70	degree
c	1	-
ρ	998.2	kg/m^3

Figure 6. Meniscus height versus flow time for different values of polytropic index 'n' (a) for 500 nm pore dia.; (b) for 2 μm pore dia.

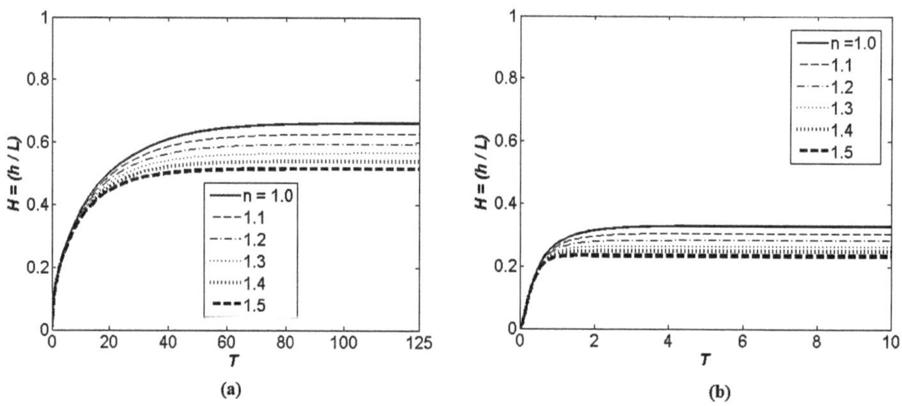

Numerical Results

The schematic diagram of the geometry and the boundary conditions that are used in the simulation are shown in Figure 4. After running the simulations of the model for both pore diameters (500nm and 2 μm) it has found that how the water-air meniscus height varies with time as shown in Figure 7. It is seen from the above figures that meniscus oscillation takes place when the diameter of the capillary tube is larger. This is because for larger diameter the inertia force of entering liquid is more. As a result, during penetration the air-liquid interface goes beyond the equilibrium height compressing the air inside the pore. Now at a certain time the compressed air pressure becomes higher and it starts opposing the penetrant liquid and finally pushes back the liquid in reverse direction. Now when the liquid starts moving in reverse direction again due to inertia the meniscus moves past the equilibrium height and again it changes the direction. This way the meniscus will oscillate and in every movement the liquid loose its energy and becomes stable after few oscillations. At stable condition two forces remain active viz. surface tension force and entrapped air pressure with equal magnitude but their directions are in opposite which makes the system equilibrium.

Comparison of Results

In the previous section (A) the need of conducting numerical simulations has been discussed and in this section a comparison has shown between analytical results and numerical result. Before making these comparison few calculations has been done for finding out the value of polytropic index ‘n’. Because without it is very difficult to solve the governing equations (Equation 15). First the value of equilibrium height from numerical solution is taken and by using Equation (17) the value of ‘n’ has been evaluated. Then using this ‘n’ Equation (15) has been solved. And finally for comparison meniscus height (H) versus time (T) is plotted in a single plot as shown in Figure 8.

From the above figures it can be inferred that the analytical and CFD results show excellent agreement with each other for 500 nm diameter pore. But a huge difference between the results is found in case of 2 μm diameter pore. It has shown in Figure 8 (b). This is the advantage of CFD approach for the solution of such problems since more accurate results cannot be obtained as compared to those obtained through analytical approach. This is simply because the analytical solution unnecessarily simplifies the problem to a large extent which has an adverse effect on the accuracy of the results.

Figure 7. Meniscus height vs. flow time for (a) 500nm pore dia. (b) 2 μm pore diameter

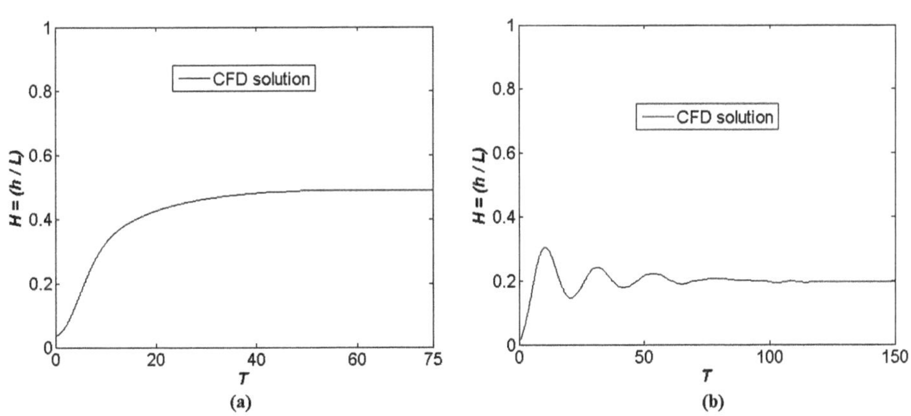

Figure 8. Comparison of results between CFD solution and analytical solution for (a) 500 nm diameter pore; (b) 2 μm diameter pore

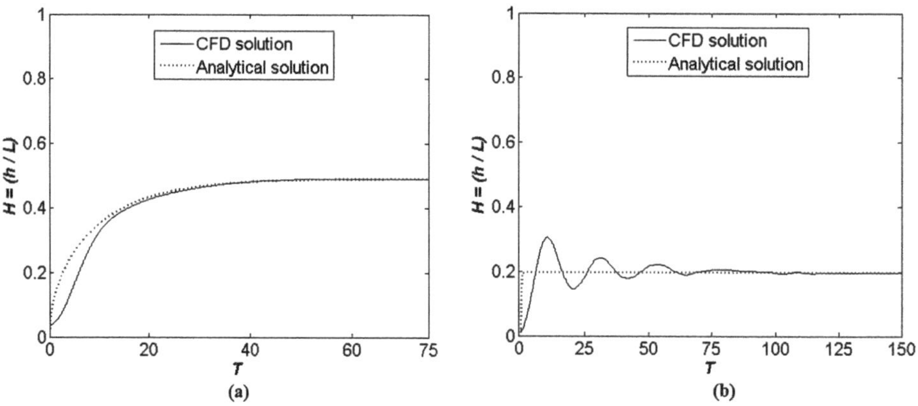

Viscosity Dependent Meniscus Dynamics

The dependence of meniscus dynamics on viscosity is shown in this section. After linearizing the governing equation it has found how the viscosity affects the meniscus dynamics as well as at what values of viscosity of liquid medium the system shows over-damped, under-damped or transition behavior.

It is seen from Figure 8 that as the viscosity value goes beyond the critical value the meniscus dynamics a monotonic behavior. In contrary if viscosity value is less than the critical value then it has found there is a rebouncing effect in meniscus dynamics or damped oscillation occurs. These behaviors are due to viscous drag. At critical point this viscous drag is just sufficient to suppress any overshoot due to liquid inertia and when viscosity goes beyond the critical viscosity then the viscous drag also becomes stronger and it makes the system over-damped. But when the viscosity is less than the critical value then viscous drag is no longer able to suppress the liquid inertia at very first rather it takes some times to suppress this overshoot and for this reason a rebouncing effect has found when viscosity goes below the critical value.

Effect of Dynamic Contact Angle on Meniscus Dynamics

The classical description of the rate of capillary rise given by Washburn equation assumes a constant value of contact angle all the times. But in some studies (Hoffman & R. L., 1975; Popescu et. al., 2008; Blake, 2006; Nichita, Zun, & Thome, 2010; Shikhmurzaev & Sprittles, 2013; Friedman, 1999; Sung, Kim, & Lee, 2012; Mourik,

Figure 9. Meniscus dynamics at different values of viscosity (dimensionless values) above and below the critical value for (a) 500 nm diameter pore; (b) 2 μm diameter pore

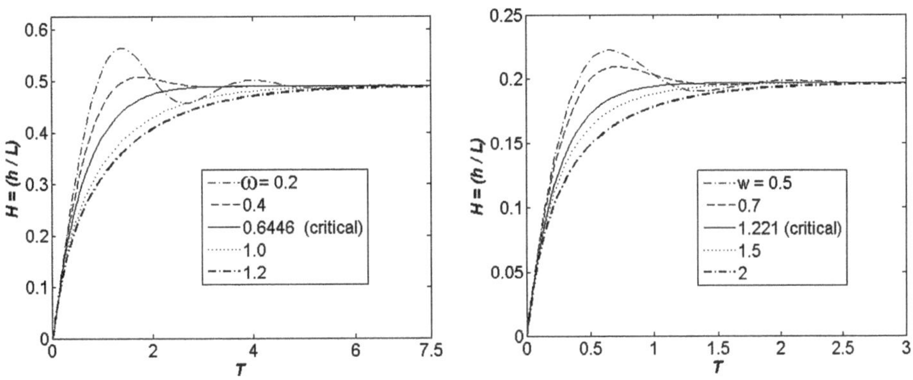

(a) (b)

Veldman, & Dreyer, 2005), it has been found that there is an influence of contact-line velocity on contact angle. This kind of phenomenon has been found in many experiments on 3-phase (solid, liquid and gas) contact line, for a wide ranges of capillary dimensions and flow velocities. Actually the magnitude of contact angle is the result of an interplay between two main forces viz. capillary forces, which tend to reduce the contact angle in case of an advancing wetting fluid, and viscous forces, which tend to increase contact angle under same condition. In this study some empirical relations have been used (given below) on dynamic contact angle in solving the governing equations to know the influence of DCA on capillary dynamics.

1. **Hoffman Model:** Under certain conditions Hoffman (Hoffman & R. L., 1975) presented a correlation for DCA as a function of capillary number (Ca), which is the ration of viscous force to interfacial force, and a shift factor, which in turn is a function of static contact angle, as:

$$\frac{\cos\theta_s - \cos\theta_m}{\cos\theta_s + 1} = \tanh\left(4.96Ca^{0.702}\right) \tag{25}$$

2. **Bracke et al. Model:** The dependence of advancing dynamic contact angle on velocity suggested by Bracke et al. is given as:

$$\cos\theta_m = \cos\theta_s - 2\left(1 + \cos\theta_s\right)Ca^{0.5} \tag{26}$$

3. **Seeberg et al. Model:** The dependence of advancing dynamic contact angle on velocity suggested by Seeberg et al. is given as:

$$\cos\theta_m = \cos\theta_s - 2.24\left(1+\cos\theta_s\right)Ca^{0.54} \tag{27}$$

Figure 10 shows the variation of contact angle with time for Hoffman's empirical model of DCA for 500 nm diameter and 2 μm diameter capillary pore. Here Equation (15) is used for getting the above results. In Equation (15) the expression of DCA has taken, replacing the static contact angle and solve the Equation (15) for different viscosity. The same procedure is followed for 2 μm diameter pore too and got the results. It is seen from above results that change in contact angle is more for 2 μm diameter capillary pore than 500 nm diameter pore and this is due to the inertia force because for 2 μm diameter capillary pore this inertia force is higher than for a 500 nm diameter capillary pore.

Now to find the effect of dynamic contact angle on meniscus dynamics equations (25 to 27) has used and Equation (15) is sloved. Here two different values of viscosity (ω) has used in non-dimensionalized form. The first one is the critical viscosity and another is less than critical viscosity. For 500 nm diameter capillary pore which is ω = 0.644 and for 2 μm diameter capillary pore ω = 1.221 (both are calculated using Equation (24). The results are shown in Figures 11 and 12 below. Here it has also shown the differences in meniscus height dynamics when dynamic contact angle has used instead of static contact angle.

Figure 10. Variation of contact angle (in degree) with time using Hoffman's empirical equation (Eq. 25); (a) for 500 nm dia. pore; (b) for 2 μm dia. pore

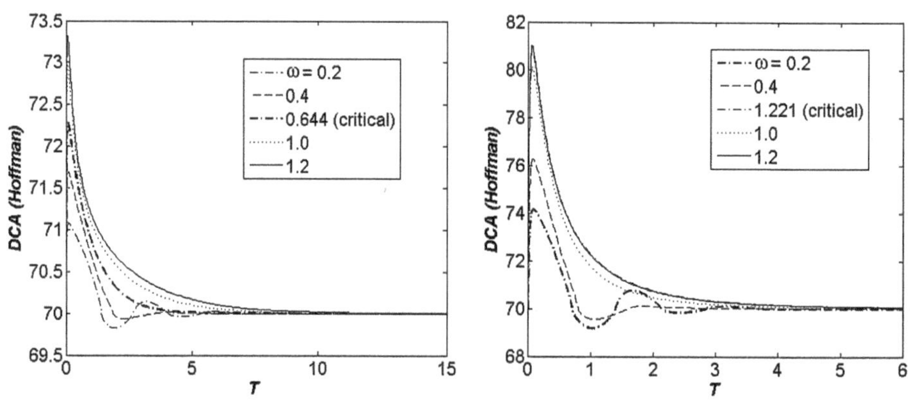

(a) (b)

Figure 11. Meniscus height vs. time at critical viscosity considering DCA. (a) for 500 nm dia. pore where $\omega_{critical} = 0.644$ *; (b) for 2 μm dia. pore where* $\omega_{critical} = 1.221$

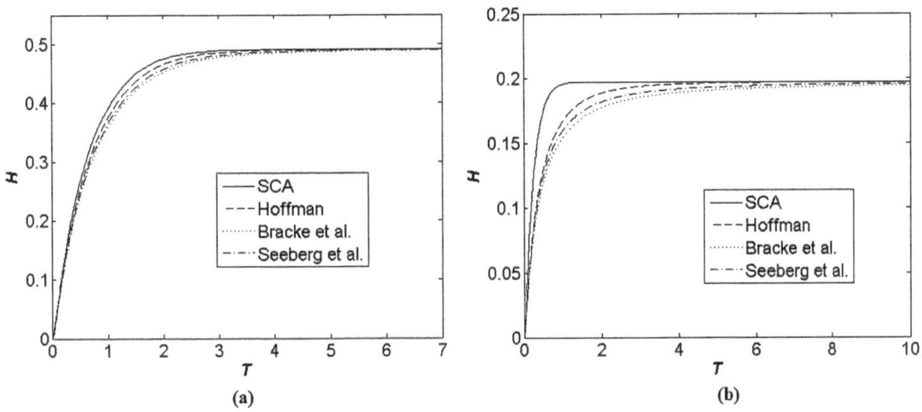

Figure 12. Meniscus height vs. time at viscosity less than critical value considering DCA. (a) for 500 nm dia. pore where $\omega_{critical} = 0.644$ *; (b) for 2 μm dia. pore where* $\omega_{critical} = 1.221$

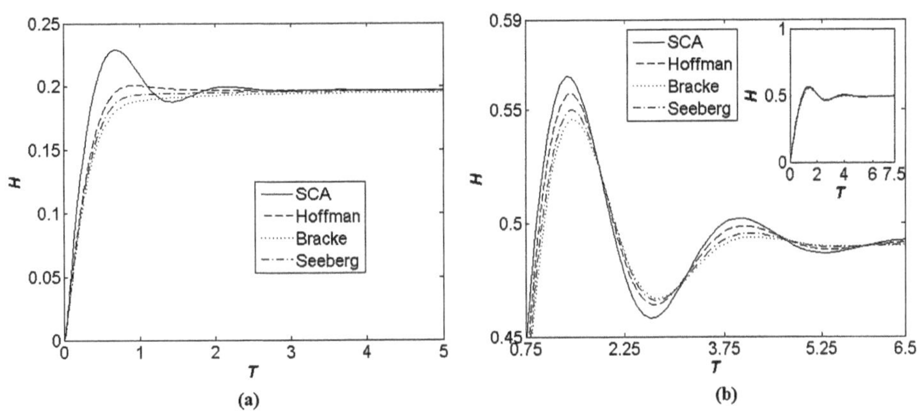

CONCLUSION

A method for deriving an analytical solution to the momentum balance of a liquid in a capillary tube is presented. Here an extended solution of the well-established Lucas-Wasburn equation has introduced. The extended solution includes the entrapped gas pressure term and excludes the gravity term. Further the governing equation is non-dimensionalized and it has found a non-dimensional viscous term. Different plots on dimensionless height versus dimensionless time are shown. Numerical approach is also included in this study where commercial CFD software is used for solving the problem numerically. Comparison between analytical and numerical solutions are also shown. During solving the governing equation Equation (15) analytically a problem has been identified on choosing the proper value of 'n' and until and unless some realistic values has been taken, the analytical approach becomes worthless. And here is the beauty of applying the numerical approach for finding out the value of 'n'.

Another important thing is in the classical description of the rate of capillary rise given by the Wasburn equation assumes that the contact angle remains constant at all times. But from many studies it has found that there is a strong dependence of dynamic contact angle on the velocity of contact line. Using some empirical co-relations the discrepancy between static contact angle (SCA) and dynamic contact angle (DCA) has been shown by means of some results and it has seen that use of DCA increases the damping effect.

By introducing a small disturbance to the equilibrium height the governing equation has linearized and from there the critical value of viscosity has calculated and by means of some figures it has presented that how the meniscus height is depending on viscosity of the penetrating liquid. Here it has seen that with the increase of viscosity as viscous drag increases, it directly effects the meniscus dynamics. More specifically meniscus oscillation decreases with increase in viscosity. However, in this case it has considered there is no interfacial mass transfer present. But if the interfacial mass transfer exists then the whole system may become unstable because evaporation always lowers the equilibrium height while condensation may results in an either lower or higher position depending on the mass flux which also depends on temperature difference across the meniscus interface and of course there should exist some stability criterion.

REFERENCES

Blake, T. D. (2006). The physics of moving wetting lines. *Journal of Colloid and Interface Science*, *299*(1), 1–13. doi:10.1016/j.jcis.2006.03.051 PMID:16631781

Brackbill, J. U., Kothe, D. B., & Zemach, C. (1992). A continuum method for modeling surface tension. *Journal of Computational Physics*, *100*(2), 335–354. doi:10.1016/0021-9991(92)90240-Y

Deutsch, S. (1979). A Preliminary study of the fluid mechanics of liquid penetrant testing. *Journal of Research of the Notionol Bureou of Standords.*, *84*(4), 287. doi:10.6028/jres.084.012

Fazio, R., & Iacono, S. (2009). Entrapped Gas Action for One-Dimensional Models of Capillary Dynamics. *World Congress on Engineering*, 2, 1-3.

Friedman, S. P. (1999). Dynamic contact angle explanation of flow rate-dependent saturation-pressure relationships during transient liquid flow in unsaturated porous medium. *Journal of Adhesion Science and Technology*, *13*(12), 1495–1518. doi:10.1163/156856199X00613

Fries, N., & Dreye, M. (2008a). The transition from inertial to viscous flow in capillary rise. *Journal of Colloid and Interface Science*, *327*(1), 125–128. doi:10.1016/j.jcis.2008.08.018 PMID:18774142

Fries, N., & Dreye, M. (2008b). An analytic solution of capillary rise restrained by gravity. *Journal of Colloid and Interface Science*, *320*(1), 259–263. doi:10.1016/j.jcis.2008.01.009 PMID:18255086

Fries, N., & Dreye, M. (2009). Dimensionless scaling methods for capillary rise. *Journal of Colloid and Interface Science*, *338*(2), 514–518. doi:10.1016/j.jcis.2009.06.036 PMID:19628213

Hoffman, R. L. (1975). A study of the advancing interface. I. Interface shape in liquid—gas systems. *Journal of Colloid and Interface Science*, *50*(2), 228–241. doi:10.1016/0021-9797(75)90225-8

Mondal, D., Pal, D., & Chaudhuri, C. R. (2015). Real-time sensing of epithelial cell-cell and cell-substrate interactions by impedance spectroscopy on porous substrates. *Journal of Applied Physics*, *118*(4), 044701. doi:10.1063/1.4927215

Mourik, S. V., Veldman, A. E. P., & Dreyer, M. E. (2005). Simulation of Capillary Flow with a Dynamic Contact Angle. *Microgravity Science and Technology*, *17*(3), 87–93. doi:10.1007/BF02872093

Nichita, B. A., Zun, I., & Thome, J. R. (2010). *A VOF method coupled with a dynamic contact angle model for simulation of two-phase flows with partial wetting*. Paper presented at the 7th International Conference on Multiphase Flow, Tampa, FL.

Popescu, M. N., Ralston, J., & Sedev, R. (2008). Capillary Rise with Velocity-Dependent Dynamic Contact Angle. *Langmuir*, *24*(21), 12710–12716. doi:10.1021/la801753t PMID:18834162

Radiom, M., Chan, W. K., & Yang, C. (2010). Capillary filling with the effect of pneumatic pressure of trapped air. *Microfluidics and Nanofluidics*, *9*(1), 65–75. doi:10.100710404-009-0527-1

Ramon, G., & Oron, A. (2008). Capillary rise of a meniscus with phase change. *Journal of Colloid and Interface Science*, *327*(1), 145–151. doi:10.1016/j.jcis.2008.08.016 PMID:18722625

Shikhmurzaev, Y. D., & Sprittles, J. E. (2013). Dynamic contact angle of a liquid spreading on an unsaturated wettable porous substrate. *Journal of Fluid Mechanics*, *715*, 273–282. doi:10.1017/jfm.2012.517

Som, S. K., Biswas, G., & Chakraborty, S. (2017). *Introduction to Fluid Mechanics and Fluid Machines* (3rd ed.). McGraw Hill Education.

Sung, J., Kim, Y. B., & Lee, M. H. (2012). *Transient Phenomena of Dynamic Contact Angle in Micro Capillary Flow*. Paper presented at the 15th International Symposium on Flow Visualization, Minsk, Belarus.

Szekely, J., Neumann, A. W., & Chuang, Y. K. (1971). The rate of capillary penetration and the applicability of the washburn equation. *Journal of Colloid and Interface Science*, *35*(2), 273–278. doi:10.1016/0021-9797(71)90120-2

Washburn, E. W. (1921). The dynamics of capillary flow. *Physical Review*, *17*(3), 273–283. doi:10.1103/PhysRev.17.273

Zhmud, B. V., Tiberg, F., & Hallstensson, K. (2000). Dynamics of Capillary Rise. *Journal of Colloid and Interface Science*, *228*(2), 263–269. doi:10.1006/jcis.2000.6951 PMID:10926465

KEY TERMS AND DEFINITIONS

Capillarity: It is a phenomenon of rise or fall of liquid level in a small passage (tube of small cross-sectional area, space between the fibers of a towel) and it does not depend on the orientation of the passage.

CFD (Computational Fluid Dynamics): It is used to visualize how a fluid (both liquid and gas) flows as well as how the flow is being affected when it flows past any object by using the applied mathematics, physics, and computational software.

Contact Angle: It is the angle that a liquid forms in contact with solids or liquids also it is the measure of wettability of a surface or material.

Impedance Spectroscopy: It is a methodology where electrochemical impedance is usually measured by applying potential to an electrochemical cell using a small excitation signal produced by the micro-motion of the cells.

Linearization: It is a mathematical process of finding the linear approximation of inputs and corresponding outputs.

Meniscus: It is a phase boundary, usually a curved surface developed due to surface tension. It may be a concave or a convex in shape depending on the types of fluids are in contact.

Surface Tension: It is a phenomenon in which the surface of a liquid acts as an elastic sheet and always try to shrink to take the shape of minimum surface area.

VOF Method: Volume of fluid (VOF) method is a numerical technique for tracking and locating the free surface of meniscus.

Chapter 5
Immiscible Two-Phase Parallel Microflow and Its Applications in Fabricating Micro- and Nanomaterials

Yujie Li
Harbin Institute of Technology at Weihai, China

Shan Song
Harbin Institute of Technology at Weihai, China

Jie Wang
Harbin Institute of Technology at Weihai, China

Peng Zhang
Harbin Institute of Technology at Weihai, China

Shijie Wang
Harbin Institute of Technology at Weihai, China

Jianguo Li
Xi'an Microelectronic Technology Institute, China

Di Li
Harbin Institute of Technology at Weihai, China

Hai Yuan
Xi'an Microelectronic Technology Institute, China

ABSTRACT

The immiscible two-phase flow behaves nonlinearly, and it is a challenging task to control and stabilize the liquid-liquid interface. Parallel flow forms under a proper balance between the driving force, the friction resistance, and the interfacial tension. The liquid-solid interaction as well as the liquid-liquid interaction plays an important role in manipulating the liquid-liquid interface. With vacuum-driven flow, long and stable parallel flow is possible to be obtained in oil-water systems and can be used for fabricating micro- and nanomaterials. Ultra-small Cu nanoparticles of 4~10

DOI: 10.4018/978-1-5225-7138-4.ch005

nm were synthesized continuously through chemical reactions taking place on the interface. This makes it possible for in situ synthesis of conductive nanoink avoiding oxidation. Well-controlled interface reactions can also be used to produce ultra-long sub-micro Cu wires up to 10 mm at room temperature. This method provided new and simple additive fabrication methods for making integrated microfluidic devices.

INTRODUCTION

The term 'two-phase flow' in fluid mechanics is applied to simultaneous flow of: (a) a single fluid flowing as two different phases, such as steam and water; (b) two different fluids with different chemical properties occurring in the same phase, for example, the oil-water liquid-liquid system; or (c) two materials with different states or phases, for instance, water flow with air bubbles or sand grains inside.

In macroscale, no well-defined interfaces between two miscible materials can be maintained and detected due to rapid mixing through convection as well as diffusion. In this case, two-phase flow generally referrers only to flow systems composed of immiscible materials.

But when the characteristic length of the flow channel is small enough, the Reynolds number is usually small and the flow typically possess a laminar feature. The mass transfer in the system is geometrically confined and dominated by diffusion. That is why in situations when a two-phase micro-flow is concerned, mixing between the two phases is largely suppressed. An interface region can be clearly found even between miscible fluids flowing side by side, as shown in Figure 1a. The interface region disappears only when the two phases are put in contact with each other for long enough time. Thus in microfluidics, both miscible and immiscible two-phase flows are studied and used for various on-chip applications like chemical synthesis, material characterization, cell culture, extraction/separation, emulsification, and even Boolean calculations. In comparison with the blurred interface developed between the two miscible phases in Figure 1a, sharp interfaces are easily built and held between immiscible micro-fluids (Figure 1b). This results in more extensive applications of the immiscible two-phase micro-flows(Atencia & Beebe, 2005).

All the applications of the immiscible micro-flow systems are related to the mass and heat exchange across the interfaces between the two phases. The interface morphology is in many cases decisive while it is largely influenced by factors like the viscosities of the fluids, interfacial tension, flow rates, shape and characteristic size of the microchannel, and the chemical and geometrical states of the channel walls. Flow patterns, such as parallel flow, wavelet flow, annular flow, droplet flow, and segmented/slug flow, are defined accordingly (Yagodnitsyna, Kovalev, & Bilsky, 2015; Zhao, Chen, & Yuan, 2006a; Azarmanesh & Farhadi, 2016; de Menech,

Figure 1. Interfaces developed in two-phase microflow systems with (a) miscible and (b) immiscible fluids flowing side by side in microchannels

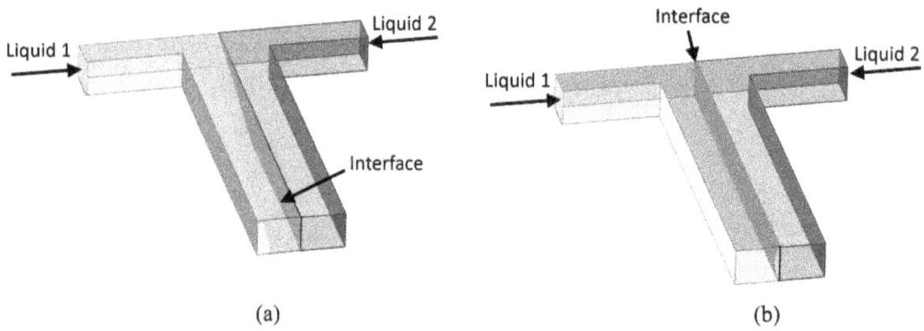

Garstecki, Jousse, & Stone, 2008; Salim, Fourar, Pironon, & Sausse, 2008). When the co-flowing two phases are running as laminar layers parallel to each other and the interface in between is even and straight, a parallel flow is formed. But if the two-phase interface is a wavy one, a wavelet pattern is built instead. Figure 2a and Figure 2b are parallel flow and wavelet flow recorded in an oil-water two-phase system. The annular flow is a pattern formed when one phase is moving continuously in the central part of the microchannel and enclosed by the other phase as if the enclosed phase is flowing in a pipe made by the outside phase. According to the above flow patterns, both phases are flowing continuously. A droplet microflow is formed when one of the immiscible phases (called the dispersed phase) is flowing in small discrete volumes while the other (called the continuous phase) is streaming continuously in a laminar way (Figure 2c). In a segment/slug flow pattern, one fluid phase is separated into segments by segments of the other phase (Figure 2d).

Figure 2. Optical microscope images of typical two-phase flow patterns. (a)A parallel flow, (b) a wavelet flow, (c) a droplet flow, and (d) a segment/slug flow

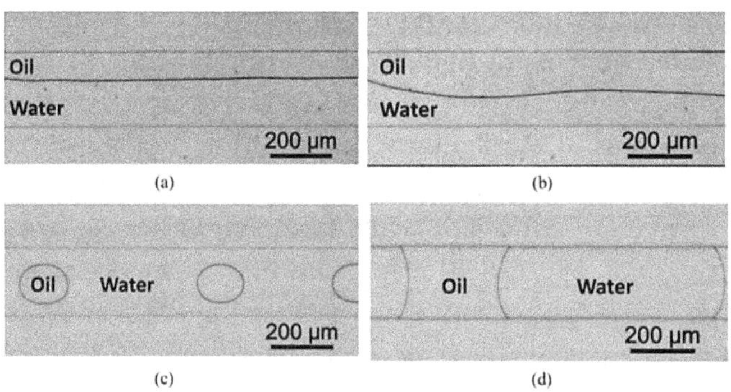

Among various flow patterns, the droplet flow is easier to obtain because of the hydrodynamic instability induced by the shearing forces. Droplet sizes can be tuned by adjusting the rate of flow of the continuous and disperse phases. Droplet based chemical and biological processes are benefitted from high and precise product selectivity, small reagent volumes, rapid reactions, efficient heat transfer, cell-like structure, cost reduction, and environmental protection.

However, for applications like continuous flow chemical processing (CFCP) (Aota, Mawatari, & Kitamori, 2009), it is then necessary to produce stable parallel microflows. The continuous interface established between the stratified co-flowing two immiscible phases enables the mass exchange or chemical reactions between the two streams occurring continuously. When multiphase parallel flow is established in a microchannel network, various chemical processes like mixing, single or multistep reaction, and separation can be combined in series and in parallel and a highly integrated microfluidic chip is functionalized.

In a certain immiscible microflow system, when the physical and chemical states of the channel and fluid phases are determined, the flow pattern is then dependent on the flow parameters. Flow rates, which can be either volumetric flow rates or mass flow rates, and phase fractions are the important ones to be considered. In order to show the influence of the flow parameters on the flow patterns, usually flow pattern diagrams are plotted against the superficial flow velocities (Salim, Fourar, Pironon, & Sausse, 2008), the flow rates (Guillot & Colin, 2005; Kashid & Kiwi-Minsker, 2011), the Reynolds numbers (Re) (Azarmanesh & Farhadi, 2016), the Weber numbers (We) (Yagodnitsyna, Kovalev, & Bilsky, 2015; Zhao, Chen, & Yuan, 2006b), or the Capillary numbers (Ca) (Azarmanesh & Farhadi, 2016) separately defined for each of the two phases. It is worth noting that not all flow patterns appear in a diagram necessarily. For example, in a microchannel with a depth of 40 μm and a width of 100 μm, when the immiscible phases were cyclohexane and water, respectively, only three patterns of flow were observed, namely the slug flow, the droplet flow, and the parallel flow, at Reynolds numbers for both phases are in the range of 0.3-10(Zhu, Fu, Gao, Liu, & Ma, 2011a).

The flow pattern diagram helps to find out proper flow parameters in controlling the desired flow pattern. The transition conditions between different patterns are identified in the diagram. However, actual control of the immiscible microflow is still quite challenging in microfluidics. The interfacial tension makes all dynamical problems nonlinear and the flow pattern stability is rather sensitive to small fluctuations in the driving and friction forces. Especially at very small dimensions of the flow channel when Reynolds number is much smaller than 1, the viscous forces of the fluid dominate the inertial forces. Shearing between the two phases easily break off the two phases flowing side-by-side into droplets of one phase dispersed in another.

To obtain long and stable parallel flow at low Reynolds numbers is rather a bottle neck technique for running multifunctional microfluidic chips based on CFCP.

Appropriate increase in the channel dimensions could help in solving such problems and most present microfluidic chips do have channels with the characteristic sizes in the range of tens to hundreds of micrometers. The Reynolds number applied can be as large as 1-100. Even in such cases, long and stable parallel flow has to be realized often with the help of novel design of the flow channels, such as selective chemical surface modifications of the microchannel walls (Zhao, Viernes, Moore, & Beebe, 2002; van der Linden, Jellema, Holwerda, & Verpoorte, 2006; Aota, Hibara, Sugii, & Kitamori, 2012) or guide structures on the solid surfaces (Aota, Hibara, & Kitamori, 2007). But in some applications, like microfluidic devices integrated in complicated multifunctional micro-electro-mechanical systems (MEMS) with miniaturization requirements or experimental simulation systems to study the flow in quasi-2D porous media (Scholz, Wirner, Li, & Bechinger, 2012) and the geometrical confinement effects (Cui, Diamant, Lin, & Rice, 2004) in micro- and nano-pores, the length of the channels may be around or even smaller than 10 μm. Correspondingly, the Reynolds number can be several orders of magnitude smaller than 1. In such channels, additional wall modification and patterning may greatly increase the difficulty of the fabrication processes or even impossible at present. To obtain and maintain a parallel flow for these applications are in a large demands but yet not available.

This chapter focuses then on control and applications of parallel microflow at very small Reynolds numbers ($< 10^{-2}$ for both phases). Co-flow of oil and water is an immiscible liquid-liquid two-phase system frequently encountered in fields like petroleum industries, chemical engineering, and biomedical science. Without loss of generality, oleic acid (analytical reagent) and water (possessed a resistivity of 18.25 MΩ·cm at 25 °C) flowing simultaneously in Y-shape channels of 240 μm in width and 8 μm in depth (the layout of a typical channel was schematically illustrated in Figure 1a) were studied and the flow pattern diagrams were mapped. The Y-shape design with the two inlets intersecting at 60° was applied since it was told to be conductive to the formation of parallel flow (Zhu, Fu, Gao, Liu, & Ma, 2011b). The channels were made in poly (dimethylsiloxane) (PDMS) by replication molding following a standard soft lithography process (Zhu, Fu, Gao, Liu, & Ma, 2011b; Xia & Whitesides, 1998a), with which the relief structures of the channels was produced with the SU-8 2005 photoresist on the surfaces of silicon wafers. A top and a bottom glass/quartz plate were bonded irreversibly to the PDMS replica to form a closed channel as shown in Figure 3b. Before bonding, the plates and the PDMS replica were treated by O_2 plasma to obtain hydrophilic channel surfaces. After being bonded the channel was tightly sealed around. The withstand pressure of the sealed flow system was larger than 2 atm (Wang, Wang, Zhang, & Li, 2017).

In order to control the temperature in the channel, glass sheets with ITO films sputtered on their backside surfaces were used as the bottom plates. The square resistance of the ITO films was about 13Ω/□.The two-phase flow were formed by first saturating the whole channel by water injected from one of the inlet through the capillary effect and then partially displaced by oleic acid injected from the other inlet. Both pressure higher and lower than the atmospheric pressure were imposed to drive the two-phase flow (Li, Wang, Li, & Liu, 2018).

Proper balance between the driving force, the friction resistance, and the interfacial tension necessary for stabilizing the oil-water parallel microflow was discussed. The liquid-solid interaction as well as the liquid-liquid interaction was found to play an important role in manipulating the liquid-liquid interface. By precise control of the flow, long and stable parallel flow was obtained. Dissolving metal salts and reducing agents into the oil and aqueous phases, respectively, the parallel flow was used as a novel method for fabricating micro- and nanomaterials. Continuous synthesis of ultra-small Cu grains of 4-6 nm in sizes as well as in situ preparation of ultra-long submicron Cu wires of up to 10 mm in length inside the microchannels were obtained. Both applications provide new and simple solutions for low-temperature interconnect for making integrated devices additively.

PRESSURE BALANCE AT THE LIQUID-LIQUID INTERFACE

The pressure balance at the liquid-liquid interface determines the interface morphology (Aota, Hibara, & Kitamori, 2007; Aota, Hibara, Sugii, & Kitamori, 2012). The pressure that drives the fluids decreases along the downstream direction in each phase because of the friction in the microchannel. For laminar flow, the friction factor f for a single-phase flow in a microchannel is usually expressed as:

Figure 3. A typical structure of the microfluidic devices. (a) The layout of the microchannels and (b) an exploded view of a device

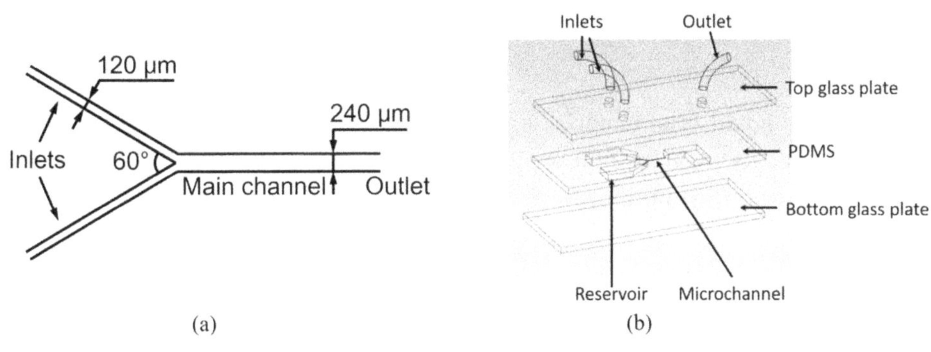

$$f = \frac{16}{\mathrm{Re}} = \frac{16\eta}{\rho v D_h} \tag{1}$$

where η, ρ, and v are the dynamic viscosity, the density, and the velocity of the fluid, respectively. D_h=4*S*/*L*, is the mean hydraulic diameter of the channel with *S* and *L* to be the cross section and the perimeter of the channel occupied by the corresponding phase.

The pressure loss in a certain phase can be then estimated by:

$$P_{loss} = \frac{2f\rho v^2 l}{D_h} = \frac{32\eta v l}{D_h^2} \tag{2}$$

here *l* is the length that the fluid passes through in the channel along the flow direction. However, this equation only sounds when non-slip boundaries are established under homogeneous wall conditions for a single fluid phase.

For an oil-water microflow system as shown in Figure 4, the pressure balance at a certain point on the interface under steady state is determined by both the hydrostatic pressure difference ΔP across the interface and the Laplace pressure P_L caused by the interfacial tension between the two phases. At a certain point on the interface, the hydrostatic pressure difference between the two liquid phases is expressed by:

$$\Delta P = P_o - P_w = \left(P_{in,o} - P_{loss,o}\right) - \left(P_{in,w} + P_{loss,w}\right) \tag{3}$$

The subscripts o and w stand for oleic acid and water, respectively. The representations are used throughout the whole chapter unless it is announced otherwise. P_{in} is the inlet pressure for either oil or water, with the outlet pressure P_{out} as a reference.

The Laplace pressure P_L is calculated according to the Young-Laplace equation:

$$P_L = \gamma\left(r_a^{-1} + r_r^{-1}\right) \tag{4}$$

where γ is the liquid-liquid interfacial tension; $r_{a,}$ and r_r the curvature radius of the interface in the xy plane and the xz plane in Figure 4, respectively. For a parallel flow, $r_a \to \infty$ and P_L simplifies to:

$$P_L = \gamma r_r^{-1} = \frac{2\gamma \sin\left(\theta - 90^o\right)}{h} \tag{5}$$

with h the height of channel, θ the contact angle of oil on the solid surface in balance with water. The value of θ is restricted to the values between the largest advancing contact angle θ_{adv} and the smallest receding contact angle θ_{rec}.

When the oil-water interface is pinned at a certain position, at each point on the interface:

$$P_L = \Delta P \tag{6}$$

$$\frac{2\gamma \sin\left(\theta_{rec} - 90^o\right)}{h} < \Delta P < \frac{2\gamma \sin\left(\theta_{adv} - 90^o\right)}{h} \tag{7}$$

The oleic acid-water interfacial tension γ measured and calculated by the pendant drop method (Adamson & Gast, 1997) at room temperature is about 10 mN/m. The sessile angle θ of the oil phase can be obtained by direct capturing of the profile of an oil drop on some solid surfaces in balance with water using a microscope optical system. When oil is continuously removed and added to the drop at a rate which is small enough, the dynamic contact angles θ_{adv} and θ_{rec} are measured just before the liquid-liquid-solid contact line starts to move. For various solid surfaces, such as PDMS, glass, and quartz, θ_{adv} and θ_{rec} are in the range of 160-180° and 96-154°, respectively (Li, Wang, Li, & Liu, 2018). The liquid-liquid interface protrudes from oleic acid to water as shown in the inset of Figure 4. In the case of a stable liquid-liquid interface, the pressure balance at each point on the interface is determined by Equation 6. The interface is pinned by the liquid-liquid-solid contact line on the upper and lower channel walls. When the balance at any place on the interface is broken, the local interface moves and changes its morphology.

FLOW STABILITY AND PRECISE CONTROL OF THE DRIVING PRESSURE

Microflows are usually actuated pneumatically, mechanically (Dessimoz, Cavin, Renken, & Kiwi-Minsker, 2008), electro kinetically (Stroock, Weck, Chiu, Huck, Kenis, Ismagilov, & Whitesideset al., 2000; Gascoyne, Vykoukal, Schwartz, Anderson, Vykoukal, Current, … Andrews, 2004), by the capillary force (Guillot & Colin,

Figure 4. Pressure balance at the oil-water interface

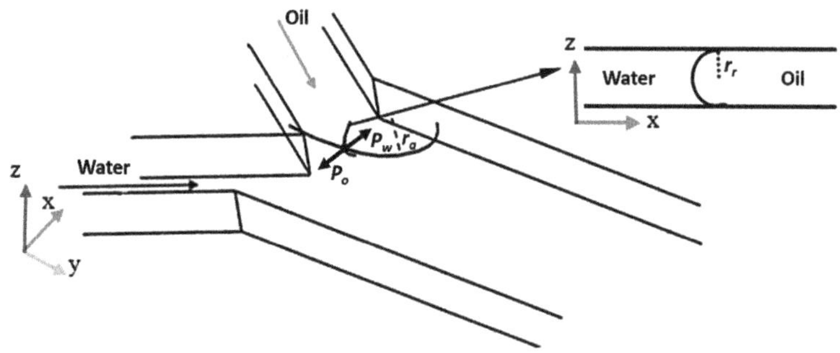

2005), or by the centrifugal force (Li, Chen, Li, Hu, & Zhao, 2010). Since immiscible two-phase microflow is sensitive to small fluctuations, the driving system has to be carefully designed to ensure stable and repeatable flow. Mechanical pumping may generate a pulsating flow. Electro kinetic pumping methods and driving by the centrifugal force can serve as continuous flow solutions. However, these are limited by the fluid concentration or design of the microfluidic chips. The capillary force needs to be combined with other forces to produce continuous flow. Pressure driven flow is commonly applied in microfluidics. The flow velocity can be adjusted in a large range simply. As long as the pressure sources are carefully designed and controlled, expected flow states can be maintained rather stable.

In the work concerned in this chapter, high pressure gas (N_2 or Ar) was supplied from high-pressure tanks and connected to both channel inlets by PTFE tubes. The gas pressure was adjusted by needle valves and measured by pressure gauges installed just before the inlets. The outlet of the main channel was simply exposed to atmospheric pressure or connected to a U-gauge loaded with water as the working media by a three-way valve to supply a certain level of vacuum. The total pressure driving system is schematically illustrated in Figure 5. As long as the gas tanks supplies a constant pressure, which is usually assured, and the valves are precise enough, the pressure control precision is determined by the pressure control at the outlet. Since the output resolution of the U-gauge used was as good as 4.9 Pa in the pressure range of -9.8~9.8 kPa (pressure in such a range was pretty enough for driving a flow at $Re < 10^{-2}$), the liquid-liquid interface morphology could be pinned at a certain position in the channel longer than 30 min. The flow interface morphology was imaged and recorded by a camera installed to a microscope focused inside the channel. When a parallel flow was built and stably hold, two interface images taken half an hour apart were overlapped to each other and shown in Figure 6. No visible change in the interface morphology and position was observable. To test the reproducibility of a flow pattern in such a system, the outlet pressure and the pressure at one of the inlets were kept constant while the pressure at the other inlet was first increased and then decreased back to its original value. The original and the final interface morphologies were also imaged and compared. No observable change was found as well. The driving system based on simple pressure control by a hydraulic head was proved to be quite efficient and precise. The pressure control precision was higher than most of the commercial systems. Another big advantage in using the U-gauge in the driving system was that it supplies pressure not only higher than but also lower than the atmospheric pressure. Vacuum realized in such a way fully avoided the pulsation effect induced by the pumps. The only shortcoming of the system was its difficulties in becoming portable and miniaturized.

Figure 5. A schematic picture of the pressure control system

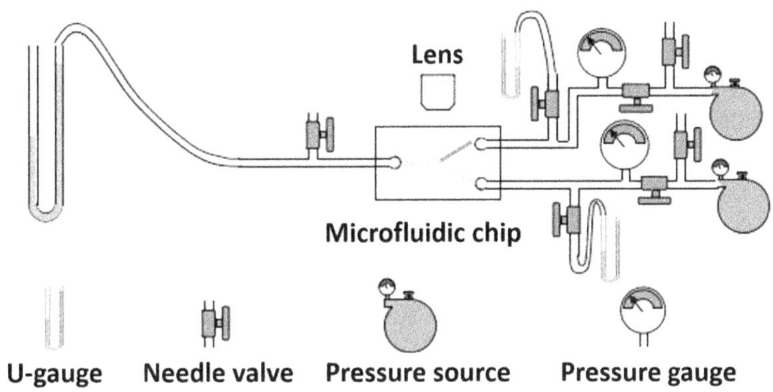

Figure 6. A picture made by overlapping two optical microscope images of a parallel flow taken half an hour apart. Both the morphology and the position of the interface were found unchanged

FLOW PATTERN CONTROL FOR OIL-WATER TWO-PHASE MICROFLOWS

With precise and repeatable control of the driving pressure, the flow parameters were definitely determined and the flow pattern diagram was possible to be plotted.

Estimation of the Flow Velocities and the Reynolds Numbers for Both Phases

In such small microchannels, it was very difficult to measure the flow rate or flow velocity directly. For dispersed flows (either droplet or slug flows) the flow velocities are estimated by the travelling time of each drop or slug between the inlet and outlet of a channel (Salim, Fourar, Pironon, & Sausse, 2008). This method can be modified for dispersed flows formed in thin channels. Since the channel height h was the characteristic dimension of the channel and the width and length of a drop/slug were tens of times larger than h, the drops/slugs were believed to fill the entire channel in the direction of height. Therefor the volume of the drop/slug V was estimated to be:

$$V = C \cdot h \tag{8}$$

where C was the cross-sectional area of each drop. In a certain time interval t, the drop/slug numbers N and the distance between a drop/slug and its follower L were determined from the flow video recorded. In an oleid acid-water system, only flow with oil as the dispersed phase was detected. The flow rate of the oil drops/slugs was then estimated by:

$$Q_o = \sum_N V_i / t \tag{9}$$

and the flow rate of water flowing continuously is:

$$Q_w = \left(\sum_N L_i \cdot W \cdot h - \sum_N V_i \right) / t \tag{10}$$

where W was the width of the main channel.

In order to find out the transition conditions between the dispersed flow and the parallel flow, a critical state was defined. At the critical state, when the driving pressure of either phase was changed by a very small value (according to the control accuracy of the pressure driven system) the dispersed flow would turn into a parallel flow or vice versa. The flow rates of both water and oil for a dispersed flow at the critical state could be calculated by Equations (9) and (10).

After turning into a parallel flow (also in the critical state), the widths of oil (W_o) and water (W_w) phases were measured. The corresponding velocities of oil and water were then estimated by:

$$u_o = \frac{Q_o}{h \cdot W_o} \tag{11}$$

and

$$u_w = \frac{Q_w}{h \cdot W_w} \tag{12}$$

The Reynolds numbers for both phases were then estimated by:

$$\mathrm{Re}_j = \frac{\rho_j u_j D_{h,j}}{\eta_j} \tag{13}$$

where the subscript j could either stand for 'o' or 'w' and accordingly ρ, u, η, and D_h were the density, velocity, dynamic viscosity, and mean hydraulic diameter of the corresponding phase. For both the oil and water phases, their densities and the dynamic viscosities are largely influenced by temperature. The values under a certain temperature were referenced from related literatures (Noureddini, Teoh, & Davis Clements, 1992; Noureddini, Teoh, & Davis Clements, 1992; Al-Shemmeri, 2012). The mean hydraulic diameter of a certain phase was calculated by the height of the main channel and the actual channel width occupied by the very phase measured from the microscopic images.

By using the expressions given above, the velocities calculated for the oil and water phases were in the range of 40-400 μm/s and 100-3000 μm/s, respectively. And the Reynolds numbers were 1×10^{-5}-1×10^{-3} for the oil phase and 1×10^{-3}-2×10^{-1} for the water phase.

Flow Pattern Control of the Oleid Acid-Water System

The diagrams were plotted against the driving pressures instead of the dimensionless numbers. This helped in understanding the pressure balance in such flow systems.

The flow velocities for oil and water were in the range of 40-400 μm/s and 100-3000 μm/s, respectively. And *Re* was in the range of 1×10^{-5} - 1×10^{-3} for oil and 1×10^{-3} - 2×10^{-1} for water, respectively, when a parallel flow was obtained.

When the characteristic dimension of the flow channel is smaller than 10 μm and the original channel wall is hydrophilic, merely displacement of water by oleic acid is applicable, instead of the other way round. The main channel was first saturated by water injected from one of the inlets and then partially displaced by oleic acid injected from the other inlet to form the two-phase flow.

At room temperature, four different flow patterns, namely the single-phase flow of water (SW), the droplet flow with the oil droplets dispersed in water (DF), the parallel flow (PF), and the single-phase flow of oleic acid (SO) appeared in turn in the measured flow pattern diagram when $P_{in,\,o}$ is increasing. No slug flow or droplet flow with water as the dispersed phase were found under the given geometric and material conditions. An annular flow with oleic acid surrounded by water only appears when the channel walls were pretreated to be superhydrophilic. This indicated that the oil phase was actually flowing on top of a very thin layer of water molecules absorbed by the channel wall before oil was introduced inside. Consequently, a faster pressure loss existed in the water phase despite the higher viscosity of the oil phase. A typical flow pattern diagram measured at the so-called homogeneous wall condition, which means that the top, bottom, and side walls of the corresponding microchannels were indistinctively made of the same material, was plotted in Figure 7. At low inlet pressures for both the aqueous and organic phases (P_w and P_o), the parallel flow regime is rather small. This primarily causes the difficulties in obtaining long and stable parallel flow at low Reynolds numbers.

According to Equation 7, lowering the liquid-liquid interfacial tension may be in favor of the parallel flow. When SDS was added as a surfactant in the aqueous solution with its concentration increased from 0 to 0.012 mol/L, the interfacial tension decreased monotonically by an order of magnitude. However, the parallel flow regime in the flow pattern diagram, especially at the low driving pressure corner, did not expand as expected. On the contrary, it first shrank into a narrow band and then gradually became larger at SDS concentrations higher than 0.002 mol/L. The parallel flow obtained with SDS added in the water phase had a rather unstable liquid-liquid interface lied in between. Small disturbance like mild perturbations in the fluid flow or tiny pinning points existing on the surface of the channel walls caused localized vortexes and led to interface fluctuations.

Raising the temperature can also lower the liquid-liquid interfacial tension. But the experimental results were also unexpected. When the temperature was raised from 25 °C, the room temperature, to 58 °C, the coverage of the parallel flow regime did not change obviously at the low pressure end. Instead, the single phase flow

Figure 7. A typical flow pattern diagram measured in a PDMS channel. The points in the figure were the transition points between different patterns measured in the experiments while the lines were just used to guide the eye

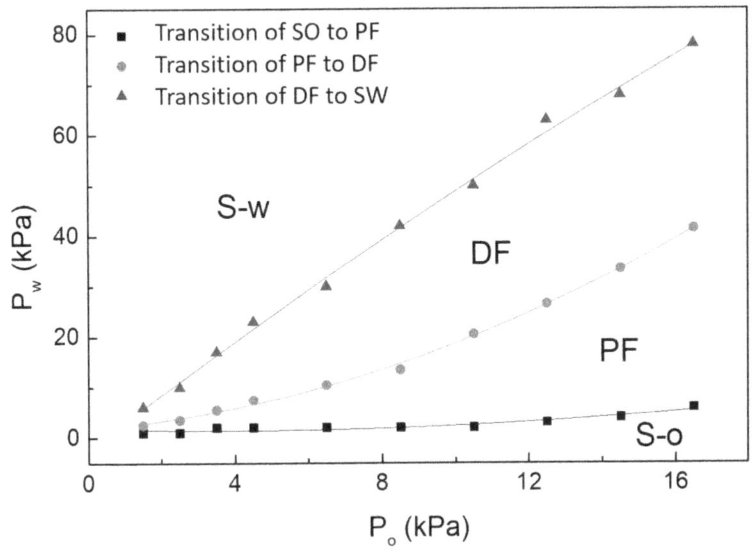

regime for water was enlarged and the parallel flow regime shifted tremendously towards the high $P_{in,o}$ side.

Parallel flow forms under a proper balance between the driving force, the friction resistance, and the Laplace pressure on the liquid-liquid interface. To lower the liquid-liquid interfacial tension γ does help to reach the balance at low Re. But on the other hand, the smaller γ is, the smaller the range of ΔP is allowed for stabilizing a parallel flow according to Equation 7. Moreover, the lower the liquid-liquid interfacial tension is, the more unstable the liquid-liquid interface becomes. These explain that why lowering γ was not an effective way in obtaining the parallel flow in the experiments.

In fact, not only the liquid-liquid interfacial tension but also the solid-liquid interfacial tension played an important role in controlling the flow pattern. The solid wall properties has two effects on the flow: (1) the uniformity of the solid surface largely influenced the dynamic contact angles of the liquid-liquid-solid three-phase contact line on the interface and as a result the interface balance is affected, and; (2) more importantly, the liquid-solid interaction determines the friction resistance exerted by the solid wall to the flow. The weaker the solid-liquid interaction, the smaller the friction resistance. When large amount of SDS was added to the fluid, the solid walls were covered by the surfactants. The friction resistance, especially on the side of the water phase, was remarkably lowered. This led to the expansion of the parallel flow regime on the flow pattern diagram. The solid-liquid interaction also accounted for the shifting of the parallel flow regime at high temperature. The channel walls were verified to have better lipophilicity at higher temperatures and the friction resistance exerted on the oil phase turned to be much higher than that at room temperature. And as a result much higher $P_{in,o}$ was necessary for balancing the pressure on the interface.

The importance of the liquid-solid interaction in controlling the flow pattern was also directly proved by comparing the patterns formed by oleic acid-water two-phase flow under the same geometric and driving conditions on various solid channel walls. The parallel flow regime was compressed when the solid surface was better wetted by the oil phase. And when the channel walls were made by two or more different kinds of materials (the so called inhomogeneous wall condition), the two-phase interfaces were largely disturbed and it was more difficult to maintain a stable parallel flow.

Long and Stable Parallel Flow Obtained at Low Re

When the liquid-liquid interaction keeps unchanged, strong interaction between the solid wall and the liquid phase increases the flow resistance for the corresponding phase and resulted in a narrowed parallel flow regime. On the contrary, lowering the flow resistance helps for expanding the parallel flow regime and make it easier for obtaining stable parallel flows, especially at low Re. A simple and convenient method to lower the flow resistance for both phases without changing the liquid-liquid interfacial tension was to supply a certain level of vacuum at the channel outlet. The water molecule layer covered on the solid wall surface become thicker geometrically and thinner in density. Friction resistance for both the water and oil phases flowing on top of such a layer was decreased remarkably while more pronounced lowering of the friction for oil took place. The flow pattern diagram measured with vacuum indeed had a significantly larger parallel flow regime. Under a condition that the vacuum level is high enough, the droplet flow regime can be completely depressed at the low $P_{in,w}$ end. Merely three flow regimes, namely the single phase flow of water, the parallel flow, and the single phase flow of oil were identified. The parallel flow turned into the single phase flow of water directly as both $P_{in,w}$ and $P_{in,o}$ were small enough. This is clearly shown in Figure 8a. In such a case, stable parallel flow as long as 3-10 mm could be obtained at $Re < 10^{-2}$ for both phases. The straight interface lay between the two liquid fluids imaged in Figure 8b remained unchanged for hours.

Figure 8. Vacuum controlled microflow. (a) A typical flow pattern diagram measured when the two-phase flow is driven by $P_{in,o} = P_{in,w} = 1.013\times10^5$ Pa and $P_{out} = -4000$ Pa and (b) the stable parallel flow formed as long as 3 mm.

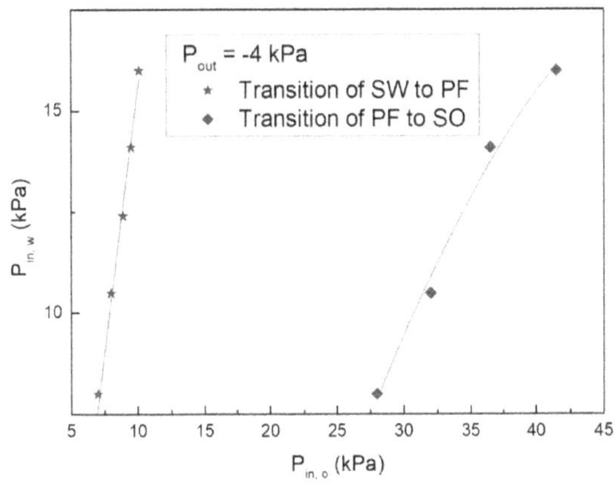

(a)

(b)

MICRO- AND NANO-MATERIALS FABRICATED CONTINUOUSLY

One of the purpose for microfluidics is to fabricate micro- and nano-materials with high and precise product selectivity, small reagent volumes, efficient heat transfer, reduced cost, environmental protection, and easy integration. Various flow patterns find their own superiorities in being used as microreactors. For example, multiple reagents can be transported in small droplets without dispersion (Song, Chen, & Ismagilov, 2006) and the droplet flow are widely used for cell/drag encapsulation (wiki, Flow focusing, 4,5), particle labeling, core-shell or even multi-core particle synthesis (wiki, Droplet-based microfluidics, 1,4). While intensified mixing is realized by the internal circulation within the liquid segments or slugs (Günther, Khan, Thalmann, Trachsel, & Jensen, 2004), segment/slug flow-based reactions can be as fast as in kilohertz rates. As for the parallel flow, when reactive solutions are flowing in the microchannels in well-controlled parallel manner, various materials, including metal, non-metal inorganic materials, and organics, in shapes of particles, tubes, or fibers, are synthesized continuously (Kenis, Ismagilov, & Whitesides, 1999). When metal nanoparticles, including gold, sliver, cobalt, palladium, and copper nanoparticles were synthesized in micro-reactors through flow chemistry, smaller particle sizes and narrower size distributions were usually obtained (Badilescu & Packirisamy, 2012; Marre & Jensen, 2010; Lohse, Eller, Sivapalan, Plews, & Murphy, 2013). For chemically active materials like copper or copper based complexes, one of the most important advantages for continuous synthesis in microchannels is the improved stability to oxidation due to the sealed and liquid saturated environment (Badilescu & Packirisamy, 2012; McQuade & Seeberger, 2013). If the products can be deposited on the inner walls of the microchannels to generate desired structures, such a method is then a special CFCP application called multiphase laminar flow patterning (Zhao Viernes, Moore, & Beebe, 2002; Kenis, Ismagilov, & Whitesides, 1999). It has been applied to pattern both inorganic and organic materials in capillaries.

No doubt that the droplet and segment flow used for material synthesizing are immiscible multiphase systems. However, in most cases as for parallel microflows miscible systems are employed (Badilescu & Packirisamy, 2012; Kenis, Ismagilov, & Whitesides, 1999) due to the difficulties in operating the immiscible parallel flow. Under such conditions, the feature sizes of the products are relatively large, usually in 5-100 μm range, and not uniform because of the diffusional mixing across the liquid-liquid interfaces. If stable immiscible parallel flow is practice, the sharp and clear interfaces without the defocusing effect caused by the radial diffusion may act as a strong geometric confinement for reactions. The feature sizes of the products can be much smaller.

Ultra-Small Cu Nanoparticles Continuously Synthesized

Metal salts dissolved in oleic acid and reducing agents dissolved in water or vice versa have been used as the oil and aqueous phases, respectively, for establishing a stable reactive parallel flow. According to the molecular dynamics (MD) simulation, the aqueous-oil interface was established within several tens of picosecond after the two phases were put in contact with each other. The interface thickness was related to the mutual solubility between the two fluids. If pure oleic acid and water phases were considered, the liquid-liquid interface was as thin as about 15 Å. If reactants were present and the reactants originally dissolved in one phase was slightly soluble in the other, the interface became thicker. To control the interface morphology in microchannels, the flow pattern diagram for such reaction immiscible systems was explored. The main features of the diagram were not significantly changed by the solvents. And a parallel flow has been achieved successfully in a microchannel as long as 6 mm when the Reynolds numbers for both phases were smaller than 10^{-2}.

When copper oleate and ascorbic acid were used to react with each other through the oil-water interface, ultra-small Cu nanoparticles were synthesized through reactions taking place at the water-oil interface. Figure 9a is a typical XRD spectrum of the nanoparticles collected at the outlet of the microchannel. Peaks centered at $2\theta = 43.5°$, 50.6°, and 74.3°, corresponding to (111), (200), and (220) of Cu were detected. The full width at half maximum (FWHM) of each peak was obviously broadened. Figure 9b is a typical high-resolution TEM (HRTEM) image of the Cu nanoparticals. Each particle was a grain of 4~8 nm in size. The distances between two adjacent lattice planes were in the range of 2.06-2.09 Å. The values only deviate slightly (< 1%) to 2.08 Å, the distance between two adjacent fcc {111} of copper. No copper oxides were detected in neither the XRD nor the TEM results. The particles possessed lipophilic properties. As long as the parallel flow was running, the particles were produced continuously and could be collected at the outlet reservoir of the microfluidic chip.

The particles were too small to be separated by centrifuging. Such small metal particles are seldom obtained. They are in principle easily agglomerated to form larger grains. TEM measurements were also made for copper nanoparticles stored in air for as long as 27 days. No oxidation was detected and the particles showed no apparent aggregation or coarsening. The average particle sizes were measured to be still smaller than 10 nm. Though no capping layers were observed on the particle surfaces in the HRTEM images, octadecylene groups originated from copper oleate were believed to cover the particle surface and stabilize them. Supposing that this process can take place in the nozzle of an inkjet printer, it is then possible for in situ synthesis of the conductive nanoink composed of ultra-small metal particles. This provides a solution for the nozzle clogging problem of the particle containing

Figure 9. Ultra-small Cu nanoparticles synthesized continuously by parallel flow. (a) A typical XRD spectrum and (b) a typical HRTEM image of the nanoparticles collected at the outlet of the microchannel

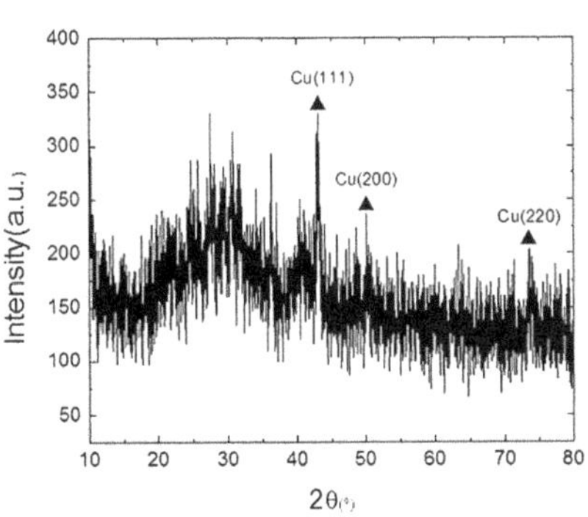

(a)

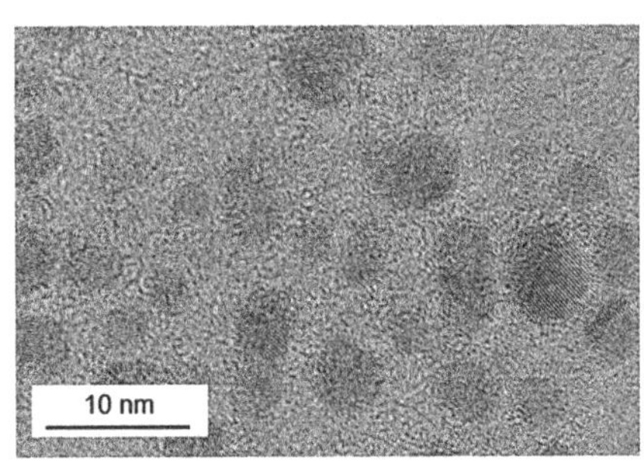

(b)

ink available on the market and the oxidation problem of active metal ink during storage and applications.

The ultra-small sizes of the copper particles resulted from the confined diffusion and reaction on the oil-water interface. If the reactants solubility in both phases were carefully matched, the interface thickness was broadened. The particles obtained then had sizes controlled from ten to several tens of nanometers due to faster mass transferring processes.

Ultra-Long Submicron Cu Wires Fabricated in Situ in Microchannels

As microfluidics is becoming a more and more important branch in the field of Micro-electro-mechanical systems (MEMS), microfluidic chips are integrated with microelectric parts to constitute multifunctional system-on-chip (SoC) or system-in-a-package (SiP) modules. However, the fabrication of the microchannels has always been a tough task (Kumar, 2017). Microchannels, which are microstructures in three dimensions (3D), are essential elements for all kinds of microfluidic systems. Materials far beyond silicon are involved in preparing the microchannels. Non-conventional microfabrication methods such as micro-machining, stereolithography, soft lithography, embossing processes, and LIGA usually have to be used. Yet these are not all the difficulties. In order to exchange information and energy with the external signal and power sources, electrical interconnect patterns, including discrete co-planar or non-coplanar electrodes (Zhang, Hu, Chen, Fan, Wang, Zheng, ... Qian, 2017; Zhang, Li, Xu, Chencheng, Hao, Meng, & Gui, 2017)and microelectrode arrays (Kim, Li, & Park, 2011; Jin, Li, Williams, Lee, & Park, 2011; Choi, Park, Hao, Holman, & Pisano, 2012), have to be fabricated and integrated directly inside the microchannels. Submicron interconnect wires are in most cases required. The transport, connection, and packaging of the fluids in the microchannels have to be taken into account during the electrical interconnect processes. And locating the electrical interconnect patterns precisely in the three dimensional microchannels also takes great efforts. It is therefore very difficult for traditional micro-patterning principles and processes to be adopted for the construction of interior electrical or other functional patterns inside the microchannels.

When the flow rate of the reactive two-phase parallel flow, the reaction rate, and the channel wall properties are suitably controlled, materials produced through the interface reactions can deposit on or be removed from the wall surfaces to conduct the multiphase laminar flow patterning as stated before. This technique is quite flexible and useful in patterning the inner walls of the microchannels. And it is a low cost patterning technique compatible with the microfluidic technique and 3D micro-fabrication processes. Metal wires were synthesized in this manner to make interconnect structures like a three-electrode system inside a 200-μm-wide channel (Atencia & Beebe, 2005). But due to the difficulties in handling immiscible parallel flows, such pattering technique mainly employs miscible multiphase microflows at the moment. The feature sizes of the patterns obtained by miscible laminar flows are only down to 5 micrometers and not uniform. Moreover, the special localization of the patterns is also limited to several micrometers. If immiscible fluid systems are used, sharp fluid interfaces can be pinned at desired positions and smaller patterns and better localization can be expected.

Based on the methods shown in this chapter, long and stable parallel flow have been realized in oil-water systems at low Reynolds numbers. The interface formed between the immiscible two liquid phases ran through the full length of the main channel. It located just in the center of the main channel as expected and did not show any movement throughout the reaction process. Copper oleate was serving as the metal source from the oil phase side and ascorbic acid as the reduction reagent from the aqueous phase side. The reaction was kept lasting for 10-60 min.

A wire as long as 3-10 mm was deposited on the bottom surface of the channel. Figure 10a is an SEM image of the whole wire taken at low magnification. When enlarged, the wire was found to be composed of submicron copper particles with dimensions of about 500-1000 nm. The width of the wire measured at different positions along the flow direction was plotted in Figure 10b. The average value of the wire width was about 730 nm and the standard deviation about 180 nm. A slight tendency for the width to become larger from the beginning to the end was detected. But this phenomenon is caused by deposition of multi-particles at the same spot on the liquid-liquid interface at the downstream end of the channel instead of broadening of the reactive region due to diffusion across the interface. The length-to-radius ratio of the wire was as large as 10^{-4}.

The shape and anisotropy of materials may largely influence their optical, mechanical, magnetic, and electrical properties, especially in nanoscale. And the geometry of the materials in many cases determines the design and characters of the devices which can be made accordingly. Researchers in the field of nanotechnology have put a lot of energy in fabricating one dimensional nanomaterials with large aspect ratios. When synthesized with hard templates (usually alumina membranes were preferred) (Murphy & Jana, 2002; Yang, Shi, Zhang, Hao, Liu, Feng, & Cui, 2018) as the reactor, the aspect ratio of the products can be varied by growth parameters but limited by the pore size of the membrane. Soft templates, such as rodlike micelles, microemulsions, and surfactants, can also lead to anisotropic growth of nanomaterials. A multistep surfactant-directed growth of gold and silver nanorods and nanowires (Yang, Shi, Zhang, Hao, Liu, Feng, & Cui, 2018; Murphy & Jana, 2002) from seeds was developed to control the aspect ratio by varying the ratio of seed to metal salt and the time interval between synthesis steps. With optimized pH values of the reacting

Figure 10. A submicron Cu wire obtained. (a) A low magnification SEM image; and (b) the width of the wire at spots with different distances to the inlet of the main channel. The lines are just used to guide the eye

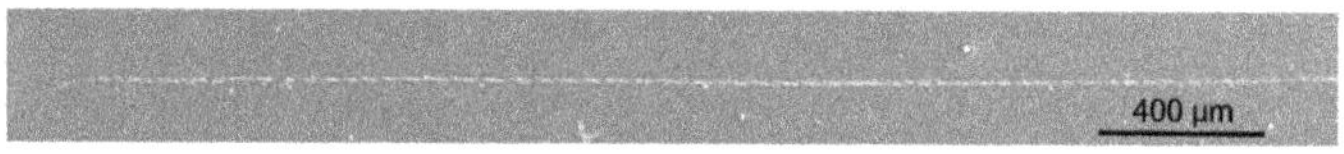

(a)

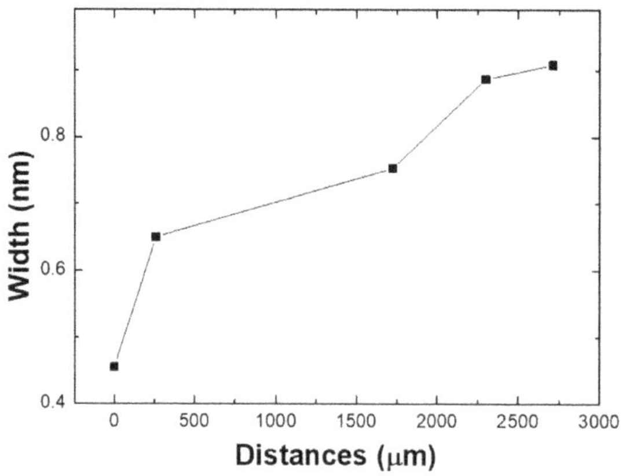

(b)

solution, the aspect ratio of up to 100 was gained. A most recent work made with the help of graphene templates showed the possibility of making gold nanowires of very high aspect ratio (Xin, De Rosa, Cao, Yin, Yu, Ye, … Yang, 2018). In the absence of ultrasound during the growth process, the average aspect ratio valued around 1100 while in the presence of ultrasound the ratio rose up to 5000. But for wires synthesized by the parallel flow method, especially with immiscible flow systems, the aspect ratio actually does not show an obvious limitation. As long as the microchannel is long enough and the flow control is precise and stable enough, a wire with nano- to submicron diameter and length in minimeters are available.

Another important issue in applications of low dimensional materials is lacking of an effective method to assemble them to 2D or 3D structures in devices and chips with high spatial precision. Radom dispersion (Adib, Eckstein, Hernandez-Sosa, Sommer, & Lemmer, 2018), catalyst-induced localized growth (Thelander et al., 2006), hydrodynamic assembly (Xing & Dittrich, 2018; S. Zhang, Yang, & Nie, 2018), and optical trapping technologies (Pauzauskie, Radenovic, Trepagnier, Shroff, Yang, & Liphardt, 2006) have been introduced. Optical traps adopted in transferring nanowires with aspect ratios of more than 100 could achieve a rather high operating accuracy in three dimensions (Pauzauskie, Radenovic, Trepagnier, Shroff, Yang, &

Liphardt, 2006). Positioning of extremely long nanowires in predetermined complex structures like microchannels of various layouts are still a rarely surveyed technique. According to the content in this chapter, in well-controlled immiscible parallel microflow, the localization of the liquid-liquid interface between the two phases was easily adjusted by varying the driving pressure of the liquid phases. The flow reacted sensitively and reproducibly to the driving pressure. And the wire fabricated was positioned at places accordingly in the microchannel irrespective to its length.

The Electrical Property of the Ultra-Long Copper Wire

Since the microchannel was saturated entirely by reductive liquids during the whole process, the main constituent of the wire was copper and little oxides were detected. Electrodes located at both ends of the main channel was deposited on the bottom plate of the channel by magnetron sputtering before fabricating the microfluidic device. Each electrode composed of metals like Ti and/or Cu several to tens of nanometers in thickness. With the thin design of the electrodes, only slight influence of the step imposed on the flow control was recorded. The parallel flow interface remained to be straight and stable. The copper wire synthesized in situ by the interface reaction connected the two electrodes and could act as an interconnect structure. Figure 11a and Figure 11b are respectively the schematic and real pictures of the microfluidic device to show the relative positions of the microchannel, the electrodes, and the interconnect wire fabricated. I-V curves were measured to show the electrical property of the wire. A typical one was plotted in Figure 11c. An ohmic-like behavior was clearly demonstrated. Good electrical conductivity of the interconnect wire and acceptable contact between the wire and the electrodes have been gained without further heat treatment. The synthesis method based on immiscible parallel flow is useful not only in patterning the channels but also in producing functioning parts in situ as the interconnect shown here.

FUTURE RESEARCH DIRECTIONS

Here only synthesizing of copper materials was presented. However, the application of immiscible liquid-liquid two-phase parallel microflow in continuous fabrication of micro- and nanomaterials is far beyond this. In principle, as long as proper immiscible fluid systems can be found and the reactants' solubility is suitable for the desired thickness of the parallel flow interface, nanoparticles and wires of metals, semiconductors, and insulators can all be made in such a manner. What's more, an important feature for CFCP is that multistep reactions are possible without disturbing

Figure 11. The electrical property of the interconnect wire. (a) A schematic picture of the interconnect structure within the microchannel. (b) An image of the microfluidic device aligned with the electrodes. And (c) a typical I-V curve

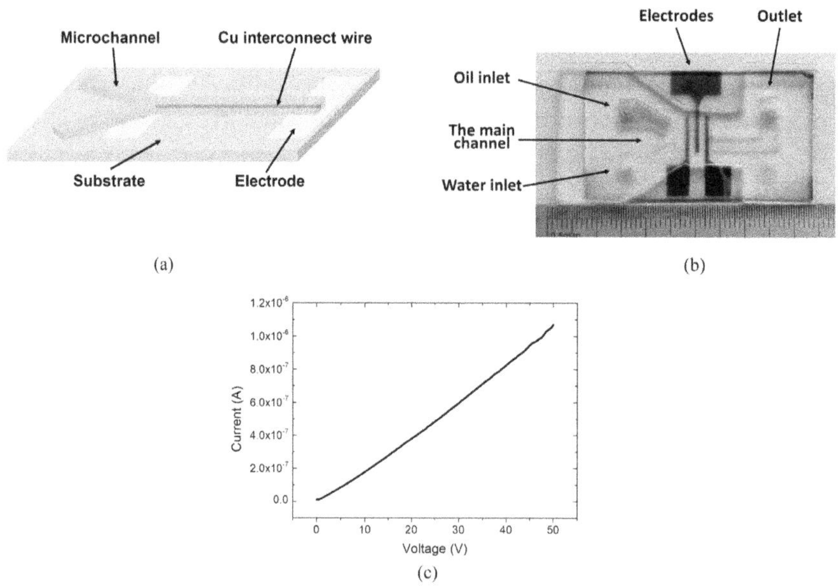

the continuity and stability of the flow. If reactants are changed during flowing of the parallel system, the products can then be modulated temporarily. Therefore, alloys, core-shell wires, and even hetero-junctions are expectable.

Except for the novelty of the fabrication methods itself, practical commercial applications have always been one of the biggest challenges for microflow chemistry (Whitesides, 2006). Though microfluidic devices and components fabricated based on PDMS and/or similar organic materials make it possible for cheap manufacturing of the devices, real-time optical observation and analysis, and good biological compatibility (Xia & Whitesides, 1998b), integration of multifunctional MEMS with microfluidics asks for materials like silicon, glass, and steel. Higher mechanical strength, better chemical and thermal stability of the devices are then assured and synthesizing conditions can be investigated in wider ranges. Easy scale-up automation is another important concern (McQuade & Seeberger, 2013; Lohse, Eller, Sivapalan, Plews, & Murphy, 2013). Multiple microreactors simply running in parallel may not be an effective solution due to the small dimensions and flow rates necessary to pin the liquid-liquid interface vertical to the substrate surfaces. New setup or assembly of the devices and components as well as new flow control or reaction strategies may be necessary.

Almost any reactions can be performed continuously in microchannels (McQuade & Seeberger, 2013). However, in small scales, especially in dimensions down to the ones applied in this work or even smaller, effects like competition and coupling between the mass transport in the liquid phases and the chemical reaction taking place on the interface between the two phases, boundary influences due to the long-range nature of viscous flows and so on are to be studied in details for better manipulation of the flow and the reaction processes (Squires & Quake, 2005). The physics behind the fabrication provides clear synthesis mechanisms and kinetics. This can help us find out the key controlling factors in obtaining materials with desired sizes, morphologies and functionalized surface properties.

CONCLUSION

In this chapter, the control of immiscible liquid-liquid two-phase flow has been discussed and tested thoroughly. Oleic acid-water flow in microchannels of 240 μm in width and 8 μm in height was studied as a microfluidic demo system for building blocks of future multifunctional integrated SoC or SiP with rather strong geometrical confinements of the microchannels due to the miniaturization demands. Under such conditions the nonlinearity of the flow made the control of flow patterns, especially the parallel flow, very difficult.

The flow pattern diagrams were measured in order to investigate the influences of surfactants, temperature, and channel wall properties. A rather complicated transition behavior between the parallel flow and the droplet flow was found. Parallel flow formed under a proper balance between the driving force, the friction resistance, and the interfacial tension. Not only the liquid-liquid interaction but also the liquid-solid interaction played an important role in manipulating the liquid-liquid interface. By driving the flow with carefully chosen and designed pressure sources, various flow patterns were obtained with very good stability and reproducibility. In addition to high pressure driving forces, certain levels of vacuum were also be applied. The first layer of water attached to the solid walls became lower in density and thicker in thickness under evacuation. The flow resistance for both phases were remarkably lowered. Long and stable parallel flows at the Reynolds numbers smaller than 10^{-2} for both phases have been obtained without any chemical modifications of the solid surfaces, additional channel structures, or surfactants.

The stable parallel flow was used as a continuous reactor for fabricating nanosized copper particles. The sizes of the particles were determined by the solubility of reactants in the liquid phases. When the metal salts and the reduction agent only dissolve in one of the phases, the interface was sharp enough to produce ultra-small

Cu nanoparticles with sizes smaller than 10 nm. This is a potential method for in situ synthesizing of conductive nanoink for printable electronics.

With well-adjusted flow rates, reaction rates, and channel wall properties, Cu interconnect wires with their widths smaller than 1 μm was deposited in situ in microchannels by the interface reactions taking place between the two immiscible liquid phases. The spatial localization of the wires were easily controlled and the wire length was up to 10 mm. Easy positioning and ultra-large aspect ratio ($\geq 10^4$) of the produced wire make the method itself an important extension of the multiphase laminar flow patterning technique for employing immiscible liquid-liquid phases. Ohmic behavior and good conductivity of the Cu interconnect wire were obtained without any heat treatment. A new low-temperature interconnect method which is simple, green, and compatible with the microfluidic techniques has been proposed.

ACKNOWLEDGMENT

This research was supported by the National Natural Science Foundation of China [grant number 51205082]; the Natural Science Foundation of Shandong Province, China [grant number ZR2017MEE002]; and Pre-research Project on Equipment in the 13th-Five-Year-Plan [JZX2017-1501/Y395].

REFERENCES

Adamson, A. W., & Gast, A. P. (1997). *Physical Chemistry of Surfaces* (6th ed.). New York: John Wiley & Sons, Inc.

Adib, M., Eckstein, R., Hernandez-Sosa, G., Sommer, M., & Lemmer, U. (2018). SnO 2 Nanowire-Based Aerosol Jet Printed Electronic Nose as Fire Detector. *IEEE Sensors Journal*, *18*(2), 494–500. doi:10.1109/JSEN.2017.2777178

Al-Shemmeri, T. (2012). *Engineering Fluid Mechanics*. Ventus Publishing ApS.

Aota, A., Hibara, A., & Kitamori, T. (2007). Pressure Balance at the Liquid–Liquid Interface of Micro Countercurrent Flows in Microchips. *Analytical Chemistry*, *79*(10), 3919–3924. doi:10.1021/ac070031d PMID:17439241

Aota, A., Hibara, A., Sugii, Y., & Kitamori, T. (2012). Shape of the Liquid–Liquid Interface in Micro Counter-Current Flows. *Analytical Sciences*, *28*(1), 9. doi:10.2116/analsci.28.9 PMID:22232217

Aota, A., Mawatari, K., & Kitamori, T. (2009). Parallel multiphase microflows: Fundamental physics, stabilization methods and applications. *Lab on a Chip*, *9*(17), 2470–2476. doi:10.1039/b904430m PMID:19680572

Atencia, J., & Beebe, D. J. (2005). Controlled microfluidic interfaces. *Nature*, *437*(7059), 648–655. doi:10.1038/nature04163 PMID:16193039

Azarmanesh, M., & Farhadi, M. (2016). The effect of weak-inertia on droplet formation phenomena in T-junction microchannel. *Meccanica*, *51*(4), 819–834. doi:10.100711012-015-0245-6

Badilescu, S., & Packirisamy, M. (2012). Microfluidics-Nano-Integration for Synthesis and Sensing. *Polymers*, *4*(2), 1278–1310. doi:10.3390/polym4021278

Choi, S., Park, I., Hao, Z., Holman, H.-Y. N., & Pisano, A. P. (2012). Quantitative studies of long-term stable, top-down fabricated silicon nanowire pH sensors. *Applied Physics. A, Materials Science & Processing*, *107*(2), 421–428. doi:10.100700339-011-6754-9

Cui, B., Diamant, H., Lin, B., & Rice, S. A. (2004). Anomalous Hydrodynamic Interaction in a Quasi-Two-Dimensional Suspension. *Physical Review Letters*, *92*(25), 258301. doi:10.1103/PhysRevLett.92.258301 PMID:15245065

de Menech, M., Garstecki, P., Jousse, F., & Stone, H. A. (2008). Transition from squeezing to dripping in a microfluidic T-shaped junction. *Journal of Fluid Mechanics*, *595*. doi:10.1017/S002211200700910X

Dessimoz, A.-L., Cavin, L., Renken, A., & Kiwi-Minsker, L. (2008). Liquid–liquid two-phase flow patterns and mass transfer characteristics in rectangular glass microreactors. *Chemical Engineering Science*, *63*(16), 4035–4044. doi:10.1016/j.ces.2008.05.005

Gascoyne, P. R. C., Vykoukal, J. V., Schwartz, J. A., Anderson, T. J., Vykoukal, D. M., Current, K. W., ... Andrews, C. (2004). Dielectrophoresis-based programmable fluidic processors. *Lab on a Chip*, *4*(4), 299–309. doi:10.1039/b404130e PMID:15269795

Guillot, P., & Colin, A. (2005). Stability of parallel flows in a microchannel after a T junction. *Physical Review. E*, *72*(6), 66301. doi:10.1103/PhysRevE.72.066301 PMID:16486053

Günther, A., Khan, S. A., Thalmann, M., Trachsel, F., & Jensen, K. F. (2004). Transport and reaction in microscale segmented gas–liquid flow. *Lab on a Chip*, *4*(4), 278–286. doi:10.1039/B403982C PMID:15269792

Jin, C. Y., Li, Z., Williams, R. S., Lee, K.-C., & Park, I. (2011). Localized Temperature and Chemical Reaction Control in Nanoscale Space by Nanowire Array. *Nano Letters*, *11*(11), 4818–4825. doi:10.1021/nl2026585 PMID:21967343

Kashid, M., & Kiwi-Minsker, L. (2011). Quantitative prediction of flow patterns in liquid–liquid flow in micro-capillaries. *Chemical Engineering and Processing: Process Intensification*, *50*(10), 972–978. doi:10.1016/j.cep.2011.07.003

Kenis, P., Ismagilov, R., & Whitesides, G. (1999, July 2). Microfabrication inside capillaries using multiphase laminar flow patterning. *Science (New York, N.Y.)*. Retrieved from http://www.ncbi.nlm.nih.gov/pubmed/10390366

Kim, J., Li, Z., & Park, I. (2011). Direct synthesis and integration of functional nanostructures in microfluidic devices. *Lab on a Chip*, *11*(11), 1946. doi:10.1039/c1lc20079h PMID:21499615

Kumar, A. (2017). Micro-channel fabrication using Cu-sheets as electrode in EDM process and optimization of process parameters using Taguchi Method. *International Journal of Current Engineering and Technology*, *7*(1), 15–18.

Li, G., Chen, Q., Li, J., Hu, X., & Zhao, J. (2010). A Compact Disk-Like Centrifugal Microfluidic System for High-Throughput Nanoliter-Scale Protein Crystallization Screening. *Analytical Chemistry*, *82*(11), 4362–4369. doi:10.1021/ac902904m PMID:20459060

Li, Y., Wang, J., Li, D., & Liu, R. (2018). Flow pattern diagrams of oil-water two-phase microflows and stable parallel flows obtained at low Reynolds numbers. *International Journal of Multiphase Flow*, *98*, 139–146. doi:10.1016/j.ijmultiphaseflow.2017.09.008

Lohse, S. E., Eller, J. R., Sivapalan, S. T., Plews, M. R., & Murphy, C. J. (2013). A Simple Millifluidic Benchtop Reactor System for the High-Throughput Synthesis and Functionalization of Gold Nanoparticles with Different Sizes and Shapes. *ACS Nano*, *7*(5), 4135–4150. doi:10.1021/nn4005022 PMID:23634842

Marre, S., & Jensen, K. F. (2010). Synthesis of micro and nanostructures in microfluidic systems. *Chemical Society Reviews*, *39*(3), 1183. doi:10.1039/b821324k PMID:20179831

McQuade, D. T., & Seeberger, P. H. (2013). Applying Flow Chemistry: Methods, Materials, and Multistep Synthesis. *The Journal of Organic Chemistry*, *78*(13), 6384–6389. doi:10.1021/jo400583m PMID:23750988

Murphy, C. J., & Jana, N. R. (2002). Controlling the Aspect Ratio of Inorganic Nanorods and Nanowires. *Advanced Materials*, *14*(1), 80–82. doi:10.1002/1521-4095(20020104)14:1<80::AID-ADMA80>3.0.CO;2-#

Noureddini, H., Teoh, B. C., & Davis Clements, L. (1992). Viscosities of vegetable oils and fatty acids. *Journal of the American Oil Chemists' Society*, *69*(12), 1189–1191. doi:10.1007/BF02637678

Pauzauskie, P. J., Radenovic, A., Trepagnier, E., Shroff, H., Yang, P., & Liphardt, J. (2006). Optical trapping and integration of semiconductor nanowire assemblies in water. *Nature Materials*, *5*(2), 97–101. doi:10.1038/nmat1563 PMID:16429143

Salim, A., Fourar, M., Pironon, J., & Sausse, J. (2008). Oil-water two-phase flow in microchannels: Flow patterns and pressure drop measurements. *Canadian Journal of Chemical Engineering*, *86*(6), 978–988. doi:10.1002/cjce.20108

Scholz, C., Wirner, F., Li, Y., & Bechinger, C. (2012). Measurement of permeability of microfluidic porous media with finite-sized colloidal tracers. *Experiments in Fluids*, *53*(5), 1327–1333. doi:10.100700348-012-1362-9

Song, H., Chen, D. L., & Ismagilov, R. F. (2006). Reactions in Droplets in Microfluidic Channels. *Angewandte Chemie International Edition*, *45*(44), 7336–7356. doi:10.1002/anie.200601554 PMID:17086584

Squires, T. M., & Quake, S. R. (2005). Microfluidics: Fluid physics at the nanoliter scale. *Reviews of Modern Physics*, *77*(3), 977–1026. doi:10.1103/RevModPhys.77.977

Stroock, A., Weck, M., Chiu, D., Huck, W., Kenis, P., Ismagilov, R., & Whitesides, G. (2000). Patterning Electro-osmotic Flow with Patterned Surface Charge. *Physical Review Letters*, *84*(15), 3314–3317. doi:10.1103/PhysRevLett.84.3314 PMID:11019078

Thelander, C., Agarwal, P., Brongersma, S., Eymery, J., Feiner, L. F., Forchel, A., ... Samuelson, L. (2006). Nanowire-based one-dimensional electronics. *Materials Today*, *9*(10), 28–35. doi:10.1016/S1369-7021(06)71651-0

van der Linden, H. J., Jellema, L. C., Holwerda, M., & Verpoorte, E. (2006). Stabilization of two-phase octanol/water flows inside poly(dimethylsiloxane) microchannels using polymer coatings. *Analytical and Bioanalytical Chemistry*, *385*(8), 1376–1383. doi:10.100700216-006-0526-y PMID:16773299

Wang, J., Wang, S., Zhang, P., & Li, Y. (2017). Easy-disassembly bonding of PDMS used for leak-tight encapsulation of microfluidic devices. In *2017 18th International Conference on Electronic Packaging Technology (ICEPT)* (pp. 1051–1055). IEEE. 10.1109/ICEPT.2017.8046623

Whitesides, G. M. (2006). The origins and the future of microfluidics. *Nature*, *442*(7101), 368–373. doi:10.1038/nature05058 PMID:16871203

Xia, Y., & Whitesides, G. M. (1998a). Soft Lithography. *Angewandte Chemie International Edition*, *37*(5), 550–575. doi:10.1002/(SICI)1521-3773(19980316)37:5<550::AID-ANIE550>3.0.CO;2-G PMID:29711088

Xia, Y., & Whitesides, G. M. (1998b). Soft lithography. *Annual Review of Materials Science*, *28*(1), 153–184. doi:10.1146/annurev.matsci.28.1.153 PMID:29711088

Xin, W., De Rosa, I. M., Cao, Y., Yin, X., Yu, H., & Ye, P. ... Yang, J.-M. (2018). Ultrasonication-assisted Synthesis of High Aspect Ratio Gold Nanowires on Graphene Template and Investigation of Their Growth Mechanism. *Chemical Communications*. doi:10.1039/C8CC01727A

Xing, Y., & Dittrich, P. (2018). One-Dimensional Nanostructures: Microfluidic-Based Synthesis, Alignment and Integration towards Functional Sensing Devices. *Sensors (Basel)*, *18*(2), 134. doi:10.339018010134 PMID:29303990

Yagodnitsyna, A., Kovalev, A., & Bilsky, A. (2015). Experimental investigation of immiscible liquids flow in a T-shaped microchannel. *10th Pacific Symposium on Flow Visualization and Image Processing*.

Yang, F., Shi, J., Zhang, X., Hao, S., Liu, Y., Feng, C., & Cui, L. (2018). The microstructure and magnetic properties of Cu/CuO/Ni core/multi-shell nanowire arrays. *Chemical Physics Letters*, *697*, 43–47. doi:10.1016/j.cplett.2018.02.064

Zhang, S., Yang, Y., & Nie, Z. (2018). Controllable Self-Assembly of One-Dimensional Nanocrystals. In Functional Organic and Hybrid Nanostructured Materials (pp. 1–38). Weinheim, Germany: Wiley-VCH Verlag GmbH & Co. KGaA. doi:10.1002/9783527807369.ch1

Zhang, X., Hu, N., Chen, X., Fan, T., Wang, Z., Zheng, X., ... Qian, S. (2017). Controllable cell electroporation using microcavity electrodes. *Sensors and Actuators. B, Chemical*, *240*, 434–442. doi:10.1016/j.snb.2016.08.172

Zhang, Y., Li, Z., Xu, K., Chencheng, G., Hao, Y., Meng, F., & Gui, Y. (2017). A low-cost flexible electrochemical accelerometer using graphene-integrated microchannel. In *Micro Electro Mechanical Systems (MEMS), 2017 IEEE 30th International Conference on* (pp. 1268–1271). Las Vegas, NV: IEEE. 10.1109/MEMSYS.2017.7863648

Zhao, B., Viernes, N. O. L., Moore, J. S., & Beebe, D. J. (2002). Control and Applications of Immiscible Liquids in Microchannels. *Journal of the American Chemical Society*, *124*(19), 5284–5285. doi:10.1021/ja025835j PMID:11996566

Zhao, Y., Chen, G., & Yuan, Q. (2006). Liquid-liquid two-phase flow patterns in a rectangular microchannel. *AIChE Journal. American Institute of Chemical Engineers*, *52*(12), 4052–4060. doi:10.1002/aic.11029

Zhu, C., Fu, T., Gao, X., Liu, W., & Ma, Y. (2011). Study on the liquid-liquid two-phase flow patterns in microchannels. *Chemical Industry and Engineering Progress*, *30*(s1), 65–69.

ADDITIONAL READING

Adamson, A. W., & Gast, A. P. (1997). *Physical Chemistry of Surfaces* (6th ed.). New York: John Wiley & Sons, Inc.

Berthier, J. (Ed.). (2013). *Micro-Drops and Digital Microfluidics* (2nd ed.). Waltham: Elsevier Inc.

Betchov, R., & Criminale, W. O. J. (1967). Stability Of Parallel Flows. London: Academic Press Inc. (London) Ltd.

Drop (liquid). (n.d.). Retrieved from https://en.wikipedia.org/wiki/Drop_(liquid)#Pendant_drop_test

Ghosh, U. U., Subramanian, S. G., Suman, C., & Sunando, E. (2018). Droplets in Microfluidics. In S. Basu, A. K. Agarwal, A. Mukhopadhyay, & C. Patel (Eds.), *Applications Paradigms of Droplet and Spray Transport: Paradigms and Applications* (pp. 347–379). Singapore: Springer Nature Singapore Pte Ltd. doi:10.1007/978-981-10-7233-8_13

Lo Faro, M. J. (2016). *Silicon Nanowires: the Route from Synthesis towards Applications*. Università degli Studi di Palermo.

Zhang, S., Yang, Y., & Nie, Z. (2018). Controllable Self-Assembly of One-Dimensional Nanocrystals. In Functional Organic and Hybrid Nanostructured Materials (pp. 1–38). Weinheim, Germany: Wiley-VCH Verlag GmbH & Co. KGaA. doi:10.1002/9783527807369.ch1

KEY TERMS AND DEFINITIONS

Annular Flow: A two-phase flow pattern formed when one phase is moving continuously in the central part of the microchannel and enclosed by the other phase as if the enclosed phase is flowing in a pipe made by the outside phase.

Droplet Flow: A two-phase flow pattern formed when one of the immiscible phases, called the dispersed phase, is flowing in small discrete volumes while the other, called the continuous phase, is streaming continuously in a laminar way.

Homogeneous Wall Condition: This is the condition when the top, bottom, and side walls of a microchannel are fully made of the same material.

Inhomogeneous Wall Condition: This is the condition when the top, bottom, and side walls of a microchannel are made by two or more different kinds of materials.

Parallel Flow: The co-flowing two phases are running as laminar layers parallel to each other and the interface in between is even and straight.

Segmented/Slug Flow: A two-phase flow pattern formed when one fluid phase is separated into segments/slugs by segments/slugs of the other phase.

Two Phase Flow: A simultaneous flow of a single fluid flowing as two different phases, two different fluids with different chemical properties occurring in the same phase, or two different fluids with different states or phases.

Wavelet Flow: The co-flowing two phases are running as laminar layers side by side with the two-phase interface in between possesses a wavy morphology.

Chapter 6
Experimental and Numerical Analyses of a Micro-Heat Exchanger for Ethanol Excess Recovery From Biodiesel

João Lameu da Silva Júnior
Federal Institute of Education, Science, and Technology of South of Minas Gerais, Brazil

Harrson Silva Santana
University of Campinas, Brazil

ABSTRACT

The use of micro-heat exchangers increased with the advancement of microfluidics. These microdevices present some advantages like elevated surface area-to-volume ratio resulting in high heat transfer rates. Micro-heat exchanger with phase change is a new application of such devices. The simultaneous momentum, heat, and mass transfer at microscale still require investigations due to the inherent complexity. The main goal of the chapter is to demonstrate experimentally and numerically the capability of the micro-heat exchanger use in the continuous process of ethanol excess recovery from the biodiesel. The influence of flow rate, ethanol/biodiesel molar ratio, and temperature on the ethanol evaporation performance was evaluated. The flow rate and the ethanol/biodiesel molar ratio influenced negatively the evaporation. In contrast, the temperature was affected positively. The mathematical model was able to capture the main features of the continuous evaporation; however, further improvements must be performed in order to consider the thermodynamics characteristics.

DOI: 10.4018/978-1-5225-7138-4.ch006

INTRODUCTION

The use of microchannels in heat and mass transfer operations has received great attention in the last decades. The advantages of the microdevices regarding the conventional macroscale equipment, especially the high surface area-to-volume ratio, enhances the transport phenomena. In heat transfer applications, very high heat fluxes are observed, resulting in fast heating or cooling of the fluids. Other advantages of using microdevices for heat transfer operations are compactness, simple manufacturing, high heat dissipation per volume unit and low amount of refrigerant fluid requirements (Kim & Mudawar, 2013a). In mass transfer applications, the main advantage is obtained from the diffusion path reduction, improving the molecular diffusion. Smart device of the design can also induce split and recombination of streams and vortex generation, even at relative low Reynolds numbers, enhancing the mass transfer rates due to chaotic advection mechanism. The hydraulic diameter decrease from macro to micro scale was experimentally demonstrated as an efficient way to improve the heat and mass transfer. However, the pressure drop restriction bounds the lower dimension of the channel. Accordingly, the microdevice design should always take into account the heat transfer jointly with the hydraulic characteristics (Wang & Wang, 2014). The microdevices applications in heat and mass transfer include cooling of microelectronic devices and high power laser diode arrays, chemical reactors, fuel cells and micro-heat exchangers (Santana, Sanchez, & Taranto, 2017a; Santana, Tortola, Silva Jr., & Taranto, 2017b; Deng, Wan, Zhang, & Chu, 2017; Bakhtiary-Davijany, Hayer, Phanam, Myrstad, Venvik, Pfeifer, & Holmen, 2011; Fazeli & Behnam, 2010).

Flow boiling in micro- and milli-channels (mini-channels) can present distinct behavior concerning the well-established observed from macroscale due to the changes in the dominant forces (Wang & Wang, 2014). Several studies were carried out in the investigation of boiling flow performance (vapor quality) dependence on the main operating parameters: wall heat flux, degree of feedstream subcooling, pressure and mass velocity (Mahmoud & Karayiannis, 2013; Sun & Mishima, 2009; Kandlikar & Balasubramanian, 2004; Huo, Chen, Tian, & Karayiannis, 2004; Kew & Cornell, 1996; Liu & Winterton, 1991). Most of the studies were performed using pure substances, once the heat transfer parameters of the boiling flow were the main goal of these studies. In contrast, the use of microchannel devices for simultaneous heat and mass transfer in mixtures, i.e., evaporation of the species A from a binary or multicomponent mixture (e.g., excess alcohol from a biodiesel synthesis downstream) still requires further investigation. In this context, the present chapter aims to describe the main characteristics of boiling flow in micro- and milli-channels and to present experimental and numerical results from a microchannel heat exchanger with phase change (micro evaporator) applied to the recovery of excess ethanol from biodiesel.

Recently, the biodiesel has become an interesting source of renewable fuel. The main chemical route to synthesize biodiesel is the vegetable oil transesterification using alcoholic solutions of alkali catalyst (sodium or potassium hydroxide) in mild conditions (temperature below the alcohol saturation temperature and ambient pressure). Commonly, short-chain alcohols are employed, preferably methanol or ethanol. The ethanol presents advantages over the methanol, since it is renewable and nontoxic. In the biodiesel transesterification, the alcohol is used in excess in order to ensure the complete conversion of the vegetable oil. Recently, some studies of biodiesel synthesis in microreactors have been developed (Chueluecha, Amaraporn, & Attasak, 2017; Santana, Sanchez, & Taranto, 2017b; Santana, Tortola, Reis, Silva Jr., & Taranto, 2016; Bhoi, Sen, Singh, Mahajani, Shenoy, Rao, & Ghosh, 2014; Rahimi, Aghel, Alitabar, Sepahvand, & Ghasempour, 2014; Santacesaria, Di Serio, Tesser, Turco, Tortorelli, & Russo, 2012). In order to become the microdevices an interesting tool for the continuous production, the excess alcohol has to be recovered. Currently, this is performed using batch rotary evaporators, and consequently, the process cannot be operated continuously. To overcome this issue, the development of micro-evaporators, i.e., micro-heat exchangers with phase change, is a fundamental key. Based on this, the present chapter aims to present experimental results of a micro-heat exchanger application for ethanol recovery based on its vaporization from biodiesel mixture. Moreover, a mathematical model was proposed to simulate the main features of this process.

BOILING FLOW IN MICROSCALE

The boiling flow heat transfer pattern can be classified in two mechanisms: nucleate boiling and convective boiling. The first one is marked by a significant portion of the channel length occupied by bubbles or slugs. The convective pattern presents slugs intermission with liquid portions along the channel. Gradual evaporation promotes the liquid film thinning, increasing the heat transfer coefficient along the channel (Kim & Mudawar, 2013a). Wang and Wang (2014) noticed that the boiling flow undergoes different flow patterns along the channel depending on the local heat and flow characteristics. Dryout incipience, i.e., the beginning of vapor dryout, occurs due to the uneven evaporation or asymmetry on the liquid film. At the dryout incipience the heat transfer coefficient experiences a considerably drop. Finally, a full dryout can occur in a forward position, where the liquid film was totally evaporated (Kim & Mudawar, 2013a).

The main operating variables of boiling in microchannels are mass velocity, heat flux and channel hydraulic diameter. The saturation temperature of the liquid also affects the boiling performance, given by the obtained vapor quality. Once the boiling flow regime changes along the channel, the prediction of the incipient dryout is a key-factor in the design and operation of these devices. Several researches were developed in order to correlate the main operating variables and system properties in the prediction of the incipient dryout vapor quality. Experimental results showed that the increment of mass velocity decreased the incipient dryout vapor quality for most of cases. However, the mass velocity effect on incipient dryout vapor quality can be positive in the specific conditions of low heat fluxes or low saturation temperature (Kim & Mudawar, 2013a). The incipient dryout vapor quality also was decreased with the increment on heat flux and with the hydraulic diameter reduction. Due to the potential occurrence of two heat transfer regimes (nucleate and convective), the boiling heat transfer coefficient, h_b, can be described by the summation of these two contributions.

$$h_b = h_n + h_c \tag{1}$$

Where h_n is the heat transfer coefficient for nucleate boiling and h_c is the heat transfer coefficient for the convective boiling. Several correlations for both nucleate and convective boiling heat transfer coefficients have been proposed and can be found in (Kim & Mudawar, 2013a,b; Mahmoud & Karayiannis, 2013; Kandlikar & Balasubramanian, 2004; Kew & Cornwell, 1997). The boiling heat transfer coefficient, h_b, presents a direct relationship with the vapor quality and depends on the predominant boiling regime. For the nucleate dominant boiling, the boiling heat transfer coefficient decreases with the vapor quality for water flow, presenting an inverse behavior for the convective dominant boiling regime. Moreover, the nucleate boiling contribution to the boiling heat transfer coefficient increases with the increment of liquid saturation temperature and the heat flux, reducing with the channel hydraulic diameter (Kim & Mudawar, 2014). Wang and Wang (2014) studied the ethanol boiling in minichannels and they reported a decrease in the boiling heat transfer coefficient with the increment of vapor quality for low heat flux conditions, and an increase in h_b with the reduction of the vapor quality for high heat flux conditions.

The incipient vapor quality is directly proportional to the heat transfer coefficient in the convective dominant boiling regime. In contrast, for the nucleate dominant regime, the inverse behavior is noticed. As expected, the increment in mass velocity favors the predominance of convective boiling regime. The nucleate boiling contribution increases with increment in the saturation temperature and in the heat flux. The channel hydraulic diameter increment results in decrease of the heat transfer coefficient for both boiling regimes, proving the advantage of the use of reduced size scales.

In order to design and optimize a microchannel for heat transfer service, the macro- to micro-scale transition must be defined. Several definitions were stated; however a universal criterion is still not established due to lack of data in different microdevice geometries (Wang & Wang, 2014). The main macro-to-micro scale threshold definitions are provided following.

A definition based on bubble confinement criteria were proposed, relating the bubble detachment diameter and the channel dimensions, as the Confinement number, *Co*:

$$Co = \left[\frac{\sigma}{g\left(\rho_l - \rho_v\right) d_h^2} \right]^{0.5} \tag{2}$$

where σ is the surface tension coefficient, N.m^{-1}, g is the gravity acceleration, m.s^{-2}, ρ is the specific mass, kg.m^{-3}, the subscripts l and v stand for liquid and vapor phases and d_h is the hydraulic diameter, m. The microchannel is defined for $Co > 0.5$ (Kew & Cornwell, 1997) or $Co > 1.0$ (Triplett, Ghiaasiaan, Abdel-Khalik, & Sadowski, 1999).

Cheng et al. (2006) proposed the scale transition classification by the Bond number, *Bo*:

$$Bo = \left(\frac{d_h}{l_c} \right)$$

with

$$l_c = \left[\frac{\sigma}{g\left(\rho_l - \rho_v\right)} \right]^{0.5} \tag{3}$$

where l_c is the capillary length. The classification is defined as:

- $Bo < 0.05$**:** Microchannel (effects of gravitational field are negligible);
- $0.05 < Bo < 3$**:** Minichannel (surface tension dominates over small effect of gravitational field);
- $Bo > 3$**:** Macrochannel (gravitational field dominates over surface tension effects).

Li and Wu (2010) defined a nondimensional correlation considering the effects of surface tension, inertial, gravitational and viscous forces. The channel is classified as a micro/mini channel for the condition: $Bo.Re_l^{0.5} < 200$, where Re_l is the Reynolds number of the liquid phase.

Harirchian and Garimella (2010) proposed the macro-to-micro transition classification based on the convective confinement number, $Bo^{0.5}Re_l$, as given by Equation 4 (Tibiriça & Ribatski, 2013):

$$Bo^{0.5}\,\mathrm{Re}_l = \frac{Gd_h^{\,2}}{\mu_l}\left[\frac{g\left(\rho_l - \rho_v\right)}{\sigma}\right]^{0.5} < 160 \tag{4}$$

where G is the vapor phase mass velocity, kg m^{-2} s^{-1}, μ_l is the dynamic viscosity of the liquid phase, kg m^{-1} s^{-1}.

The heat transfer and the pressure drop of two phase boiling flow are directly related to the liquid and vapor phases distribution inside the microchannels. According to Tibiriça and Ribatski (2013), the flow pattern in microchannels includes:

1. **Disperse Flows:** Including bubbly flow composed by small bubbles in continuous liquids and mist flows composed by small liquid droplets inside the continuous vapor core;
2. **Annular Flow:** A continuous liquid film flows in the wall zone around the continuous vapor phase core;
3. **Intermittent Flow:** Including churn flow and slug flow.

Experimental observation noticed that there is not a constant flow pattern. The flow pattern depends locally of the heat and flow characteristics, resulting in transitions of the flow pattern along the channel.

The physico-chemical properties present a fundamental role in the heat transfer performance. Fluids with higher liquid/vapor density ratios and high latent heat of vaporization require great amount of heat flux to initiate the boiling (Peng, Hu, & Wang, 1998). After the beginning of the boiling, liquids with smaller surface tension coefficients and latent heats of vaporization provide higher heat transfer coefficients (Greco & Vanoli, 2005).

The geometric characteristics also influence the boiling performance. In general, a decrease on the hydraulic diameter results in heat transfer performance enhancement, however, the cross section aspect ratio also has to be considered. Tibiriça et al. (2012) demonstrated that the decrease in the cross section aspect ratio reduced the heat transfer coefficient. According to Tuckerman and Pease, (1981) microchannel with higher cross section aspect ratios are expected to reduce the thermal resistance (Wang & Wang, 2014) being more attractive due to its elevated surface area to cross sectional area ratio and reduced vapor phase pressure drop, resulting in superior thermal-hydraulic performance.

Sempértegui-Tapia and Ribatski (2017), Tibiriça and Ribatski (2013) and Revellin and Thome (2007) evaluated the two-phase boiling flow pressure drop behavior dependence on several factors in micro and millichannels, including: working fluid, channel diameter and length, channel cross section geometry, fluid saturation temperature and mass velocity. The working fluid influence was related to the specific volume of the vapor phase. Fluids with higher specific volumes resulted in higher superficial velocities of the two-phase mixture, subsequently, higher pressure drops were noticed. The increment in the saturation temperature resulted in the pressure drop reduction. This behavior was attributed to the increase of the vapor density, decreasing the vapor-phase velocity, and consequently, the pressure drop. As expected and already observed in the analyses of the working fluid and the saturation temperature influence in the pressure drop, the increment in mass velocity resulted in higher pressure drops, since it is directly related to the flow superficial velocity.

Revellin and Thome (2007) observed no influence of the channel length in the pressure drop. In contrast, the cross section geometry influenced the pressure drop, as found by Sempértegui-Tapia and Ribatski (2017), by experimental measurements of pressure drop in circular and non-circular (triangular and square) channels. The results showed, in descending order of pressure drop: triangular > square > circular. The pressure drop behavior was explained by the equivalent diameter provided by the cross section shapes. The diameter has an inverse relationship with the pressure drop. The increment of the diameter resulted in lower pressure drop.

Revellin and Thome (2007) described the friction factor behavior for the two-phase boiling flow in a Moody's diagram-like charts. The authors divided the friction factor in three zones, depending on the two-phase Reynolds number, Re_{TP}. For $Re_{TP} < 2000$, the flow presents laminar regime characteristics, and the friction factor decreases linearly with the two-phase Reynolds number. From $Re_{TP} = 2000$ to 8000, the two-phase flow exhibited a transitional behavior, and the friction factor could increase or decrease, depending on other factors, as for example, the fluid saturation temperature and the degree of subcooling. For $Re_{TP} > 8000$, the flow is in turbulent regime, and the friction factor decreases with the increment of two-phase Reynolds number.

EXPERIMENTAL STUDY OF THE A MICRO-HEAT EXCHANGER

Materials and Methods

The experimental evaluation of the excess alcohol evaporation from biodiesel was performed using ethyl alcohol (Dinâmica, > 99.5%) and pure biodiesel (B100) donated by the LAPEC (Laboratory of Fuels) of Federal University of Amazonas. The experimental microdevice is detailed in Figure 1. The device presents the dimensions: 30 mm of total longitudinal length, 210 μm of channel height. It was composed by 15 channels of 0.5 mm of width and 20 mm of length (Figure 1.a). The separation between each channel was 0.5 mm. The microdevice was manufactured in polydimethylsiloxane (PDMS) by soft lithography at the Microfabrication Laboratory (LMF) of Brazilian National Laboratory of Nanotechnology (LNNano – Campinas/ SP, Brazil). The micro heat exchanger (Figure 1b) presents one inlet and two outlets (superior and inferior), located at the same longitudinal position (Figure 1c). The proposed design aims to allow the separation of vapor phase from the liquid phase from the outlet streams, providing in an ideal condition, the flow of vapor and liquid phases through superior and inferior outlets, respectively.

Figure 1. Micro heat exchanger: (a) details of the microdevice geometry, in mm; (b) top view of the manufactured microdevice; (c) details of inlet and outlets.
Reprinted from Santana, H.S., Sanchez, G.B, & Taranto, O.P. (2017a). Evaporation of excess alcohol in biodiesel in a microchannel heat exchanger with Peltier module, Chemical Engineering Research and Design, 124, 20-28, with permission from Elsevier

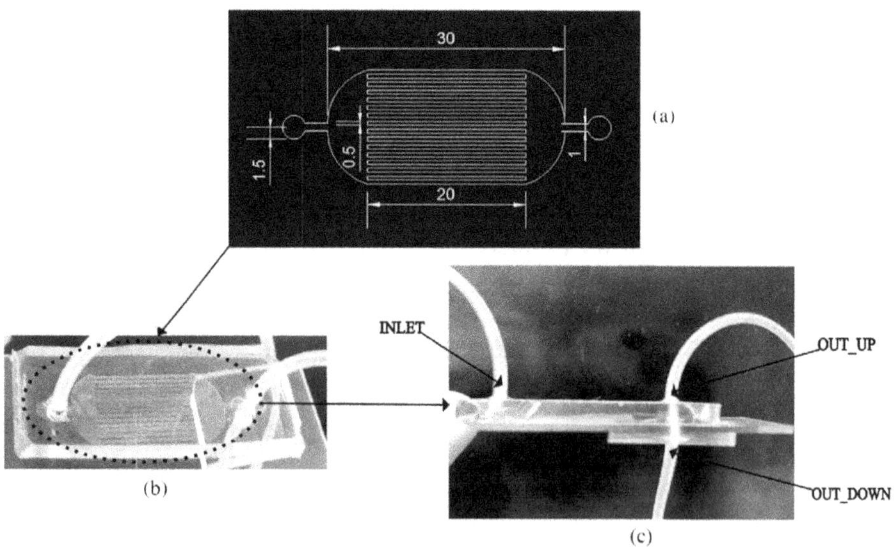

The experimental heating apparatus employs a Peltier plate (model HTC-50-12-15-4). The Peltier plate generates the required heating by the use of a potential difference provided from an electric tension source. Moreover, a temperature sensor located at the upper surface of the microdevice, a digital temperature controller (model MTB48, LOTI) and a syringe pump (model 74905-54, Cole-Parmer) integrate the experimental apparatus. The temperature control was carried out by the control system, composed the digital temperature controller, the power source and the source outlet from the Peltier plate. During the experimental tests, an aluminum plate was used under the microdevice/Peltier plate assembly in order to remove the heat from the cold side of the Peltier element. In high power supplies, the Peltier plate performance decreased due to the temperature uniformity through the system.

EXPERIMENTAL PROCEDURE

In the experimental runs, 50 mL of the biodiesel-ethanol mixture were pumped by the syringe pump at a constant flow rate to the microdevice previously heated at the specified temperature. Samples were collected at the microdevice outlet and were immediately weighed in an analytical balance. The evaporation efficiency was determined from the material balance over the microdevice, as shown in Figure 2.

The global material balance (Figure 2) is given by:

$$m = m_v + m' \tag{5}$$

Figure 2. Schematic diagram of the mass balance of the ethanol evaporation process. *Reprinted from Santana, H.S., Sanchez, G.B, & Taranto, O.P. (2017a). Evaporation of excess alcohol in biodiesel in a microchannel heat exchanger with Peltier module, Chemical Engineering Research and Design, 124, 20-28, with permission from Elsevier*

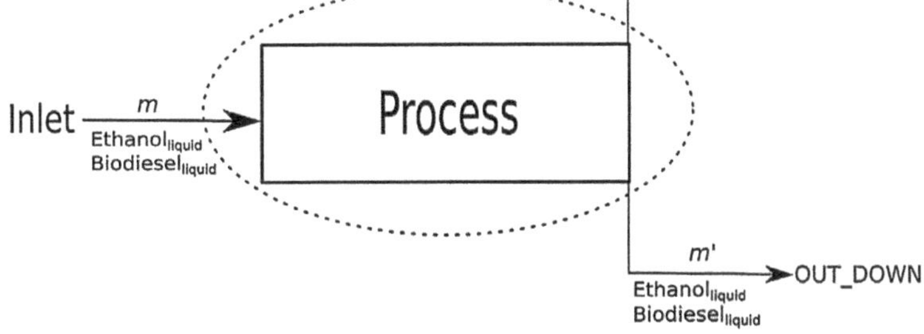

where m is the total mass entering the microdevice, m_v is the ethanol mass from vapor stream (considering that only the ethanol evaporates along the microdevice at the evaluated operating conditions, the biodiesel evaporation was neglected based on the biodiesel saturation temperature (340 °C – 375 °C, at room pressure (Goodrum, 2002) and m' is the ethanol-biodiesel mass (liquid stream). The total mass m is given by the summation of the ethanol and biodiesel amounts in the feed mixture:

$$m = m_e + m_b \tag{6}$$

At the inferior outlet, the material balance provides:

$$m' = m_{e,l} + m_{b,l} \tag{7}$$

where $m_{e,l}$ is the liquid ethanol mass and $m_{b,l}$ is the liquid biodiesel mass. Once the biodiesel evaporation was not considered:

$$m_b = m_{b,l} \tag{8}$$

At the superior outlet, the material balance provides:

$$m_v = m_e - m_{e,l} \tag{9}$$

After rearrangements and manipulation of Equations 6-9, the global mass balance is satisfied, and then the Equation 11 can be used to evaluate the evaporation performance, defining the Mass Loss:

$$m_v = m - m' \tag{10}$$

$$\text{Mass Loss}(\%) = \frac{m_v}{m} = \frac{m - m'}{m} \tag{11}$$

A Central Composite Rotatable Design (CCDR) was used to evaluate the process variables effect on the ethanol evaporation performance. The variable boundaries were defined according to the process limitations and are summarized in Table 1.

Table 1. Evaluated range of the process variables

Variable	Factor	Levels				
		-1,68	**-1**	**0**	**1**	**1,68**
Temperature (°C)	x_1	80	88	100	112	120
Flow rate (ml min^{-1})	x_2	0.1	0.3	0.6	0.9	1.2
Ethanol/biodiesel molar ratio	x_3	2	4	6.5	9	11

The experimental data of evaporation performance were statistically modeled by the second order equation:

$$Y' = \beta_0 + \beta_1 x_1 + \beta_2 x_2 + \beta_3 x_3 + \beta_{12} x_1 x_2 + \beta_{13} x_1 x_3 + \beta_{23} x_2 x_3 + \beta_{11} x_1^2 + \beta_{22} x_2^2 + \beta_{33} x_3^2 \quad (12)$$

where Y' is the response (Mass Loss), β_0 is the data average, β_i and β_{ii} and are the linear and quadratic regression coefficients, respectively, β_{ij} are the second interaction coefficients and x_1, x_2 and x_3 are the analyzed variables (operating variables). The statistical data analysis was carried out in the software Statistica 7. The potential difference applied in the Peltier plate was 10 V.

MATHEMATICAL MODELING OF A MICRO-HEAT EXCHANGER

The mathematical model used was based on Eulerian multiphase approach. Three phases were considered: liquid and vapor ethanol and biodiesel. The continuity (Equation 13) was solved for each phase, considering incompressible and steady-state conditions:

$$\nabla \cdot \left(\alpha_k \rho_k \vec{U} \right) = S_{m,k} \quad (13)$$

where α is the volumetric fraction of the phase, ρ is the specific mass (kg m^{-3}), U is the velocity vector (m s^{-1}), the subscript k indicates the k-th phase ($k = ve, le, b$ stand for ethanol vapor, ethanol liquid and biodiesel), and $S_{m,k}$ represents the mass sources (kg m^{-3} s^{-1}) due to the ethanol evaporation. The biodiesel evaporation was neglected based on the biodiesel saturation temperature (340 °C – 375 °C, Goodrum, 2002). The mass sources are given by:

$$S_{m,ve} = m_{vap} \tag{14}$$

$$S_{m,le} = -m_{vap} \tag{15}$$

$$S_{m,b} = 0 \tag{16}$$

where m_{vap} is the ethanol evaporation volumetric rate (kg m^{-3} s^{-1}), modeled in accordance with Equation 17.

The evaporation was modeled based on the phase change model by saturation temperature (Rahmat & Hubert, 2010), as described in Equation 7:

$$m_{vap} = \left(\frac{h_{le}\left(T_{sat} - T_{le}\right) + h_{ve}\left(T_{sat} - T_{ve}\right)}{-\Delta H_{v}} a_{le-ve} \right) \tag{17}$$

$$\Delta H_{v} = 1.254 \times 10^{6} \left(1 - \frac{T_{le}}{T_{c}}\right)^{0.31245} \tag{18}$$

where h is the convective heat transfer coefficient (W m^{-2} K^{-1}), $a_{el\text{-}ev}$ is the ethanol liquid-vapor specific area (m^{-1}) (Equation 27), ΔH_v is the latent heat of vaporization at the temperature T_{le} (J kg^{-1}), obtained from Poling et al. (2008), T_{sat} is the ethanol saturation temperature (78.4 ºC), T_{le} and T_{ve} are the absolute temperature of liquid and vapor ethanol (K), T_c is the critical temperature of the ethanol. The physical properties of ethanol were defined according to the operating temperature, estimated from expression given by Poling et al. (2008). The biodiesel physical properties were taken from literature (Esteban, Riba, Baquero, Rius, & Puig, 2012; Fasina & Colley, 2008). The convective heat transfer coefficients, h_{le} and h_{ve}, were estimated by Nusselt number, Nu, flow conditions and geometric parameters, based on the empirical correlation for microchannels given by Kalaivanan and Rathnasamy (2011) (Equation 19):

$$Nu = \frac{h\,D_h}{\lambda} = 0.949\,Re^{0.06} Pr^{0.4} \tag{19}$$

where D_h is the hydraulic diameter of the microchannels (m).

The velocity and pressure fields where shared by the three phases, namely, a homogeneous modeling was used in the solution of momentum balance. For multiphase flows at steady-state conditions, the homogeneous modeling provides:

$$\nabla \cdot \left(\rho \vec{U} \vec{U}\right) = -\nabla P + \mu \nabla^2 \vec{U} + \rho \vec{g} \tag{20}$$

where P is the pressure, μ is the dynamic viscosity (kg m^{-1} s^{-1}) and g is the gravity acceleration (m s^{-2}). The fluid properties, ρ and μ, were weighted by the volumetric fraction of each individual phase.

$$\rho = \sum_{k=1}^{n_p} \alpha_k \rho_k$$

and

$$\mu = \sum_{k=1}^{n_p} \alpha_k \mu_k \tag{21}$$

The energy balance was solved for each phase, and is given by Equation 22:

$$\nabla \cdot \left(\alpha_k \rho_k \vec{U}_k H_k\right) = \nabla \cdot \left(\alpha_k \lambda_k \nabla T_k\right) + S_{h,k} \tag{22}$$

where H is the specific enthalpy (J kg^{-1}), λ is the thermal conductivity (W m^{-1} K^{-1}), T is the thermodynamic temperature (K), S_h represents extra sources due to the interphase heat transfer, given by:

$$S_{h,le} = \left(Q_{le-ve} + Q_{le-b}\right) \tag{23}$$

$$S_{h,ve} = -\left(Q_{b-ve} + Q_{le-ve}\right) \tag{24}$$

$$S_{h,b} = \left(Q_{b-ve} - Q_{le-b}\right) \tag{25}$$

where Q_{k-i} indicates the heat rate transfer due to interphase convection, modeled by Equation 26:

$$Q_{k-i} = h_{k-i}\, a_{k-i} \left(T_i - T_k\right) \tag{26}$$

where *k-i* denotes the interaction between *k*-th and *i*-th phases, h_{k-i} is the convective heat transfer coefficient (W m^{-2} K^{-1}) obtained from Equation 19 using the respective fluid properties, and a_{k-i} is the specific interfacial area between the *k*-th and *i*-th phases (m^{-1}), given by Equation 27:

$$a_{k-i} = \frac{2\left|\nabla \alpha_k\right|\left|\nabla \alpha_i\right|}{\left|\nabla \alpha_k\right| + \left|\nabla \alpha_i\right|} \tag{27}$$

The discretization process was carried out using the Ansys ICEM, resulting in about 255,000 tetrahedral control volumes. The Ansys CFX 16 was used in the pre-processing, solver and post-processing steps. Steady-state simulations were performed. High resolution interpolation schemes were employed, and the number of iterations was defined between 100 and 1000, with the root mean square residual target of 10^{-4}.

A Central Composite Rotatable Design (CCRD) was employed to analyze the effects of three operating variables on ethanol evaporation performance: temperature (80-120 °C), mixture flow rate (0.1-1.2 mL/min), ethanol/biodiesel molar ratio, E:B (2-11). The evaporation process performance was evaluated using the Vaporized Ethanol Percentage - VEP (%), given as the ratio of mass flow rate of vapor ethanol at superior outlet by the mass flow rate of liquid ethanol entering the microdevice, according to Equation 28:

$$VEP(\%) = 100 \cdot \left(\frac{\dot{m}_{ve,outlet\ vapor}}{\dot{m}_{le,inlet}} \right) \tag{28}$$

The boundary conditions were:

- **Inlet:** Velocity, temperature and chemical species mass fractions prescribed in accordance with the case;
- **Outlets:** Opening to relative pressure zero, prescribed temperature in accordance with the case;

- **Heat Plate and Walls:** No-slip boundary conditions for all phases, prescribed temperature in accordance with the case.

RESULTS AND DISCUSSION

Experimental Study

The operating condition effects on the ethanol evaporation performance was experimentally evaluated. Table 2 summarizes the Mass Loss response to the evaluated variables using the CCDR with 6 axial points and 3 repetitions at the central point.

The Mass Loss (%) ranged from 9.98% to 82%. The effects of each operating variable on the ethanol evaporation were estimated and are listed on Table 3.

From the data summarized in Table 3, the linear terms of temperature (x_1), ethanol/biodiesel molar ratio (x_3) and temperature-molar ratio interaction (x_1x_3) were statistically significant (p-value < 0.05). The flow rate (x_2), the quadratic

Table 2. Uncoded values and Mass Loss (%) response

Test	T (°C)	Flow Rate (mL min^{-1})	Ethanol/Biodiesel Molar Ratio	Mass Loss (%)
1	88	0.3	4	31.30
2	112	0.3	4	76.76
3	88	0.9	4	30.44
4	112	0.9	4	54.89
5	**88**	**0.3**	**9**	**9.89**
6	112	0.3	9	18.16
7	88	0.9	9	12.72
8	112	0.9	9	17.23
9	80	0.6	6.5	39.34
10	120	0.6	6.5	50.55
11	100	0.1	6.5	55.43
12	100	1.2	6.5	42.38
13	**100**	**0.6**	**2**	**82.16**
14	100	0.6	11	17.74
15	100	0.6	6.5	38.81
16	100	0.6	6.5	46.48
17	100	0.6	6.5	38.88

Table 3. Effects of the variables on the Mass Loss (%)

Factors	Regression Coefficient	Standard Error	t(2)	p-Value	Estimates per Interval (95%)	
					Lower Limit	Upper Limit
Average	**42.47**	**2.54**	**16.71**	**0.004**	**31.54**	**53.41**
x_1 (L)	**14.87**	**2.38**	**6.22**	**0.024**	**4.59**	**25.14**
x_1 (Q)	-4.99	2.62	-1.89	0.197	-16.29	6.31
x_2 (L)	-6.26	2.38	-2.62	0.12	-16.53	4.00
x_2 (Q)	-2.19	2.62	-0.83	0.49	-13.50	9.11
x_3 (L)	**-35.69**	**2.38**	**-14.95**	**0.004**	**-45.96**	**-25.42**
x_3 (Q)	-1.45	2.62	-0.55	0.63	-12.76	9.85
x_1x_2	-6.19	3.11	-1.98	0.18	-19.61	7.22
x_1x_3	**-14.28**	**3.11**	**-4.57**	**0.04**	**-27.70**	**-0.85**
x_2x_3	6.15	3.11	1.97	0.18	-7.26	19.57

(L) – linear; (Q) – quadratic.

Table 4. ANOVA for Mass Loss

Source of Variation	Sum of Squares	Degrees of Freedom	Mean Square	F_{calc}
Regression	5646.53	4	1411.63	3.68
Residues	1396.15	12	116.34	
Total	7042.68	16	440.16	

% explained variation (R^2) = 80.59; $F_{4;12;0.05}$ = 3.2

terms and the interaction terms x_1x_2 and x_2x_3 did not present statistically significant effects. The ANOVA results are listed in Table 4. The statistically non-significant terms were coupled into the residues term of Equation 29, except the flow rate that was maintained.

The F_{calc} from the regression (3.68) was greater than the 3.2 obtained from the statistical table. The explained variation of the model, considering the linear terms of temperature, flow rate and ethanol/biodiesel molar ratio and the temperature-molar ratio interaction term, was 80.59%. The adjusted model in coded terms (x1, x2 and x3) is given by Equation 29:

$$Y' = 39 + 14.87x_1 - 6.26x_2 - 35.69x_3 - 14.28x_1x_3 \quad (29)$$

The temperature and the ethanol/biodiesel molar ratio presented strong effect on the mass loss. The increment in temperature resulted in higher ethanol mass loss. In contrast, the increment in the molar ratio resulted in the decrease of the ethanol mass loss, reducing the evaporation efficiency. Similar behavior was observed for the flow rate, however with a small effect regarding the other two operating variables.

The temperature effect on the ethanol evaporation can be analyzed by the relationship between the mass and the latent heat of vaporization:

$$Q = m\Delta H_v \rightarrow m = \frac{Q}{\Delta H_v} \tag{30}$$

where Q (J), is the heat supplied to the liquid mass m (kg), and ΔHv is the latent heat of vaporization (J.kg-1) (Equation 18). Accordingly, the vaporized mass is inversely proportional to the latent heat of vaporization. The increment in temperature provides a decrease in the latent heat of vaporization, resulting in the increase of the vaporized mass, explaining the positive effect of temperature observed for the vaporization efficiency.

The relationship between the flow rate and the ethanol vaporization efficiency can be explored by the Newton's law of cooling:

$$q''_s = h\Delta T \rightarrow \Delta T = \frac{q''_s}{h} \tag{31}$$

where q''_s is the convective heat flux from the solid surface ($W.m^{-2}$), h is the convective heat transfer coefficient ($W.m^{-2}.K^{-1}$) and ΔT is the temperature difference between the solid surface and the fluid stream. From Newton's law of cooling, the heat flux from the surface is proportional to the heat transfer coefficient. The temperature difference between the solid surface and the fluid stream is inversely proportional to the heat transfer coefficient. Usually, the relationship between the convective heat transfer coefficient and the fluid flow and geometric properties are obtained from experimental data and are given in terms of the nondimensional numbers of Nusselt, Reynolds and Prandlt, as the empirical correlation proposed by Kalaivanan and Rathnasamy (2011) (Equation 19).

The convective heat transfer coefficient, *h,* can be estimated for the evaluated cases. The Reynolds number ranged from 1 to 100, the vapor phase Prandtl number about 1, the thermal conductivity of 0.18 W m^{-1} K^{-1} and the hydraulic diameter of 2.96 x 10^{-4} m (smaller channels), the convective heat transfer coefficient ranged about 577 – 760 W m^{-2} K^{-1}. This relative low variation of *h* can be attributed to the small influence of the flow rate on the evaporation performance.

The evaporated mass is directly proportional to the rate of energy supplied (Equation 30) to the system. The flow rate increment resulted in the decrease of the residence time, reducing the amount of energy received by the fluid flowing along the microdevice, and consequently affecting the ethanol evaporation process.

The most influencing variable on the evaporation performance was the ethanol/biodiesel molar ratio. The increment of this variable reduced considerably the ethanol evaporation. Moharana et al. (2013) studied the liquid-to-gas conversion of ethanol/water mixture for the production of hydrogen in microchannels. The authors found that the heat transfer coefficient varied with the ethanol volumetric fraction or holdup for a given temperature difference. They showed that the heat transfer coefficient decreased with the increment of the ethanol holdup.

The reduction in the heat transfer coefficient with the increment of the ethanol/biodiesel molar ratio can be explained considering that the ethanol/biodiesel mixture present similar thermodynamic behavior of an ethanol/water system. The ethanol/water (and many others, as for example ethanol/benzene and ethanol/chloroform) presents a positive deviation from the Raoult's law (Ball, 2014), i.e., it is a non-ideal liquid mixture. The Raoult's law establishes that the partial vapor pressure of one compound in a mixture is equal to the product of the liquid-phase molar fraction of the compound and the vapor pressure of the pure compound at a given temperature. A positive deviation of the Raoult's law stands that the mixture real vapor pressure is greater than the expected for an ideal mixture. This behavior means that the molecules present an easy separation from the mixture regarding the pure liquid.

For the case of ethanol/water mixture, from the phase diagram analysis for a given temperature and pressure, the increment on the liquid ethanol concentration is in equilibrium with a greater amount of ethanol in vapor phase. However, the increment of liquid ethanol above 95.6% w/w (azeotrope point) results in an inverse behavior. Guo et al. (2007) studied the influence of the ethanol addition on the biodiesel vapor pressure. They concluded that the biodiesel/ethanol mixture presented a positive deviation of the Raoult's law. Accordingly, the ethanol/biodiesel mixture exhibited a similar behavior of the ethanol/water mixture related by Moharana et al. (2013). The convective heat transfer coefficient decreased with the increment of ethanol amount on the stream, subsequently, the rate of energy supplied to the mixture stream should be larger in order to increase the evaporation efficiency. Once the energy rate remained constant, the ethanol vapor phase concentration decreased for liquid feeds with higher ethanol/biodiesel molar ratio.

In order to prove the relationship between the ethanol mass loss (or even, the evaporation rate) and ethanol/biodiesel molar ratio, experimental runs were carried out for the conditions of constant temperature (100 °C) and flow rate (0.6 mL min^{-1}), ranging the molar ratio from 2 to 11, and the results are presented in Figure 6.10.

From the Figure 6.10, the inverse relationship between the ethanol mass loss and the ethanol/biodiesel molar ratio can be noticed proving the above mentioned hypothesis.

NUMERICAL STUDY

The evaporation performance was evaluated using a Central Composite Rotatable Design (CCRD), analyzing the effect of three operating variables (temperature, mixture flow rate, ethanol/biodiesel molar ratio), using six axial points and one central point, resulting in 15 runs. Table 5 presents the VEP (%) predictions. For the studied conditions, VEP (%) ranged from 6.90% to 63.74%. Using the simulation results, the effects of each one of the three process variables in the evaporation performance were calculated using the software Statistica 7, as summarized in Table 6. Considering a statistical significance level of 95% (p-value < 0.05), the linear terms of temperature (x_1), flow rate (x_2) and E:B molar ratio (x_3) were statistically significant. Accordingly, the numerical predictions showed that the evaporation

Figure 3. Ethanol mass loss profile in function of the ethanol/biodiesel molar ratio at 100 °C and 0.6 mL min^{-1}.
Reprinted from Santana, H.S., Sanchez, G.B, & Taranto, O.P. (2017a). Evaporation of excess alcohol in biodiesel in a microchannel heat exchanger with Peltier module, Chemical Engineering Research and Design, 124, 20-28, with permission from Elsevier

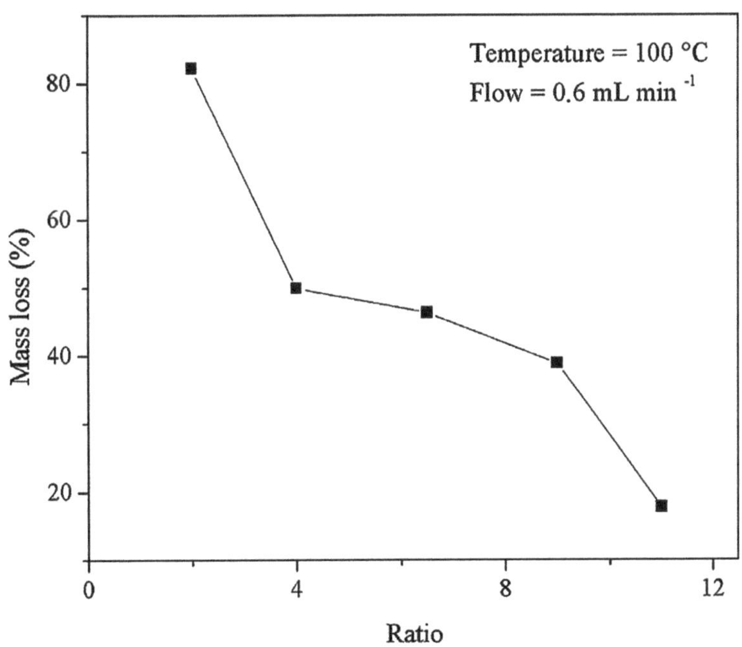

efficiency (vaporization of ethanol liquid molecules to vapor state) is strongly affected by the evaluated process variables. The increment of temperature provided an increase in the vaporized ethanol percentage, while the decrease of both, flow rate and E:B molar ratio, resulted in higher evaporation efficiencies. This behavior is illustrated in the Pareto diagram (Figure 4). Figure 5 shows the good agreement by the mathematical model in data prediction, explaining 94.97% of data variability.

The temperature influence on ethanol evaporation can be analyzed by the relation between vaporized mass and heat of vaporization ($m = Q\ H_v^{-1}$), where an inverse relationship between the mass and the energy amount is observed. The increment of temperature promotes a decrease in the heat of vaporization, and thus, increasing the vaporized mass.

The flow rate effect also could be explained by the relationship between vaporized mass and heat of vaporization. The increment of flow rate results in a decrease of fluid residence time inside the micro heat-exchanger, resulting in lower rates of heat received by the fluid mixture, consequently, influencing the ethanol evaporation rate.

The ethanol/biodiesel molar ratio resulted in a negative effect in ethanol evaporation performance, namely, the increment of E:B molar ratio decrease the VEP (%).This behavior could be related to the interphase heat transfer phenomena between the liquid phases (ethanol and biodiesel). Smaller amount of biodiesel (higher E:B

Table 5. Description of operating conditions and VEP(%) results for the cases of study

Case	Q (mL/min)	T (ºC)	E:B Molar Ratio	VEP(%)
1	0.3	88	4	32.79
2	0.3	112	4	58.59
3	0.9	88	4	16.56
4	0.9	112	4	36.23
5	0.3	88	9	25.32
6	0.3	112	9	45.19
7	0.9	88	9	25.31
8	0.9	112	9	27.40
9	0.6	80	6.5	6.90
10	0.6	120	6.5	39.71
11	0.1	100	6.5	63.74
12	1.2	100	6.5	21.71
13	0.6	100	2	46.54
14	0.6	100	11	25.95
15	0.6	100	6.5	30.39

Table 6. Variables effects on vaporized ethanol percentage, VEP (%)

Factor	Effects	Std. Deviation	t(5)	p-Value	Interval Estimative (95%)	
					Inferior Limit	Superior Limit
Average	30.56	5.69	5.36	0.003	15.91	45.21
x_1 (L)	17.96	3.10	5.79	0.002	9.98	25.93
x_1 (Q)	-5.48	4.66	-1.17	0.292	-17.46	6.49
x_2 (L)	-18.61	3.10	-5.99	0.002	-26.58	-10.64
x_2 (Q)	8.24	4.66	1.77	0.137	-3.73	20.22
x_3 (L)	-8.13	3.10	-2.62	0.047	-16.11	-0.17
x_3 (Q)	3.66	4.66	0.77	0.467	-8.31	15.64
x_1x_2	-5.98	4.05	-1.47	0.200	-16.39	4.44
x_1x_3	-5.88	4.05	-1.45	0.207	-16.29	4.54
x_2x_3	5.19	4.05	1.28	0.256	-5.22	15.61

% explained variation (R^2) = 94.97

Figure 4. Pareto diagram for the operating variables effects on vaporized ethanol percentage, VEP (%)

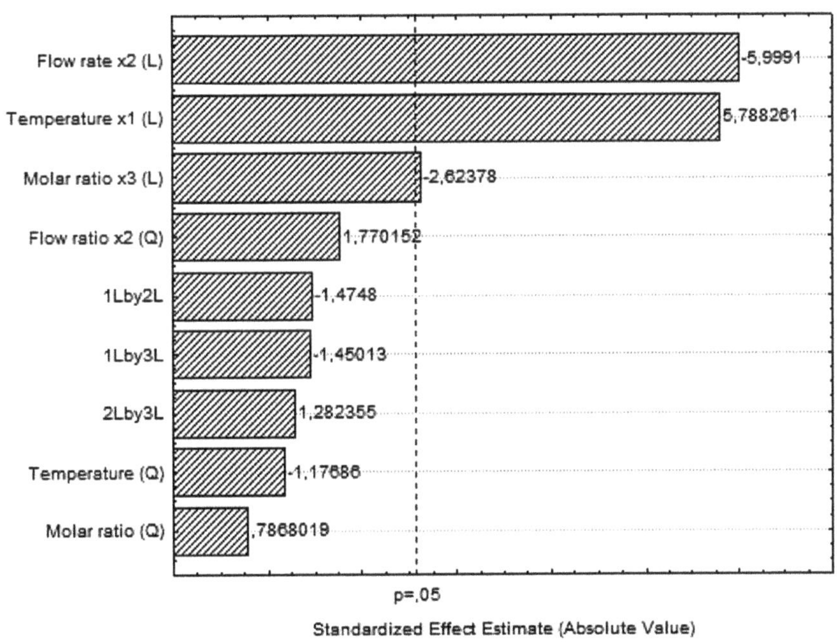

Figure 5. Observed and predicted values for vaporized ethanol percentage, VEP (%)

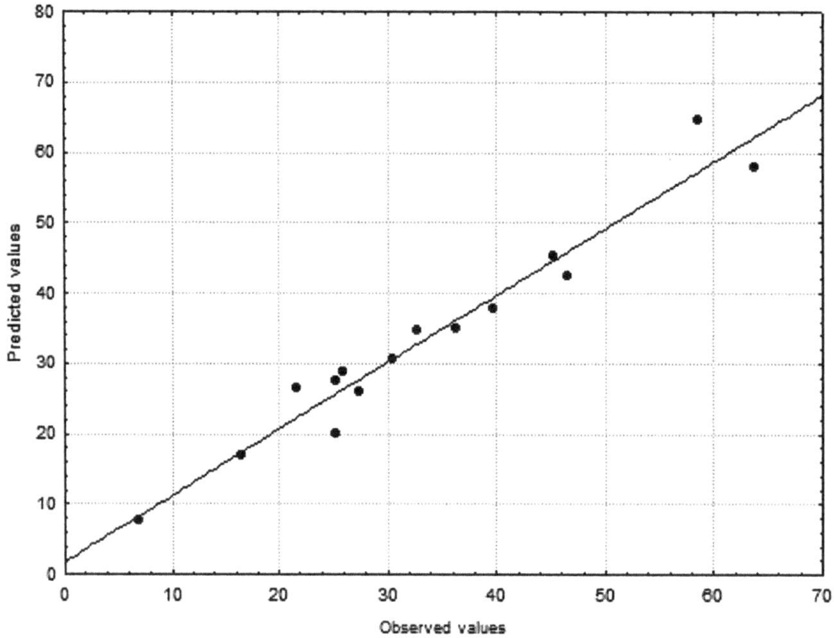

values) resulted in lower amount of heat rates received by the liquid ethanol due to the smaller specific interfacial area, resulting in the decrease of the vaporized ethanol.

CONCLUSION

The experimental and numerical studies can be compared in order to verify the performance of the proposed mathematical model, and then, the application of the numerical simulation in further geometry optimization procedure, as well, the process optimization. The temperature influenced positively the evaporation process, as observed in both approaches, experimental and numerical. The flow rate and the ethanol/biodiesel molar ratio affected negatively the evaporation process, as predicted by the numerical results and verified in the experimental tests. Accordingly, it can be concluded that the mathematical model was able to represent ethanol evaporation process from the biodiesel-ethanol mixture in the microheat exchanger.

Despite the capability of the proposed mathematical model to predict the main features of the ethanol evaporation process, the most influent variable in both approaches was different, being the flow rate (numerical simulations) and the ethanol/biodiesel molar ratio (experimental runs). The influence of the ethanol/biodiesel molar ratio on the evaporation performance can be experimentally explained by the Raoult's law positive deviation of the biodiesel-ethanol mixture. The experimental results proved the potential capability of the use of microdevices for the continuous ethanol recovery from biodiesel synthesis process. The deviations observed of the numerical model regarding the experimental results can be attributed to the no considerations concerning the thermodynamics of the studied system, since it was based on the evaporation rate considering the latent heat of the pure ethanol. Therefore, further details should be considered in the mathematical modeling in future researches.

REFERENCES

Bakhtiary-Davijany, H., Hayer, F., Phanam, X. K., Myrstad, R., Venvik, H. J., Pfeifer, P., & Holmen, A. (2011). Characteristics of an Integrated Micro Packed Bed Reactor-Heat Exchanger for methanol synthesis from syngas. *Chemical Engineering Journal*, *167*(2-3), 496–503. doi:10.1016/j.cej.2010.08.074

Ball, D. W. (2014). *Physical Chemistry* (2nd ed.). Stamford, CT: Cengage Learning.

Bertsch, S. S., Groll, E. A., & Garimella, S. V. (2009). Effects of heat flux, mass flux, vapor quality, and saturation temperature on flow boiling heat transfer in microchannels. *International Journal of Multiphase Flow*, *35*(2), 142–154. doi:10.1016/j.ijmultiphaseflow.2008.10.004

Bhoi, R., Sen, N., Singh, K. K., Mahajani, S. M., Shenoy, K. T., Rao, H., & Ghosh, S. K. (2014). Transesterification of Sunflower oil in microreactors. *International Journal of Chemical Reactor Engineering*, *12*(1), 47–62. doi:10.1515/ijcre-2013-0105

Cheng, P., & Wu, H. Y. (2006). Mesoscale and microscale phase-change heat transfer. *Advances in Heat Transfer*, *39*, 461–563. doi:10.1016/S0065-2717(06)39005-3

Chueluecha, N., Amaraporn, K., & Attasak, J. (2017). Biodiesel synthesis using heterogeneous catalyst in a packed-microchannel. *Energy Conversion and Management*, *141*, 145–154. doi:10.1016/j.enconman.2016.07.020

Deng, D., Wan, W., Zhang, J., & Chu, X. (2017). Flow boiling enhancement of structured microchannels with micro pin fins. *International Journal of Heat and Mass Transfer*, *105*, 338–349. doi:10.1016/j.ijheatmasstransfer.2016.09.086

Esteban, B., Riba, J. R., Baquero, G., Rius, A., & Puig, R. (2012). Temperature dependence of density and viscosity of vegetable oils. *Biomass and Bioenergy*, *42*, 164–171. doi:10.1016/j.biombioe.2012.03.007

Fasina, O. O., & Colley, Z. (2008). Viscosity and specific heat of vegetable oils as a function of temperature: 35 °C to 180 °C. *International Journal of Food Properties*, *11*(4), 738–746. doi:10.1080/10942910701586273

Fazeli, A., & Benham, M. (2010). Hydrogen production in a zigzag and straight catalytic wall coated micro channel reactor by CFD modeling. *International Journal of Hydrogen Energy*, *35*(17), 9496–9503. doi:10.1016/j.ijhydene.2010.05.052

Goodrum, J. W. (2002). Volatility and boiling points of biodiesel from vegetable oils and tallow. *Biomass and Bioenergy*, *22*(3), 205–211. doi:10.1016/S0961-9534(01)00074-5

Greco, A., & Vanoli, G. P. (2005). Flow boiling heat transfer with HFC mixtures in a smooth horizontal tube. Part I: Experimental investigations. *Experimental Thermal and Fluid Science*, *29*(2), 189–198. doi:10.1016/j.expthermflusci.2004.03.003

Guo, Y., Zhong, J., Xing, Y., Li, D., & Lin, R. (2007). Volatility of Blended Fuel of Biodiesel and Ethanol. *Energy & Fuels*, *21*(2), 1188–1192. doi:10.1021/ef0605807

Harirchian, T., & Garimella, S. V. (2010). A comprehensive flow regime map for microchannel flow boiling with quantitative transition criteria. *International Journal of Heat and Mass Transfer*, *53*(13-14), 2694–2702. doi:10.1016/j.ijheatmasstransfer.2010.02.039

Huo, X., Chen, L., Tian, Y. S., & Karayiannis, T. G. (2004). Flow boiling and flow regimes in small diameter tubes. *Applied Thermal Engineering*, *24*(8-9), 1225–1239. doi:10.1016/j.applthermaleng.2003.11.027

Kalaivanan, R., & Rathnasamy, R. (2011). Experimental Investigation of Forced Convective Heat Transfer in Rectangular Micro-channels. *Jordan Journal of Mechanical and Industrial Engineering*, *5*, 383–387.

Kandlikar, S. G., & Balasubramanian, P. (2004). An extension of the flow boiling correlation to transition, laminar, and deep laminar flows in minichannels and microchannels. *Heat Transfer Engineering*, *25*(3), 86–93. doi:10.1080/01457630490280425

Kew, P. A., & Cornwell, K. (1997). Correlations for the prediction of boiling heat transfer in small-diameter channels. *Applied Thermal Energy.*, *17*(8-10), 705–715. doi:10.1016/S1359-4311(96)00071-3

Kim, S. M., & Mudawar, I. (2013a). Universal approach to predicting saturated flow boiling heat transfer in mini/micro-channels – Part I. Dryout incipience quality. *International Journal of Heat and Mass Transfer*, *64*, 1226–1238. doi:10.1016/j.ijheatmasstransfer.2013.04.016

Kim, S. M., & Mudawar, I. (2013b). Universal approach to predicting saturated flow boiling heat transfer in mini/micro-channels – Part II. Two-phase heat transfer coefficient. *International Journal of Heat and Mass Transfer*, *64*, 1239–1256. doi:10.1016/j.ijheatmasstransfer.2013.04.014

Kim, S. M., & Mudawar, I. (2014). Review of databases and predictive methods for heat transfer in condensing and boiling mini/micro-channel flows. *International Journal of Heat and Mass Transfer*, *77*, 627–652. doi:10.1016/j.ijheatmasstransfer.2014.05.036

Li, W., & Wu, Z. (2010). A general correlation for evaporative heat transfer in micro/mini-channels. *International Journal of Heat and Mass Transfer*, *53*(9-10), 1778–1787. doi:10.1016/j.ijheatmasstransfer.2010.01.012

Liu, Z., & Winterton, R. H. S. (1991). A general correlation for saturated and subcooled flow boiling in tubes and annuli based on a nucleate pool boiling equation. *International Journal of Heat and Mass Transfer*, *34*(11), 2759–2766. doi:10.1016/0017-9310(91)90234-6

Mahmouda, M. M., & Karayiannis, T. G. (2013). Heat transfer correlation for flow boiling in small to micro tubes. *International Journal of Heat and Mass Transfer*, *66*, 553–574. doi:10.1016/j.ijheatmasstransfer.2013.07.042

Moharana, M. K., Nemade, R. M., & Khandekar, S. (2013). Phase-Change Heat Transfer of Ethanol-Water Mixtures: Towards Development of a Distributed Hydrogen Generator. *Proceedings of the ASME 2013 Summer Heat Transfer Conference*.

Peng, X. F., Hu, H. Y., & Wang, B. X. (1998). Boiling nucleation during liquid flow in Microchannels. *International Journal of Heat and Mass Transfer*, *41*(1), 101–106. doi:10.1016/S0017-9310(97)00096-3

Poling, B. E., Thomson, G. H., Friend, D. G., Rowley, R. L., & Wilding, W. V. (2008). Physical and Chemical Data. In *Perry's Handbook of Chemical Engineering* (8th ed.). New York: McGraw-Hill Education.

Rahimi, M., Aghel, B., Alitabar, M., Sepahvand, A., & Ghasempour, H. R. (2014). Optimization of biodiesel production from soybean oil in a microreactor. *Energy Conversion and Management*, *79*, 599–605. doi:10.1016/j.enconman.2013.12.065

Rahmat, M., & Hubert, P. (2010). Two-phase simulations of micro heat pipes. *Computers & Fluids*, *39*(3), 451–460. doi:10.1016/j.compfluid.2009.09.014

Revellin, R., & Thome, J. R. (2007). Adiabatic two-phase frictional pressure drops in microchannels. *Experimental Thermal and Fluid Science*, *31*(7), 673–685. doi:10.1016/j.expthermflusci.2006.07.001

Santacesaria, E., Di Serio, M., Tesser, R., Turco, R., Tortorelli, M., & Russo, V. (2012). Biodiesel process intensification in a very simple microchannel device. *Chemical Engineering and Processing*, *52*, 47–54. doi:10.1016/j.cep.2011.12.001

Santana, H. S., Sanchez, G. B., & Taranto, O. P. (2017a). Evaporation of excess alcohol in biodiesel in a microchannel heat exchanger with Peltier module. *Chemical Engineering Research & Design*, *124*, 20–28. doi:10.1016/j.cherd.2017.05.022

Santana, H. S., Tortola, D. S., Reis, E. M., Silva, J. L. Jr, & Taranto, O. P. (2016). Transesterification reaction of sunflower oil and ethanol for biodiesel synthesis in microchannel reactor: Experimental and simulation studies. *Chemical Engineering Journal*, *302*, 752–762. doi:10.1016/j.cej.2016.05.122

Santana, H. S., Tortola, D. S., Silva, J. L. Jr, & Taranto, O. P. (2017b). Biodiesel synthesis in micromixer with static elements. *Energy Conversion and Management*, *141*, 28–39. doi:10.1016/j.enconman.2016.03.089

Sempértegui-Tapia, D. F., & Ribatski, G. (2017). Two-phase frictional pressure drop in horizontal micro-scale channels: Experimental data analysis and prediction method development. *International Journal of Refrigeration*, *79*, 143–163. doi:10.1016/j.ijrefrig.2017.03.024

Sun, L., & Mishima, K. (2009). An evaluation of prediction methods for saturated flow boiling heat transfer in mini-channels. *International Journal of Heat and Mass Transfer*, *52*(23-24), 5323–5329. doi:10.1016/j.ijheatmasstransfer.2009.06.041

Tibiriçá, C. B., & Ribatski, G. (2013). Flow boiling in micro-scale channels - Synthesized literature review. *International Journal of Refrigeration*, *36*(2), 301–324. doi:10.1016/j.ijrefrig.2012.11.019

Tibiriçá, C. B., Ribatski, G., & Thome, J. R. (2012). Saturated flow boiling heat transfer and critical heat flux in small horizontal flattened tubes. *International Journal of Heat and Mass Transfer*, *55*(25-26), 7873–7883. doi:10.1016/j.ijheatmasstransfer.2012.08.017

Triplett, K. A., Ghiaasiaan, S. M., Abdel-Khalik, S. I., & Sadowski, D. L. (1999). Gas-liquid two-phase flow in microchannels part I: Two-phase flow patterns. *International Journal of Multiphase Flow*, *25*(3), 377–394. doi:10.1016/S0301-9322(98)00054-8

Tuckerman, D. B., & Pease, R. F. W. (1981). High-performance heat sinking for VLSI. Electron Device Lett., 2, 126-129.

Wang, Y., & Wang, Z. G. (2014). An overview of liquidevapor phase change, flow and heat transfer in mini- and micro-channels. *International Journal of Thermal Sciences*, *86*, 227–245. doi:10.1016/j.ijthermalsci.2014.07.005

ADDITIONAL READING

Abed, W. M., Whalley, R. D., Dennis, D. J. C., & Poole, R. J. (2015). Numerical and experimental investigation of heat transfer and fluid flow characteristics in a micro-scale serpentine channel. *International Journal of Heat and Mass Transfer*, *88*, 790–802. doi:10.1016/j.ijheatmasstransfer.2015.04.062

Al-Neama, A. F., Kapur, N., Summers, J., & Thompson, H. M. (2017). An experimental and numerical investigation of the use of liquid flow in serpentine microchannels for microelectronics cooling. *Applied Thermal Engineering*, *116*, 709–723. doi:10.1016/j.applthermaleng.2017.02.001

Caney, N., Louahlia-Gualous, H., & Bontemps, A. (2016). Two-Phase Flow in Microchannels. In M. Rebay, S. Kakaç, & R. M. Cotta (Eds.), *Microscale and Nanoscale Heat Transfer: Analysis, Design, and Applications* (pp. 445–478). Boca Raton, FL, USA: CRC Press. doi:10.1201/b19261-19

Leal, L. G. (2007). *Advanced Transport Phenomena: Fluid Mechanics and Convective Transport*. New York, NY, USA: Cambridge University Press. doi:10.1017/CBO9780511800245

Rebay, M., Kabar, Y., & Kakaç, S. (2016). Numerical Simulation of Combined Microscale Effects on Convective Heat Transfer in Single-Phase Flows. In M. Rebay, S. Kakaç, & R. M. Cotta (Eds.), *Microscale and Nanoscale Heat Transfer: Analysis, Design, and Applications* (pp. 283–307). Boca Raton, FL, USA: CRC Press. doi:10.1201/b19261-12

Serth, R. W., & Lestina, T. G. (2014). *Process Heat Transfer: Principles and Applications and Rules of Thumb* (2nd ed.). Oxford, UK: Elsevier Academic Press.

Yang, D., Wang, Y., Ding, G., Jin, Z., Zhao, J., & Wang, G. (2017). Numerical and experimental analysis of cooling performance of single-phase array microchannel heat sinks with different pin-fin configurations. *Applied Thermal Engineering*, *112*, 1547–1556. doi:10.1016/j.applthermaleng.2016.08.211

Zotin, J. L. Z., Knupp, D. C., & Cotta, R. M. (2016) Analytical–Numerical Solutions for Conjugated Heat Transfer in Multistream Microsystems. 349-367. In Rebay, M., Kakaç, S., & Cotta, R.M., Microscale and Nanoscale Heat Transfer: Analysis, Design, and Applications. (pp. 349-367). Boca Raton, FL, USA, CRC Press.

KEY TERMS AND DEFINITIONS

Boiling Heat Transfer Coefficient: Global heat transfer coefficient given by the contribution of nucleate boiling and convective boiling mechanisms.

Conjugate Heat Transfer Flow: Fluid flow with simultaneous heat and momentum transfer (i.e., fluid flow with heating or cooling).

Eulerian Multiphase Approach: Mathematical framework treating all phases of the system as a continuous and interpenetrating phase.

Microevaporator: Microdevice for heat transfer service promoting liquid-to-vapor phase change.

Phase Holdup: Volumetric fraction occupied by a phase in a multiphase system (i.e., the ratio of the volume of the phase to the total volume of mixture).

Subcooling Degree: Temperature difference between the liquid and the liquid saturation temperature.

Thermal-Hydraulic Performance: Performance of the equipment based on the balance between the heat transfer coefficient (or the heat transfer capacity) and the pressure drop.

Vapor Quality: Ratio of the vapor mass to the vapor-liquid mixture mass.

Chapter 7
Design Techniques for Microfluidic Devices Implementation Applicable to Chemical Analysis Systems:
Theoretical and Experimental Analysis of Microcomponents

Reinaldo Lucas dos Santos Rosa
University of São Paulo, Brazil

Antonio Carlos Seabra
University of São Paulo, Brazil

ABSTRACT

This chapter provides a guide for microfluidic devices development and optimization focused on chemical analysis applications, which includes medicine, biology, chemistry, and environmental monitoring, showing high-level performance associated with a specific functionality. Examples are chemical analysis, solid phase extraction, chromatography, immunoassay analysis, protein and DNA separation, cell sorting and manipulation, cellular biology, and mass spectrometry. In this chapter, most information is related to microfluidic devices design and fabrication used to perform several steps concerning chemical analysis, process preparation of reagents, samples reaction and detection, regarding water quality monitoring. These steps are especially relevant to lab-on-chip (LOC) and micro-total-analysis-systems (µTAS). µTAS devices are developed in order to simplify analytical chemist work, incorporating several analytical procedures into flow systems. In the case of miniaturized devices, the analysis time is reduced, and small volumes (nL) can be used.

DOI: 10.4018/978-1-5225-7138-4.ch007

INTRODUCTION

Through microelectronic manufacturing methods, fabrication of micro-devices capable of conducting and handling fluids in cellular scale microstructures, performing analysis and detection of very small samples become possible. In addition, microelectromechanical systems (MEMS) technology enables microstructures fabrication linked to a physical principle, adding sensitive and actuation capabilities to the microfluidic device. Essentially, microfluidic technology deals with liquids and gases manipulation within channels whose dimensions are in the range: 10 – 100 μm. Fully integrated microfluidic devices implementation do not have any standard component, whereas different technologies are used in order to fabricate components for LOC/ *μTAS* devices, such as electric, magnetic, piezoelectric, thermal, pneumatic actuation systems for microvalves fabrication (Oh & Ahn, 2006). The main research area relates microsystems development that work in continuous flow regime. Most equipment developed for in-*time* water monitoring despite having the ability to perform a broad range of water analysis in flow systems, show high installation and maintenance cost. Moreover, these systems exhibit high reagent consumption and operational problems. Regarding most routine instrumentation, combination of more than two or three different analytical steps is seldom. Therefore, they are just a portion of a whopping process. Besides that, for many microfluidic devices, the use of auxiliary equipment, such as injection systems, power supplies or acquisition devices, is mandatory, opposing miniaturization benefits. Furthermore, these types of systems require skilled employee destined to device operation and upkeep. In this scenario, integrated analytical system development, combining all analysis stage into a single on-line device becomes essential. Such a total analysis system must to be autonomous, functionally flighty, robust and reliable. This was the motivation for the development of miniaturized autonomous chemical analyzers (Martínez-Cisneros, da Rocha, Ferreira, Valdés, Seabra, Góngora-Rubio, & Alonso-Chamarro, 2009; Ibanez-Garcia, Mercader, Mendes da Rocha, Seabra, Góngora-Rubio, & Chamarro, 2006) applying Low Temperature Co-fired Ceramic (LTCC) prototyping material, capable of integrating several steps regarding analytical chemistry process applicable to water quality monitoring, through microfluidic components/devices interconnection within a single chip. Well-established microfabrication technologies typically uses glass, polymers and/or silicon substrate, intended for fully integrated microanalyzers fabrication on a single chip. LTCC technology is widely used for multilayer electronic circuits manufacturing due to its excellent electrical, mechanical and thermal properties (Derenovsek, Eberstein, Schiller, Naeini, Preu, & Wersing, 2001; Wang, Zang, & Ma, 2002), and owing to its fast prototyping process is suitable for microfluidic devices implementation. Additionally, most manufacturing process associated to other materials (unlike LTCC material), such as silicon, polymers or glass, requires additional processing steps in order to accomplish necessary surface modifications or to obtain properly device integration. This chapter aims to introduce fluidic components modeling methods, materials science related to manufacturing process and devices prototyping, integration towards fully microfluidic chemical analyzer development and experimental work for devices characterization, exploiting LTCC technology as prototype material used for microfluidic devices fabrication.

BACKGROUND

The first notorious microfluidic device developed portray a gas chromatograph performed on a silicon chip, consisting of a valve and a detached capillary column fabricated on a silicon wafer, including a thermal conductivity detector integrated into the silicon substrate (Terry, Jerman, & Angell, 1979) and also an ink-jet printer developed at IBM (Petersen, 1979, pp. 1918-1920). Terry reports the construction and theory related to a sensor operation that analyzes gaseous samples based on the gas chromatographic principles. A miniature capillary column was fabricated in a sequence of photolithography and etch steps, treated with an organosilane compound in order to promote adherence between lining liquid and capillary channel walls. Also, a miniature one-valve injection system was fabricated on the opposite side of the silicon wafer from the capillary column, allowing its integration. A completed thermal conductivity detector was fabricated using standard IC processing techniques, and mechanically clamped to the CG wafer directly over an etched gas channel. Adding a one-chip microcomputer intended to operation control and signal analysis, a complete gas analysis system were achieved in a small package comparable to conventional techniques. Manz and his group first suggested μTAS concept (Manz, Graber, & Widmer, 1990), as an extension for Total Analysis System (TAS) idea, applicable to analytical chemistry in miniaturized form. The TAS technology was first proposed as a solution for automation issues related to analytical chemistry, offering an integral incorporation of chemical analytical procedures into flow systems, whereby fluidic paths are connected in order to perform the required analysis (Widmer, 1983, VIII-X.). The μTAS technology works with nL to μL volumes, with analysis time of seconds. Besides that, these devices uncover important advantages in contrast to conventional total analysis systems: reagents consumption reduction, improved automation and reduction of manufacturing costs (Becker & Locascio, 2002). Moreover, a wide range of applications could be performed applying μTAS devices, such as chemiluminescent (Cheek, Steel, Torres, Yu, & Yang, 2001), potentiometric (Tantra & Manz, 2000), electrochemical (Martin, Gawron, Lunte, & Henry, 2000), fluorescence (Maims, Hulme, Fielden, & Goddard, 2001), optical (Chau, Osborn, Wu, & Yager, 1999), among others. Other noteworthy field is the multidisciplinary lab on a chip (LOC) technology, that require contribution from biology, chemistry, software development, physics, material science, microfabrication technologies or surface machining and engineering. Usually, LOC devices are of millimeter to centimeter size with more than one laboratorial process within a single device. More often, such devices contains microfluidic elements responsible for fluidic manipulation, control, data processing and detection. Albeit silicon and glass are the most used material nowadays for these applications, seen that they are chemically inert, versatile and have funded microscale-manufacturing processes, there is no standard material or fabrication method grounded with respect to LOC manufacturing.

Most microfluidic devices work with flow injection analysis (FIA) method, suggested by Ruzicka and Hansen (Ruzicka & Hansen, 1975). This method is suitable for environmental species determination, especially for water analysis. The FIA method is based on the fast injection of a sample aliquot within a continuous flow (carrier solution), driven by a pump carrying forward the sample along the fluidic route. After chemical mixing, the analyte is shifted to the detection system where the response signal will be recorded. This method promotes high sampling rate, automation, low sample/reagent consumption and system versatility. Electronic valves/pumps combined with FIA method allow specific volumes insertion of sample/reagents in time, improving even more system automation. Besides that, this methodology allow multiple determination of chemical species, without vary initial configuration, as a multicommutation and binary sampling approaches developed for nitrite, nitrate and ammonia determination (Rocha & Reis, 2000). Detection was carried out with a homemade photometer based on the arrangement of a dual-color LED and a photodiode. The flow network was assembled with a set of three-way solenoid valves, independently computer-controlled, allowing to achieve suitable reaction conditions for each determination. In 2006 a multiparametric system for sodium, potassium, calcium and magnesium determination was developed, establishing the necessary measurement conditions for each element in the same device (Piston, Dol, & Knochen, 2006).

Regarding sensing system used in microfluidic device, sensibility and scalability to smaller dimensions are two of the main coefficient considered in detection system choice. Since sensitive portable systems are required for microfluidic devices, electrochemical detection techniques, such as amperometric, potentiometric or conductometric detection, are not proper selections when detection limits are very low. However, output signal is dependent on electrode surface area, in contrast with available detection volume (spectrophotometric detection case), and therefore, some electrochemical methods can scale better upon miniaturization. The most common optical detection is spectrometric detection, suitable for a wide range of applications: absorbance, fluorescence and chemiluminescence. Normally, light-emitting diodes

(LEDs) or laser diodes and optical fibers are the mainly optical components used in these detectors, providing a portable detection system integrated to microfluidic devices. An UV absorption detection system using an optical fiber was developed in order to monitor peptides separation (Jindal & Cramer, 2004). The detection was executed at channel edge with no stationary phase, while the separation channel was filled with a sol-gel-immobilized stationary phase. A CCD array detector receive the collected light that was guided by a bottom placed optical fiber, initially transmitted by a top fiber connected to a deuterium-tungsten light source, whereby microchip is situated between the fibers. A microfabricated silicon flow-cell for optical monitoring of albumin in serum has been demonstrated (Chau, Osborn, Wu, & Yager, 1999), coupling a single emitting optical fiber and six receiving fibers to a silicon flow-cell allowing absorbance measurement. An integrated light collimating system was fabricated for absorbance measurement (Ro, Lim, Shim, & Hahn, 2005), containing cylindrical microlens in order to minimize the divergence of light from optical fibers, as well as the presence of a pair of slits that greatly reduce scattered light. Detection occurred on a Z-shaped flow-cell. These were some of implemented systems containing absorbance detection examples. Fluorescence detection has a great importance on microsensing systems, due to its upper selectivity and sensitivity (Reyes, Iossifidis, Auroux, & Manz, 2002). Several excitation sources for fluorescence detection are available, however the implementation of a few are recurrent. Laser-induced fluorescence source is the most adaptable regarding microchips dimensions, since the projected light from laser beam has good coherence and low divergence, small volume specimen can be easily detected (Auroux, Iossifidis, Reyes, & Manz, 2002). Lamp-based excitation systems are more flexible considering wavelength choice, whereby microscope detectors using xenon or mercury lamps have been successfully applied over a wide range of analytical determinations (Kuswandi, Huskens, & Verboom, 2007). A laser-induced fluorescence system used for glutathione detection was developed, with functional integration of cell sampling, single cell loading, docking, lysing and capillary electrophoretic separation fabricated in microchannels of a single glass chip (Gao, Yin, & Fang, 2004). Some examples of lamp-based detection using epifluorescence microscopes are (Mourzina, Kalyagin, Steffen, & Offenhausser, 2005; Varjo, Ludwig, Belder, & Riekkola, 2004; Wicks & Li, 2004). Chemiluminescence detection method has a simplest instrumentation, since external light source is unnecessary. However, this technique has a limited number of chemiluminescence reagents. Several chips layouts were implemented regarding different chemiluminescence methods including their respective evaluated performances (Liu, Ozaki, Utsumi, Hattori, & Terabe, 2003).

Prior to microchannels discussion, fluids motion has a considerable importance over microfluidic devices performance. In the case of chemical analysis systems, liquid flow is much more iterant. Liquid flow can be single-phase continuous flow within microchannels or multiphase flow of droplets flowing through microchannels or over a surface. For single-phase flow, fluid motion can be induced by a pressure gradient (pressure driven device), capillarity (capillary driven device), electrokinetic force (electrokinetic driven device) or a centrifugal force (centrifugal driven device), among others. Regarding multiphase flow, single droplets can be uninterrupted generated and manipulated within microchannels (droplet-based devices) or individual droplets can be moved and controlled over a surface by electric fields (electrowetting-based devices). When it comes to microflows, relative significance of surface to volume forces rises. Hence, fluid-channel surface interactions become relevant, wherein viscous forces on the surface can cause significant pressure drops, even for mild velocities profiles. Laminar flow represents microflows, since viscous forces are more dominant in contrast with inertial forces, whereupon the inertial to viscous forces ratio, Reynolds number (***Re***), can be defined as shown in Equation 1:

$$\mathrm{Re} = \frac{\rho VD}{\mu} \tag{1}$$

where ρ (kg/m^3) is the fluid density, V (m/s) is the velocity, μ (kg/ms) is fluid dynamic viscosity, and D (m) is the hydraulic diameter. Considering a water flow over a 100 μm hydraulic diameter channel with a typical velocity of 1 cm/s, the dimensionless Reynolds number will be 1. Normally, pressure-driven flows in microchannels have a parabolic velocity profile, since pressure promotes a uniform force against the cross-sectional area of the channel, and owing to solid boundary interactions, the momentum leaves the flow at the walls, forcing zero velocity. Microchannels fabricated using silicon as substrate, made by techniques such as wet or dry etching processes, normally has a glass cover sealing, in order to promote transparency to microfluidic channels. Field-assisted or anodic bonding is used in this particular case (Kovacs, 1998, pp. 586-587; Madou, 2002). This bonding is accomplished by voltage application, in the order of 200 to 1000 Volts, over silicon/glass junction, where the negative electrode is glass. Large electric fields are generated in the interface pulling surfaces together. Using LTCC technology, microchannels can be formed on a simpler process.

Microsystems fluids control inevitably involves valves utilization, which consequently increases the interest on microvalves development, especially in the micromechanical domain, showing advantages such as response time reduction, energy consumption and dead volume reduction. In addition, they enable high reproducibility bulk manufacturing and can be integrated into multifunctional microfluidic systems. Evaluating fully integrated microfluidic systems miniaturization and commercialization from a practical point of view, its inefficiency may be associated with the scarcity of reliable microfluidic components, such as microvalves and micropumps. Traditionally, microvalves are divided into active and passive valves (Oh & Ahn, 2006). Active microvalves have internal actuation structures that can be controlled by external elements, and passive microvalves have actuation structures that cannot be controlled externally to the valve structure. Mechanical active microvalves are manufactured using surface micromachining technology and/or MEMS technology, where a mobile mechanical structure is coupled to magnetic, electrical, piezoelectric or thermal actuation methods. Non-mechanical active microvalves are operated through the use of intelligent materials such as phase change materials or rheological materials. Moreover, active microvalves can be operated externally, using integrated modules or pneumatic systems. Regarding valve initial state, they can be classified as normally open, normally closed or bistable microvalves. In this chapter, active mechanically actuated valves are the focus due to automation capability offered to device development. In 1979 Terry and his research group (Terry, Jerman, & Angell, 1979) developed a miniaturized electromagnetic microvalve employing a solenoid piston physically connected to a micromachined silicon membrane. The microvalve consists of a base constructed from etched silicon, also containing a silicon substrate, a sealing ring, an orifice, a nickel diaphragm and a solenoid actuator connected to a piston. In 1995, Yanagisawa et al. (Yanagisawa, Kuwano, & Tago, 1995) manufactured a microvalve with a thin NiFe membrane acting as a cover inside a small tube with an externally mounted solenoid coil. The valve is constructed on a silicon substrate, and uses the valve cap and valve plate to control fluid flow within the channel. An externally installed coil, supported by a flat angular-spiral shape spring, controls the valve cap, made of soft magnetic material (NiFe). Valve cap vertical motion causes valve closure or opening. The main functionality of the spring is to prevent large initial deflection due to stress caused during the manufacturing process caused by an instability between the front and rear surfaces. In 1999, Meckes et al. (Meckes, Behrens, Kayser, Benecke, Becker, & Müller, 1999) developed a microvalve with a gold coil integrated to a movable membrane whose deflection is controlled by magnetic force produced by magnetic fields interaction between a microcoil and a permanent magnet.

Other microfluidic component is the micromixer. This chapter focus on passive micromixers: devices that works with no energy input, only by load loss mechanism. Channels performance must be optimized, improving mixing efficiency, utilizing the smallest sample/reagents volume as possible over each analysis. In order to obtain a microchannel satisfactory mixing rate, since microfluidic flows are laminar, it is necessary to introduce load losses of two types: distributed load losses caused by shear of the fluid in channel walls, and singular load losses caused by section change or flow direction change. In laminar flows, species mixing occurs due to diffusion mechanism, intrinsically sluggish process. Therefore, the main focus of micromixers methods is to enhance mixing efficiency achieved within shorter channels as possible, ensuring system miniaturization. Regarding active micromixers, normally these components are developed through incorporation of mechanical transducers into microfluidic systems applying microfabrication techniques. Besides system become more complex, energy consumption is increased, thus suitable passive micromixers are preferred. In passive micromixers, the most common method is related to fluids contact surface enhancement, with smaller diffusion path. Essentially, passive micromixers can be categorized as lamination, chaotic advection and special designs related to mixed phases adjustment (Lee, Lin, & Fu, 2015). A Y-mixer manufactured using co-fired ceramic sheets, having a 90-degree curve, was able to generate vortices at Reynolds number beyond 10 (Yi & Bau, 2003). A T-mixer was reported depicting glass/silicon manufacturing process, using relative high Reynolds numbers ($\boldsymbol{Re} = 500$), with velocities as high as 7.6 m/s and driven pressures of the order of 7 bar (Wong, Ward, & Wharton, 2004).

Microfluidic devices manufacturing methods imposes important requirements and limitations on microsystems development. Silicon, glass and more briefly polymers are the main materials used on LOCs/μTAS fabrication, however, co-fired ceramic tapes is an emerging technology showing excellent performance towards microdevices development applicable to chemical analysis. Ceramic substrate technology known as LTCC technology was developed by DuPont ™ in the 80s (Golonka, 2003, August). Acknowledged deals with a distinct material platform that allows multiple materials technological implementation for a variety of multidisciplinary applications. Originally LTCC technology was used in applications involving hybrid electronic circuits encapsulation such as multi-chip modules (MCM), in which different chips are encapsulated in the same substrate, allowing the creation of interconnection paths and passive components (resistors, capacitors and inductors) deposition with the aid of screen printing process. Currently this technology is used in automotive, aerospace, military and telecommunications applications areas. Ceramic sheets are produced using "Doctor Blade" method (Gongora-Rubio, Espinoza-Vallejos, Sola-Laguna, & Santiago-Aviles, 2001), with thicknesses in the range of 50 to 400 μm. It is possible to make cuts and holes by micro-machining and also embossing by ceramic tape patterning prior to the sintering process, however, micro-machining process is more often used. Laser machining is a widely used method due to its high repetition accuracy and ability to create shapes of very small sizes (in the order of 20-50 μm). By stacking and laminating all the processed layers, an inseparable multilayer block is obtained. After laminate sintering, a rigid block with the desired functionality (channels, mixers, sensors, actuators, etc.) is generated. The required temperature for ceramic sintering is 850 °C, relatively low in comparison to the sintering temperature of other ceramic materials, around 1400 °C. LTCC technology allows the application of embedded metal electrodes to perform fluid heating, integration of optical and microwave components for fluids characterization and analysis. It also allows the development of 2D (two-dimensional) and 3D (three-dimensional) structures such as microchannels, micromixers, electric plates, microheaters and micropumps (Groß, Thelemann, Schneider, Boskovic, & Köhler, 2008; Malecha, Pijanowska, Golonka, & Torbicz, 2009; Rebenklau, Wolter, & Howitz, 2000; Barlow, Wood, Elshabini, Stephens, Feeler, Kemner, & Junghans, 2009). Ceramic sheets/tapes, known as green ceramics due to their initial malleability, are glass-ceramic structure with micro-granular elements composed essentially of alumina (Al_2O_3) and silicon oxide (SiO_2). Main advantages related to LTCC technology are:

- LTCC tape machining using several sizes (in the range of 20 µm to millimeters);
- Application of large-scale production methods;
- Has good mechanical and excellent electrical properties;
- Thermo-physical properties such as thermal conductivity can be altered (Peterson, Knudson, Garcia, Patel, Okandan, Ho, & Zawicki, 2008, June);
- Materials can be incorporated to ceramic tapes (elastomers, piezoelectric, magnetic etc.), changing specific properties such as magnetic permeability and elasticity;
- Ease of adjacent layers interconnection (electrical or fluidic);
- Allows self-encapsulated devices manufacturing;
- Applied manufacturing techniques are simple and low cost compared to other microfabrication technologies, and environmentally friendly;
- Allows detection systems integration (optical, potentiometric, amperometric, mechanical etc.).

MAIN FOCUS OF THE CHAPTER

Issues, Controversies, Problems

Development of fully integrated microfluidic devices applicable to chemical analysis, especially water quality monitoring, with high-performance analysis, robust device with fast analysis and small reagents consumption, is the focus of this chapter. Therefore, the study of components development, such as micro-actuators, sensors, microchannels, micromixers, flow cells and electronic devices, as well as integration procedures are necessary in order to accomplish an entire µTAS. In addition, steps involved in chemistry analysis require automation considering FIA method enforcement. Regarding detection systems used in microfluidic devices applied to chemical analysis, miniaturization become an issue, as analysis volume are decreased, the quantity of analyte spare for detection will be decreased as well, making detection more difficult. Whereby this chapter discuss chemical analysis microsystems, pressure driven devices will be the focus considering the source of fluidic motion. Additionally, low temperature co-fired ceramic (LTCC) technology will be explored, in order to portray fast processing of fully integrated microfluidic systems applicable to chemical analysis systems.

Microchannel

In order to understand and work with microfluidic pressure-driven networks, one must first understand fluids hydrodynamic behavior. Then, the properties of the fluids will be reviewed, as well as the dimensional parameters that characterize them. As fluids are continuous materials, discrete quantities such as mass and force are replaced by continuous fields, such as mass per unit volume (fluid density), ρ, and force per unit volume, f. Applied forces over fluid elements originate with fluid stress exerted on the surface, given by the stress tensor, σ, added to external forces applied on the element, f. Figure 1 presents the diagram model considered in this chapter for the description of rectangular and circular channels. The rectangular cross section with length l, width w and height h and the circular cross section with length l, and radius r are the most used surfaces in hydrodynamic models and the development of microchannels for microfluidic systems.

Considering microfluidic flows, fluids (initially water and aqueous solutions) may be considered incompressible (Kirby, 2010), thus, velocity field for a Newtonian

Figure 1. Model of a (a) rectangular and a (b) circular channel over which fluid travels with velocity u

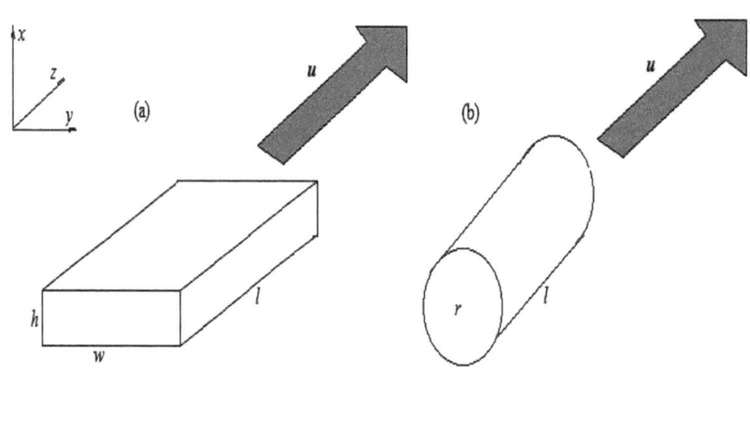

fluid follows Navier-Stokes equation, which essentially represents the continuous version of Newton's second law, in a unit-by-volume basis (Equation 2):

$$\rho\left(\frac{\partial u}{\partial t}+u\cdot\nabla u\right)=\nabla\cdot\vec{\sigma}+f=-\nabla p+\eta\nabla^{2}u+f \tag{2}$$

The terms associated with inertial acceleration are on the left side and the forces on the right side. Also with respect to Equation (2), u represents velocity field (m/s), ρ is fluid density (kg/m^3), η is fluid viscosity (Pa.s) and p is the pressure applied on the fluid (Pa). Since viscosity is a fluidic property that varies with temperature, it must be assumed that temperature does not change significantly. When inertial forces are very small compared to viscosity forces, which is usually the case in microfluidic systems, the nonlinear term can be neglected (convective forces), simplifying Navier-Stokes equation (Equation 3):

$$\rho\frac{\partial u}{\partial t}=\nabla\cdot\vec{\sigma}+f=-\nabla p+\eta\nabla^{2}u+f \tag{3}$$

Assuming no external forces acting on the fluid, considering a long channel in which the fluid moves in the steady state where the fluid flow has completely developed within the channel, velocity is unidirectional and laminar, and the fluid does not undergo acceleration. In the case of microfluidic channels this condition is reached quickly. Therefore, Equation (3) becomes:

$$\nabla p=\eta\nabla^{2}u \tag{4}$$

Considering a circular channel with length l, and radius R, the solution for Equation (4) is obtained knowing that fluid velocity is independent in relation to its angular position. As mentioned previously, microflows are incompressible; hence, the velocity is independent regarding axial direction as well. In addition, the flow is oriented to z direction, and the boundary condition is zero velocity for r equals to R (circular channel radius).

$$u\left(r\right)=\frac{\left(\frac{\partial p}{\partial z}\right)}{4\eta}\left(r^{2}-R^{2}\right) \tag{5}$$

In order to obtain total flow rate for this circular channel, Q (m³/s), it is necessary take the integral over the space adding any velocity contribution:

$$Q = \left[\frac{\left(\frac{\partial p}{\partial z}\right)}{4\eta}\right] \int_0^{2\pi}\int_0^{R} \left\{\left(r^2 - R^2\right)\right\} r dr d\theta \quad (6)$$

Is reasonable to assume that pressure gradient is uniform through channel length. Therefore:

$$-\frac{dp}{dz} = \frac{\Delta p}{l} \quad (7)$$

Where Δp is pressure drop in Pascal and flow rate is defined positive respective to a flow from inlet to outlet. Defining hydraulic resistance as the relation between pressure drop and flow rate, write as expressed in Equation (8), solving Equation (6) is possible to obtain circular channel hydraulic resistance (Equation 9).

$$\Delta p = r_H Q \quad (8)$$

$$r_c = \frac{8l\eta}{\pi r^4} \quad (9)$$

Assuming a fluid flowing on z direction within a rectangular channel (height h, width w and length l), where the ordinate axis is on the vertical, and the coordinate system origin is located over the rectangle cross-section center; Equation (4) can be rewritten as expressed in Equation (10).

$$\frac{\partial p}{\partial z} = \eta\left(\frac{\partial^2 u}{\partial x^2} + \frac{\partial^2 u}{\partial y^2}\right) \quad (10)$$

After performing specific manipulations, the solution for Equation (10) considering zero velocity on the walls ($x = \pm\frac{w}{2}$ and $y = \pm\frac{h}{2}$) is presented in Equation (11).

$$u = -\frac{8K_1h^2}{\pi^3}\left\{\frac{\cosh\frac{\pi x}{h}}{\cosh\frac{\pi w}{2h}}\cos\frac{\pi y}{h} - \frac{1}{3^3}\frac{\cosh\frac{3\pi x}{h}}{\cosh\frac{3\pi w}{2h}}\cos\frac{3\pi y}{h}\ldots\right\} - \frac{1}{2\eta}\frac{dp}{dz}\left(\frac{h^2}{4} - y^2\right) \tag{11}$$

Performing the integral of Equation (11) over superficial cross-section, Equation (12) is obtained, given the flow rate through rectangular channel.

$$Q = -\frac{1}{12}\frac{wh^3}{\eta}\frac{dp}{dz}\left\{1 - \frac{192}{\pi^5}\frac{h}{w}\tanh\frac{\grave{A}w}{2h} + \frac{1}{3^5}\tanh\frac{3\grave{A}w}{2h}\ldots\right\} \tag{12}$$

Appling Equations (7) and (8), the hydraulic resistance for a rectangular channel is as expressed on Equation (13).

$$r_{re} = \frac{12\eta l}{wh^3\left\{1 - \frac{192}{\pi^5}\frac{h}{w}\sum_{n=1,3,5}^{n=\infty}\frac{1}{n^5}\tanh\frac{n\pi w}{2h}\right\}} \tag{13}$$

Figure 2 illustrates a simulation result for a circular channel with radius r equals to 0.5 mm. For comparison effect, the rectangular channel cross-section was set πr^2 ($w = r$ and $h = \pi r$), both with same length variation. The simulation was carried out using Matlab software, solving Equations (9) and (13). It is possible to note that rectangular channel always has bigger resistance to fluid flow in comparison with circular channel. Moreover, as channels length increase, discrepancy between circular and rectangular hydraulic resistance is increased as well. Rectangular channel reaches almost 10 GPa.s/m^3 of hydraulic resistance, while circular channel reaches a wispy more than 5 GPa.s/m^3, for 13 cm channel length.

Micromixer

Graphical and numerical simulation of fluids in microchannels helps choosing and optimize micromixer geometry, in order to intensify the contact interface between liquids, improving fluidic mixing within micromixers. Using ANSYS 14.5 Software was possible to simulate fluids mixing over several distinct mixers shapes, allowing

Figure 2. Rectangular and circular channel fluidic resistance simulation for 1 - 13 cm channel length range

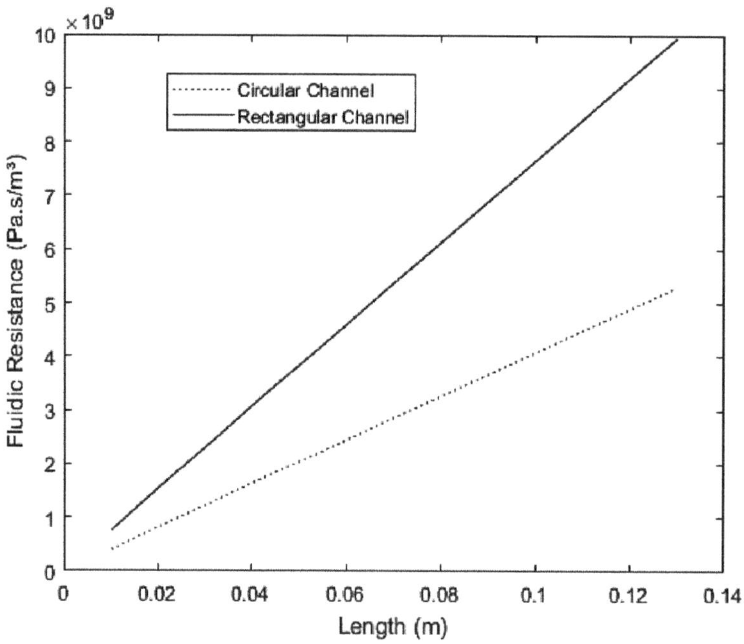

its performance evaluation for two distinct liquids mixing. Figure 3 shows an example where is possible to evaluate interface evolution between the fluids along flow axis.

Author's research group manufactured a LTCC micromixer, where a 254-μm thickness ceramic tape ware used. The first step executed in order to obtain LTCC structures is the development of cutting lines design, predicting geometric shapes required to each ceramic sheet, so that superimposition of the various tapes in the proper order allows generating the desired device. The use of a graphic software to accomplish this task is required (Autodesk AutoCAD 2004 was used in the case of Author research group). Alignment holes should be positioned within the working area of the ceramic tapes, regardless the developed device, in order to ensure complete tapes alignment during lamination process. Figure 4 presents several layers required for photometric microlaboratory construction. In figure 4 is possible to observe the layers responsible for flanges definition and electronic circuit paths, fluidic ports positioning, detection system scheme and the 3D micromixer development, all integrated in the same substrate. In order to analyze solutions behavior within microfluidic systems (Figure 5), a micromixer device was constructed without ceramic base sealed using a biocompatible plastic normally applied in PCR chambers (ABgene Adhesive Plate Seals – Thermo Scientific). Three different dyes were used

Figure 3. Concentration maps for different mixers shapes depicting cuts in the x-y plane. a. square wave; b. 3D square wave; c. Zigzag 1; d. Zigzag 2; e. 3D serpentine 1; f. 3D serpentine 2

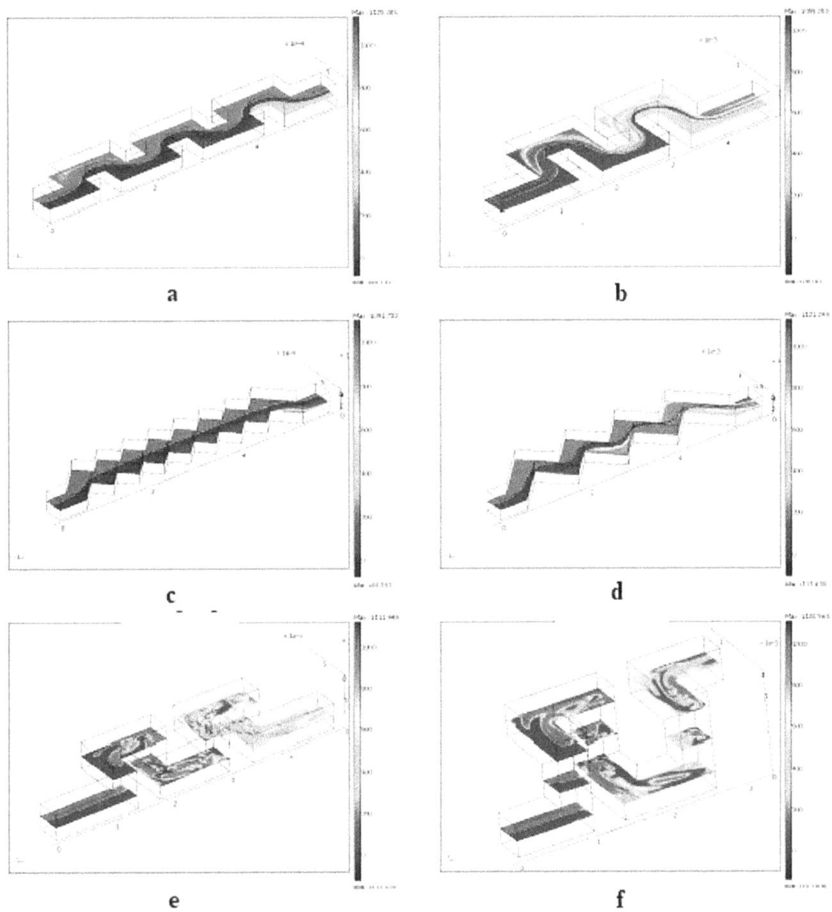

**For a more accurate representation see the electronic version.*

in order to perform a functional test of LTCC micromixer efficiency. Red, yellow and green dyes were used on experimental setup. Figure 5 shows the device completely filled by different solutions being possible to verify the different paths covered by these colorful solutions. Since red and blue dyes were much more concentrated than green dye, even with a higher flow rate for the green dye the others colors showed up dominants.

Figure 4. Several ceramic tapes cutting for microfluidic device manufacture using LTCC processing technology. A) and D) detection system and fluidic ports; B) fluidic and electrical connections; C) 3D micromixer and flow-cell

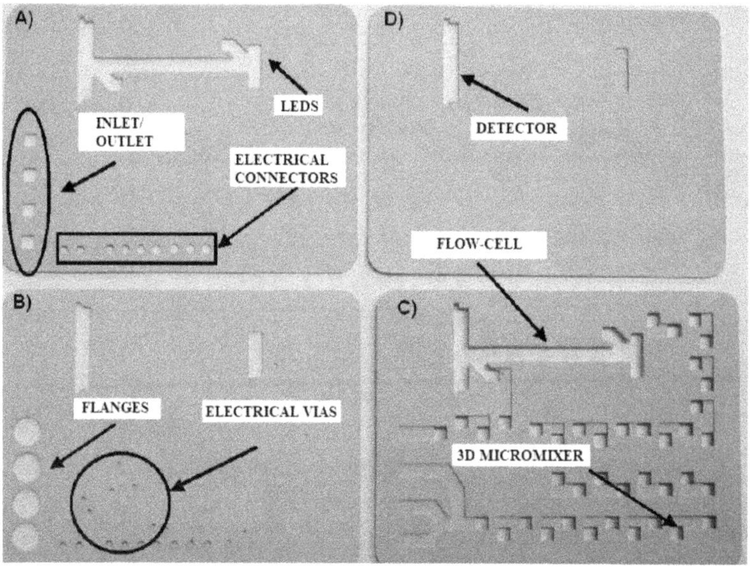

Figure 5. Device tested with different dyes to observe LTCC micromixer mixing efficiency

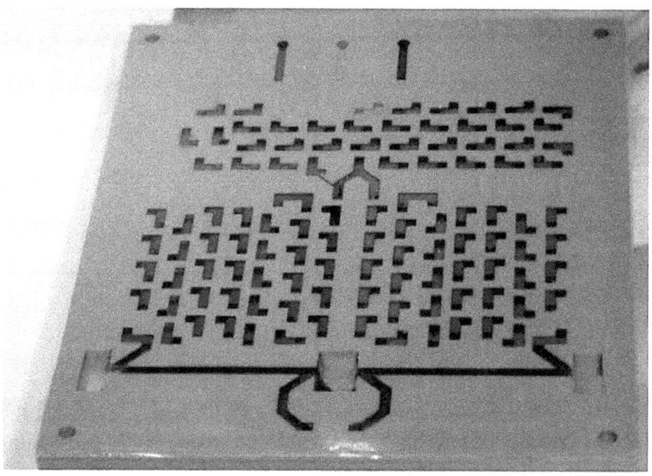

Microvalve

Design requirements associated with microvalves development are determined based mostly on the functionality application: chemical analyzes applied to water quality monitoring is very relevant application seeing also environmental impact and laboratorial automation potential, and thereat is the main case of study referring to this chapter. Considering this application area (chemical analysis), there is a wide range of possibilities related to manufacture technology, and one should prioritize some goals in order to select the best structure to compose the final system. Regarding microvalve actuation method, one must choose a type of active element among the feasible structures for the construction of a miniaturized valve. Typically, active components used in the construction of successful micro-valves are ball-type, pinch-type and membrane-type. Considering valves made using spheres, the associated manufacturing process is usually very laborious, complex and time-consuming. In relation to structured valves based on elements that strangulate the channel, through the action of pistons, due to the space occupied by the active element and the space reserved for its displacement, the final device can have dimensions that go beyond those projected for the aim of miniaturized analyzers. Several materials were used for the fabrication of active elements integrate to a microvalve device as reviewed on the background. However, PDMS is a commonly used material in microfluidic devices manufacture due to its excellent mechanical, optical and good compatibility with chemical reagents. In addition, this material combined with LTCC technology enables hybrid microdevices construction with relative ease. Figure 6 presents an image of a microfluidic system in LTCC constructed with 1.3 cm of side and 0.8 mm of height. In this system a microchannel 1 cm long, 0.3 mm wide and 0.2 mm high interconnects inlet and outlet with access to the opening in the central part.

In order to develop a functional microvalve is necessary to couple an actuation system with an active element, which is part of microvalve body. Electric, magnetic, piezoelectric, thermal, pneumatic actuation systems are the most popular methods applied in microvalves design. In order to improve device portability, low power supply is an important requirement, especially dealing with water, a polar liquid. Piezoelectric actuation normally requires high voltages for effective actuation or bulky piezoelectric stacks. In electrostatic actuation, polar liquids can be influenced by electric fields disturbing microvalve actuation, unsetting components design. Applying pneumatic actuation is difficult achieve device miniaturization. Thermal actuation have elevated response time owing to the phenomenology associated with the method in most cases. Hence, electromagnetic actuation method is probably the most suitable for automated water quality monitoring microsystems, since can be combined with electronic instrumentation, and meets actuation and miniaturization requirements. An attractive implementation proposal for an on-off microvalve applied

Figure 6. Microfluidic system constructed with a channel (1 cm long, 0.3 mm wide and 0.2 mm high), an aperture for actuation (0.4 mm x 0.6 mm cross-section and 0.4 mm height), inlets and outlets for fluid handling, for normally open LTCC microvalves

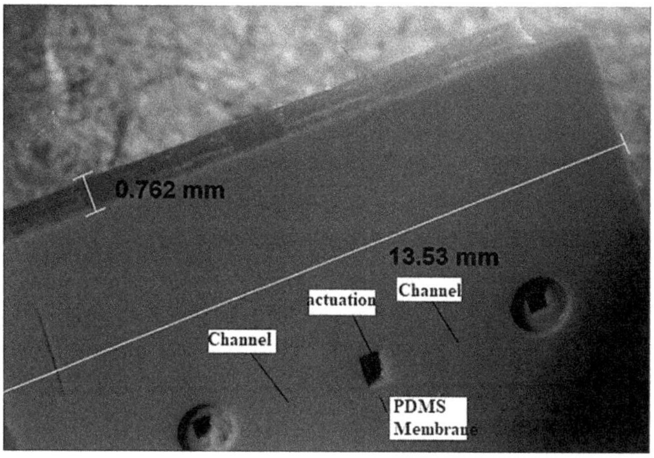

to fluidic control within microanalyzers used in water quality monitoring can be developed to be integrated to the 13.5 mm x 0.76 mm LTCC microfluidic system showed in figure 6. An electromagnetic coil coupled with a permanent magnet attached to PDMS membrane composes the proposed actuation system. Actuation is carried out from electromagnetic interaction between the electromagnet and the membrane fixed to permanent magnet. Applying a potential difference between coil terminals, a magnetic force is induced in the permanent magnet, causing membrane deflection depending on the electric current direction. Modeling of this actuation system prioritizes the magnetic force generated by the coil and the force developed in the flexible membrane.

Magnetic force produced by a coil over a body on space depends on bulk magnetization (dipole moment *m*) and magnetic flux density ***B*** (Becker & Sauter 1964), as shown on Equation (14):

$$F = \left(m \cdot \nabla\right) B \tag{14}$$

For a body, with volume *V*, a fixed magnetization M_{Sat}, oriented in *z* direction, the force created by the coil is:

$$F = M_{Sat}\left(\int_V \frac{\partial B_x}{\partial z} dVi + \int_V \frac{\partial B_y}{\partial z} dVj + \int_V \frac{\partial B_z}{\partial z} dVk\right) \tag{15}$$

The force computed on Equation (15) must be balanced with elastic membrane force, seeing that permanent magnet is completely fixed to PMDS membrane, in order to control microvalve actuation. PDMS membrane can be modeled as an elastic material, respecting Hooks law, expressed on Equation (16):

$$F_z = ku_z \tag{16}$$

Where k is PMDS microbridge elasticity, and u_z is the relative displacement. Considering a flexible membrane with both end fixed, also known as microbridge configuration, the elasticity for a material with Young modulus E, width w, length l and thickness t, considering a force applied with a distance l_x with respect to one end of the system, is given by:

$$k = \frac{3Ewt^3}{\left\{l_x\left(1 - \left[\frac{l_x}{l}\right]\right)\right\}^3} \tag{17}$$

Actuation system design for an electromagnetic method coupled to a flexible active element can be optimized using Equations (15) and (16). Moreover, the design is able to be implemented using LTCC technology, improving integration potential of the microvalve related to autonomous microlaboratories presented in this chapter.

Detection System

The current detection system, highlighted in Figure 7, is relatively simple regardless of been innovative, with the purpose to serve as proof of concept. The innovation in this system was the use of non-encapsulated LEDs, which allow illumination of the channel with different wavelengths without having to use optical fibers/switches. The detection system use seven LEDs with different wavelengths, all in the same body, improving system flexibility. Future research in order to improve this detection method is:

- Optimization of LEDs polarization circuit in order to increase light emission stability, mainly as a function of temperature.
- Use of an ultra-sensitive detection system applying avalanche photodiodes (avalanche photodiodes - APDs) in order to increase detection system sensibility and shorten flow cell size.
- Use a system capable of integrating over time (slow processes, in order of minutes) using charge coupled devices – CCDs.

Fully Integrated Microfluidic Device

Currently the research done with Autonomous Microlaboratories is intended for reagents manipulation used to perform nutrient detection of species in water. These devices allow colorimetric analytical methods application for the determination of phosphate concentration, taking reactive phosphorus in water from ammonium molybdate and phosphate ion reaction, producing an ammonium phosphomolybdate, in yellow color. Applying a reducing agent, a blue compound of intense staining is produced from ammonium molybdate reduction (Mo (VI) to Mo (V)), increasing analyzer sensitivity. Figure 8 shows an assay for phosphorus concentration determination of a sample pattern using commercial equipment, where analysis is performed in a static manner, with the analyte at rest inside of test tubes. The first six tubes are used to perform equipment calibration, relating measured absorbance

Figure 7. Detail of LEDs in silicon (non-encapsulated, 250x250um) that directly illuminates the flow cell within microfluidic colorimetric (absorbance-based) detection system

versus phosphorus concentration to standard concentration values. The last test tube contains the sample with an unknown phosphorus concentration. It can be seen that analyte coloration becomes more intense (for the first six bottles) from left to right. This is because phosphorus concentration increases in this direction, that is, the first bottle (from left to right) contains only deionized water, and successively the standard phosphorus concentrations increase (0.1 mg/L, 0.25 mg/L, 0.5 mg/L, 0.75 mg/L and 1 mg/L). With the use of valves it is possible to perform the calibration automatically, from the controlled dilution of only a 1 mg/L standard solution, through a technique known as multicommutation, which employs digitally controlled valves, allowing the injection of samples and reagents in exactly the same instants, increasing the versatility and level of automation regarding these manufactured systems. Hence, it is possible to dilute the 1 mg/L standard solution using deionized water with precision, allowing obtaining the calibration standards solutions from the control of valves opening time.

SOLUTIONS AND RECOMMENDATIONS

LTCC microfluidic devices, with micro-channels formed by laser patterning of ceramic tape, which can be designed in 3D configuration, are suitable for micromixers, microvalves, micropumps development. The LTCC technology is a fast prototyping technology and allows the application of embedded metal electrodes to perform the

Figure 8. Analytes produced in a chemistry laboratory where 5 mL of standard phosphorus water samples, 200 μL of ammonium molybdate solution, and 25 μL of stannous chloride solution were added to chemical reaction

heating of the fluid, integration of optical and microwave components for interaction, characterization and analysis of fluids on the same device.

Application of colorimetric analytical methods for the determination of reactive phosphorus concentration, for example, in water from a chemical reaction occurring whiten mixer device, improving microsensor sensibility. Using micro-valves and micro-pumps, all in the same device, it is possible to control the injection of samples and reagents at precisely predetermined moments, increasing the versatility and automation of manufactured systems. Development of electromagnetic, electric or piezoelectric microfluidic actuators, for micro-valves and micro-pumps manufacturing is also possible, related to LTCC technology. Design and simulation of fluidic, electromagnetic, electric or piezoelectric and micro-mechanic systems applicable to micro-components development.

FUTURE RESEARCH DIRECTIONS

Development of LOC, μTAS and/or bioMEMS devices applicable to medicine and biosensor fields.

CONCLUSION

Several mathematical models must be described in order to assist the design of the components and all microfluidic devices. Moreover, several experimental material related to microfluidic devices and the obtained models.

Development of autonomous and low cost laboratories capable of integrating and automating analytical procedures necessary for a given analysis, with steps of mixing, detecting, controlling and acquiring integrated data on the same substrate was demonstrated. LTCC technology was used in order to allow integration of chemical, microfluidic and electronic platforms in the same substrate, allowing autonomous microlaboratories implementation based on optical detection applied to water quality parameters evaluation.

The development of magnetically actuated membrane microvalve manufactured using LTCC/PDMS technology was demonstrated. It was seen that design consists of the correct actuation, channel, aperture and membrane structures dimensioning, which are controlled from the manufacturing processes involved, in a practical manner, and determined using the principles associated with the phenomena involved wih the systems, theoretically. Experiments need to be executed in order to evaluate real system performance.

REFERENCES

Auroux, P. A., Iossifidis, D., Reyes, D. R., & Manz, A. (2002). Micro total analysis systems. 2. Analytical standard operations and applications. *Analytical Chemistry*, *74*(12), 2637–2652. doi:10.1021/ac020239t PMID:12090654

Barlow, F., Wood, J., Elshabini, A., Stephens, E. F., Feeler, R., Kemner, G., & Junghans, J. (2009). Fabrication of precise fluidic structures in LTCC. *International Journal of Applied Ceramic Technology*, *6*(1), 18–23. doi:10.1111/j.1744-7402.2008.02315.x

Becker, H., & Locascio, L. E. (2002). Polymer microfluidic devices. *Talanta*, *56*(2), 267–287. doi:10.1016/S0039-9140(01)00594-X PMID:18968500

Becker, R., & Sauter, F. (1964). *Electromagnetic fields and interactions* (Vol. 1). Courier Corporation.

Chau, L. K., Osborn, T., Wu, C. C., & Yager, P. (1999). Microfabricated silicon flow-cell for optical monitoring of biological fluids. *Analytical Sciences*, *15*(8), 721–724. doi:10.2116/analsci.15.721

Cheek, B. J., Steel, A. B., Torres, M. P., Yu, Y. Y., & Yang, H. (2001). Chemiluminescence detection for hybridization assays on the flow-thru chip, a three-dimensional microchannel biochip. *Analytical Chemistry*, *73*(24), 5777–5783. doi:10.1021/ac0108616 PMID:11791544

Dernovsek, O., Eberstein, M., Schiller, W. A., Naeini, A., Preu, G., & Wersing, W. (2001). LTCC glass-ceramic composites for microwave application. *Journal of the European Ceramic Society*, *21*(10-11), 1693–1697. doi:10.1016/S0955-2219(01)00096-6

Gao, J., Yin, X. F., & Fang, Z. L. (2004). Integration of single cell injection, cell lysis, separation and detection of intracellular constituents on a microfluidic chip. *Lab on a Chip*, *4*(1), 47–52. doi:10.1039/b310552k PMID:15007440

Golonka, L. J. (2003, August). Low Temperature Co-fired Ceramic (LTCC) technology in microelectronics. In *Proc. Symp. Processing of Electroceramics* (pp. 313-329). Academic Press.

Gongora-Rubio, M. R., Espinoza-Vallejos, P., Sola-Laguna, L., & Santiago-Aviles, J. J. (2001). Overview of low temperature co-fired ceramics tape technology for meso-system technology (MsST). *Sensors and Actuators. A, Physical*, *89*(3), 222–241. doi:10.1016/S0924-4247(00)00554-9

Groß, G. A., Thelemann, T., Schneider, S., Boskovic, D., & Köhler, J. M. (2008). Fabrication and fluidic characterization of static micromixers made of low temperature cofired ceramic (LTCC). *Chemical Engineering Science*, *63*(10), 2773–2784. doi:10.1016/j.ces.2008.02.030

Ibanez-Garcia, N., Mercader, M. B., Mendes da Rocha, Z., Seabra, C. A., Góngora-Rubio, M. R., & Chamarro, J. A. (2006). Continuous flow analytical microsystems based on low-temperature co-fired ceramic technology. Integrated potentiometric detection based on solvent polymeric ion-selective electrodes. *Analytical Chemistry*, *78*(9), 2985–2992. doi:10.1021/ac051994k PMID:16642984

Jindal, R., & Cramer, S. M. (2004). On-chip electrochromatography using sol–gel immobilized stationary phase with UV absorbance detection. *Journal of Chromatography. A*, *1044*(1-2), 277–285. doi:10.1016/j.chroma.2004.05.065 PMID:15354449

Kirby, B. J. (2010). *Micro-and nanoscale fluid mechanics: transport in microfluidic devices*. Cambridge University Press. doi:10.1017/CBO9780511760723

Kovacs, G. T. (1998). *Micromachined transducers sourcebook*. New York: WCB/McGraw-Hill.

Kuswandi, B., Huskens, J., & Verboom, W. (2007). Optical sensing systems for microfluidic devices: A review. *Analytica Chimica Acta*, *601*(2), 141–155. doi:10.1016/j.aca.2007.08.046 PMID:17920386

Liu, B. F., Ozaki, M., Utsumi, Y., Hattori, T., & Terabe, S. (2003). Chemiluminescence detection for a microchip capillary electrophoresis system fabricated in poly (dimethylsiloxane). *Analytical Chemistry*, *75*(1), 36–41. doi:10.1021/ac026096s PMID:12530816

Madou, M. J. (2002). *Fundamentals of microfabrication: the science of miniaturization*. CRC Press. doi:10.1201/9781482274004

Maims, C., Hulme, J., Fielden, P. R., & Goddard, N. J. (2001). Grating coupled leaky waveguide micro channel sensor chips for optical analysis. *Sensors and Actuators. B, Chemical*, *77*(3), 671–678. doi:10.1016/S0925-4005(01)00773-0

Malecha, K., Pijanowska, D. G., Golonka, L. J., & Torbicz, W. (2009). LTCC microreactor for urea determination in biological fluids. *Sensors and Actuators. B, Chemical*, *141*(1), 301–308. doi:10.1016/j.snb.2009.06.026

Martin, R. S., Gawron, A. J., Lunte, S. M., & Henry, C. S. (2000). Dual-electrode electrochemical detection for poly (dimethylsiloxane)-fabricated capillary electrophoresis microchips. *Analytical Chemistry*, *72*(14), 3196–3202. doi:10.1021/ac000160t PMID:10939387

Martínez-Cisneros, C. S., da Rocha, Z., Ferreira, M., Valdés, F., Seabra, A., Góngora-Rubio, M., & Alonso-Chamarro, J. (2009). A monolithic continuous-flow microanalyzer with amperometric detection based on the green tape technology. *Analytical Chemistry*, *81*(17), 7448–7453. doi:10.1021/ac9012418 PMID:19715366

Mourzina, Y., Kalyagin, D., Steffen, A., & Offenhausser, A. (2005). Electrophoretic separations of neuromediators on microfluidic devices. *Electrophoresis*, *26*, 1849–1860. doi:10.1002/elps.200410295 PMID:15719361

Oh, K. W., & Ahn, C. H. (2006). A review of microvalves. *Journal of Micromechanics and Microengineering*, *16*(5), R13–R39. doi:10.1088/0960-1317/16/5/R01

Petersen, K. E., & Petersen, K. E. (1979). Fabrication of an integrated, planar silicon ink-jet structure. *IEEE Transactions on Electron Devices*, *26*(12), 1918–1920. doi:10.1109/T-ED.1979.19796

Peterson, K. A., Knudson, R. T., Garcia, E. J., Patel, K. D., Okandan, M., Ho, C. K., . . . Zawicki, L. R. (2008, June). LTCC in microelectronics, microsystems, and sensors. In *Mixed Design of Integrated Circuits and Systems, 2008. MIXDES 2008. 15th International Conference on* (pp. 23-37). IEEE.

Rebenklau, L., Wolter, K. J., & Howitz, S. (2000). Realization of hybrid microfluidic systems using standard LTCC process. In Electronic Components & Technology Conference, 2000. 2000 Proceedings (pp. 1696-1700). IEEE.

Reyes, D. R., Iossifidis, D., Auroux, P. A., & Manz, A. (2002). Micro total analysis systems. 1. Introduction, theory, and technology. *Analytical Chemistry*, *74*(12), 2623–2636. doi:10.1021/ac0202435 PMID:12090653

Ro, K. W., Lim, K., Shim, B. C., & Hahn, J. H. (2005). Integrated light collimating system for extended optical-path-length absorbance detection in microchip-based capillary electrophoresis. *Analytical Chemistry*, *77*(16), 5160–5166. doi:10.1021/ac050420c PMID:16097754

Rocha, F. R., & Reis, B. F. (2000). A flow system exploiting multicommutation for speciation of inorganic nitrogen in waters. *Analytica Chimica Acta*, *409*(1-2), 227–235. doi:10.1016/S0003-2670(99)00883-1

Ṙužička, J., & Hansen, E. H. (1975). Flow injection analyses: Part I. A new concept of fast continuous flow analysis. *Analytica Chimica Acta*, *78*(1), 145–157. doi:10.1016/S0003-2670(01)84761-9

Tantra, R., & Manz, A. (2000). Integrated potentiometric detector for use in chip-based flow cells. *Analytical Chemistry*, *72*(13), 2875–2878. doi:10.1021/ac000036+ PMID:10905321

Terry, S. C., Jerman, J. H., & Angell, J. B. (1979). A gas chromatographic air analyzer fabricated on a silicon wafer. *IEEE Transactions on Electron Devices*, *26*(12), 1880–1886. doi:10.1109/T-ED.1979.19791

Varjo, S. J., Ludwig, M., Belder, D., & Riekkola, M. L. (2004). Separation of fluorescein isothiocyanate-labeled amines by microchip electrophoresis in uncoated and polyvinyl alcohol-coated glass chips using water and dimethyl sulfoxide as solvents of background electrolyte. *Electrophoresis*, *25*(12), 1901–1906. doi:10.1002/elps.200405914 PMID:15213991

Wang, Y., Zhang, G., & Ma, J. (2002). Research of LTCC/Cu, Ag multilayer substrate in microelectronic packaging. *Materials Science and Engineering B*, *94*(1), 48–53. doi:10.1016/S0921-5107(02)00073-9

Wicks, D. A., & Li, P. C. (2004). Separation of fluorescent derivatives of hydroxyl-containing small molecules on a microfluidic chip. *Analytica Chimica Acta*, *507*(1), 107–114. doi:10.1016/j.aca.2003.09.067

Yanagisawa, K., Kuwano, H., & Tago, A. (1995). Electromagetically driven microvalve. *Microsystem Technologies*, *2*(1), 22–25. doi:10.1007/BF02739524

ADDITIONAL READING

Bruus, H. (2008). *Theoretical microfluidics*. Oxford: Oxford university press.

Karniadakis, G., Beskok, A., & Aluru, N. (2006). *Microflows and nanoflows: fundamentals and simulation* (Vol. 29). Springer Science & Business Media.

Kovacs, G. T. (1998). *Micromachined transducers sourcebook* (pp. 586–587). New York: WCB/McGraw-Hill.

Lee, C. Y., Lin, C. H., & Fu, L. M. (2015). Passive Micromixers. Encyclopedia of Microfluidics and Nanofluidics. 2663-2673.

Madou, M. J. (2002). *Fundamentals of microfabrication: the science of miniaturization*. CRC press. doi:10.1201/9781482274004

Manz, A., Harrison, D. J., Verpoorte, E., & Widmer, H. M. (1993). Planar chips technology for miniaturization of separation systems: A developing perspective in chemical monitoring. *Advances in Chromatography*, *33*, 1–66.

Sanders, G. H., & Manz, A. (2000). Chip-based microsystems for genomic and proteomic analysis. *TrAC Trends in Analytical Chemistry*, *19*(6), 364–378. doi:10.1016/S0165-9936(00)00011-X

Senturia, S. D. (2007). *Microsystem design*. Springer Science & Business Media.

Verpoorte, E. (2002). Microfluidic chips for clinical and forensic analysis. *Electrophoresis*, *23*(5), 677–712. doi:10.1002/1522-2683(200203)23:5<677::AID-ELPS677>3.0.CO;2-8 PMID:11891702

Wong, S. H., Ward, M. C., & Wharton, C. W. (2004). Micro T-mixer as a rapid mixing micromixer. *Sensors and Actuators. B, Chemical*, *100*(3), 359–379. doi:10.1016/j.snb.2004.02.008

Yi, M., & Bau, H. H. (2003). The kinematics of bend-induced mixing in micro-conduits. *International Journal of Heat and Fluid Flow*, *24*(5), 645–656. doi:10.1016/S0142-727X(03)00026-2

KEY TERMS AND DEFINITIONS

Autonomous Microlaboratories: A microfluidic device manufactured for chemical analysis application, integrating several steps performed in conventional laboratories autonomously.

Ball-Type Microvalve: A microvalve developed with a sphere as a mechanical actuation element.

Membrane-Type Microvalve: A microvalve developed with a membrane as a mechanical actuation element.

Multicommutation System: A flow injection analysis system that apply several sets of digital controlled valves manipulating fluid translation.

Multiparametric System: A microfluidic device capable of performing distinct chemical elements determinations with minimal change on physical configuration.

Pinch-Type Microvalve: A microvalve developed with a piston as a mechanical actuation element over a strangled channel.

Stopped Flow System: A microfluidic device capable of performing specific flow breaks in order to synchronize chemical reaction.

Chapter 8
Smart Temperature Sensors:
History, Current Practices, and Future Trends

Ashraf A. Zaher
American University of Kuwait, Kuwait

ABSTRACT

Many real-world applications depend on temperature sensing and/or control. This includes a wide range of industrial processes, chemical reactors, and SCADA systems, in addition to other physical, mechanical, and biological systems. With the advancement of technology, it became possible to produce a new generation of smart and compact temperature sensors, which are capable of providing digital outputs that are more accurate, robust, and easily interfaced and integrated into measurement and control systems. This chapter first surveys traditional analog temperature sensors, such as RTDs and thermocouples, to provide a strong motivation for the need to adopt better and smarter techniques that mainly rely on digital technology (e.g., CMOS designs). Different interfacing techniques that do not need ADCs are introduced, including the programmable Arduino microcontrollers. Different applications will be explored that include automotive accessories, weather forecast, healthcare, industrial processing, firefighting, and consumer electronics. Both wired and wireless technologies, including the IoT, will be investigated as means for transmitting the sensed data for further processing and data logging. A special case study to provide information redundancy in industrial SCADA systems will be analyzed to illustrate the advantages and limitations of smart temperature sensors. The chapter concludes with a summary of the design effort, accuracy, performance, and cost effectiveness of smart temperature sensors while highlighting future trends in this field for different applications.

DOI: 10.4018/978-1-5225-7138-4.ch008

INTRODUCTION

Temperature is perhaps the far most important physical quantity to be measured and controlled. Many industrial processes, with physical, chemical, mechanical, electrical, and biological nature, exhibit dependence, some way or another, on temperature. In the last few decades, a huge effort has been made to produce compact, cost-effective, and more accurate temperature sensors that can produce outputs in digital format for easy integration in the new computer-based and microcontroller-based measurement and control systems.

Many analog sensors are used to measure temperature. Among them are the resistance temperature detectors (RTDs), thermistors, and thermocouples. Despite the rapid advances in technology, these analog sensors are still in operation and are widely used in laboratories and many fields in the industry (Huynh, 2015). RTDs are sensors that contain resistors, which change value, according to changes in temperature, almost linearly. Besides having a long life, they are famous for being accurate, stable, and capable of offering good repeatability and interchangeability. They have a wide temperature range from -50°C to 850°C, as reported by Dames (2008). Platinum RTDs are considered the most accurate and can follow two different standards; the European standard, DIN/IEC 60751 that requires the RTD to have a nominal electrical resistance, R_0, of 100.0 Ω at 0°C and a temperature coefficient of resistance (TCR) of 3.85 mΩ/Ω/°C between 0 and 100°C. The American standard, used mostly in North America, has a resistance of 100.0 ±0.10 Ω at 0°C and a TCR of 3.92 mΩ/Ω/°C between 0 and 100°C. Equation (1) depicts the temperature dependence of the resistance.

$$R = R_0 \left(1 + TCR \times T\right) \quad (1)$$

RTDs can also have resistances of 200, 500, 1000, and 2000 Ω at 0°C. These RTDs have the same TCR as the 100.0 Ω RTD, but because of their higher resistances at 0°C, they exhibit greater resolution. RTDs are found in many different classes, e.g. A, B, C, and AA, depending on the temperature range, TCR, and the tolerance associated with it. There are many different connections for the RTDs that make use of a bridge circuit for measuring the resistance-dependent voltage, as shown in Figure 1. Usually, the measured voltage is set to zero, through a calibration process; changes in the temperature will result in equivalent changes in the resistance that will cause the voltage to linearly change, accordingly (azom.com, 2010).

RTDs have two main constructions; the thin film and the wire wound, as show in Figure 2, in (a) and (b), respectively. The thin film is available only in the DIN/IEC 60751 standard, in addition to a special TRC of 3.75 mΩ/Ω/°C for appliances

Figure 1. Different wiring configurations for RTDs

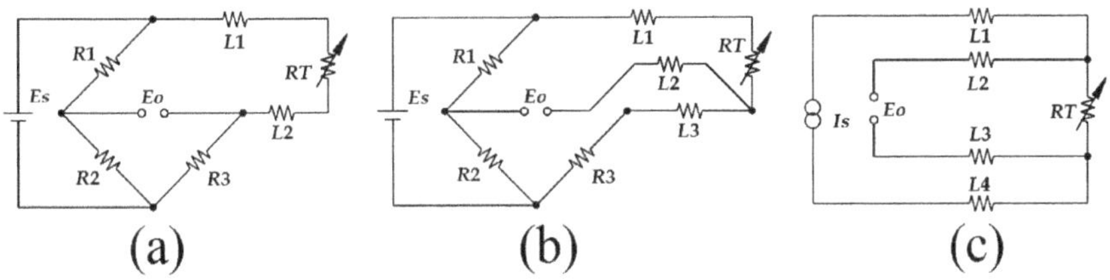

industry. The most commonly used wire wound RTDs have a wire packaged inside a ceramic or glass. RTDs can be effectively used to measure surface temperature, in addition to temperature of liquids and gases, including air.

RTDs suffer from the self-heating problem that causes deviation from the correct readings, which puts a limitation to their usage. In addition, they have a slow response time, poor sensitivity to small changes in temperature, and can't be used for very high temperatures; above 660°C (Dames, 2008; azom.com, 2010; and omega.com, 2017).

Unlike RTDs that use metals with positive temperature coefficients (PTC), thermistors use semiconductor materials such as ceramic and polymers that have both PTC and negative temperature coefficients (NTC). Increasing the temperature generates more charge carriers; thus, more current will follow, causing the resistance to effectively decrease. Thermistors are almost linear for small temperature ranges and can provide high precession measurements. They can be used for wide ranges of temperature at the expense of sacrificing linearity and the need for calibration. Equation (2) shows the mathematical relationship between the resistance and the temperature (Janata, 2009):

$$R = R_0 e^{NTC\left(\frac{1}{T_0} - \frac{1}{T}\right)}, \ T_0 \text{ is the room temperature} \qquad (2)$$

Thermistors can have two different types of coatings; Epoxy for low temperature ranges that are typically between -50 to 150°C, and Glass for higher temperature applications that have a typical range from -50 to 300°C. The base resistance of

Figure 2. Different constructions of RTDs (omega.com, 2017)

the thermistors, R_0, is usually specified at room temperature (25°C), and can take several values, depending on the range of the temperature; typically, 3, 5, 10, 30, and 50 kΩ. Figure 3 illustrates different packages for the thermistors that are smaller in size than RTDs, which include radial leaded, axial leaded, glass, threaded, and probes, in (a), (b), (c), (d), and (e), respectively.

Thermistors share the need for a bridge circuit, with RTDs, to measures the temperature, in addition to suffering from the self-heating problem, as well. Thermistors have faster response, more stability, and higher sensitivity, compared to RTDs. Thermocouples are yet another type of analog temperature sensors that offer direct voltage readings for the measurement of temperature in contrast to both RTDs and thermistors that produce indirect measurements. Thermocouples depend on the thermoelectric effect in their operation, as two dissimilar electrical conductors are joined at one end to form the so called electrical junction that acts as a temperature sensor. A voltage is produced between the two free ends, which is linearly dependent on the temperature of the electrical junction. This voltage is usually called the Seebeck voltage, after Thomas Seebeck who discovered the thermoelectric effect back in 1821 (Love, 2007). Figure 4 illustrates the typical connection of a thermocouple.

Thermocouples have a PTC like RTDs, and some types of thermistors, with a fast response time. Depending on the used material, thermocouples can be categorized into several types; J, K, T, E, and N, for base metals, and R, S, C, and GB, for noble metals. They cover a wide range of temperatures; e.g. types J, K, T, and E

Figure 3. Different packages of thermistors (resistorguide.com, 2018)

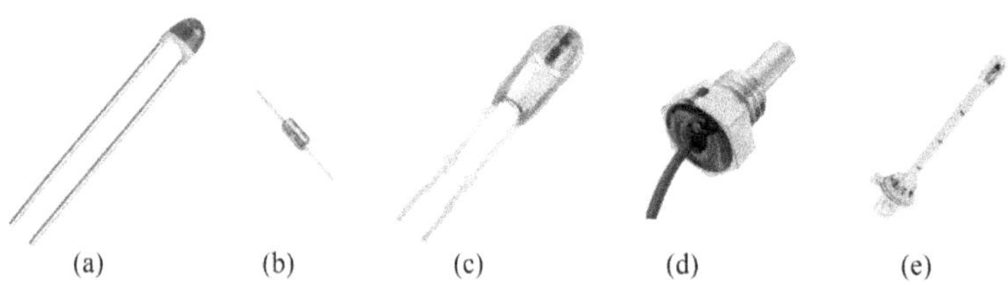

Figure 4. Temperature measurement circuit, using thermocouples

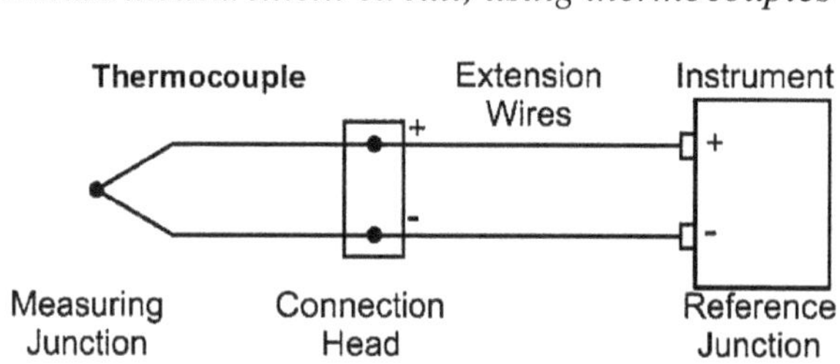

correspond to the ranges, 0° to 750°C, -200° to 1250°C, -200° to 900°C, and -250° to 350°C, respectively. Finally, thermocouples are considered more cost-effective than RTDs and more durable; however, they are considered less accurate and less stable (omega.com, 2016). Figure 5 shows typical examples for thermocouples in two different formats, beaded wire and probe, in (a) and (b), respectively.

RTDs, thermistors, and thermocouples require direct contact with the temperature source in order to measure its temperature. IR thermometry is another technique that relies on noncontact temperature measurement via estimating the infrared IR energy emitted by the heat source. This special technique, although relatively expensive, has a fast response and can be effectively used in scenarios where contact-based measurements are difficult, or even impossible. This includes moving objects, hazardous environments, and other applications that require working in vacuum or have geometry constraints. Optics and electronics are two different categories, when deciding on implementing these kinds of sensors. A major problem with IR-based temperature sensors is that it is necessary to take into consideration not only the target and its emissivity, but also the surroundings atmosphere (omega.com, 2019). For most applications, a single wavelength measurement is needed; however, for improved accuracy, multi wavelength radiometry should be considered. IR sensors are found in a variety of forms that include portable, a simple two-wire transmitter, a sophisticated ruggedized sensing unit, or a scanning device. Its electromagnetic power spectrum is shown in Figure 6.

While the analog temperature sensors, discussed in this section, are still used in plenty of applications and provide various means of connectivity, they lack the smart features that are required for current modern instrumentation systems. This

Figure 5. Different constructions of thermocouples

Figure 6. The infrared region in the electromagnetic power spectrum

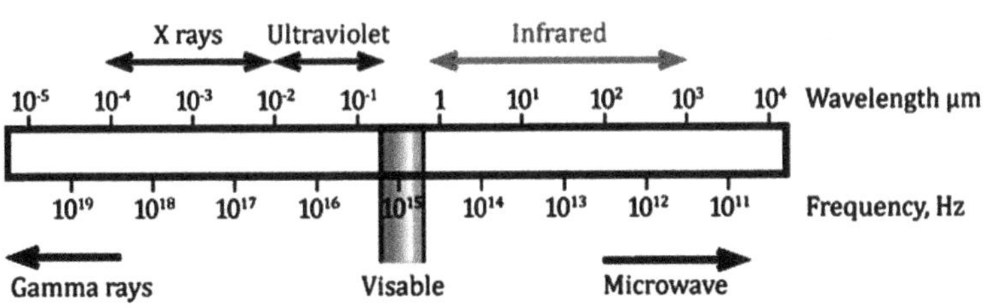

includes noise rejection, ease of interfacing with digital circuitry, and simplicity of calibration and conditioning. Remedies to these shortcomings will be explained in the subsequent sections as follows. Section 2 introduces the category of temperature sensors that rely on the relatively new CMOS technology, which employs new variables to indirectly measure temperature. These sensors are compact, more accurate, and very much suitable for integration in smart measurement systems that rely on digital technology. Sections 3 completes the introductory part via exploring different types of additional circuitries and components that are needed to implement the complete measurement system, including signal conditioning, amplification, noise rejection, and calibration. This paves the way to Section 4, the heart of this chapter that investigates different types of smart temperature sensors, along with their most important current practices. Other considerations regarding further improving the performance of smart sensors and highlighting the most important future trends are explored in Section 5. Finally, a conclusion, with a rich discussion, is given to summarize the topics, covered in this chapter.

TEMPERATURE SENSORS BASED ON CMOS TECHNOLOGY

Using CMOS technology, both bipolar junction and MOS transistors can be used to generate the required signals to measure temperature. In bipolar junction transistors (BJT), the voltage across the base-emitter junction, along with the saturation current can be used to generate a signal that is temperature-dependent, while in MOS transistors, the required temperature-dependent signal is derived from the threshold voltage and the mobility of the charge carriers. Temperature sensors, implemented in CMOS technology, are known to be both cost and energy effective. In addition, they exhibit different performances, and based on the target application, the selection of the required sensor will strongly depend on the required accuracy, resolution, power consumption, conversion time, operating supply voltage range, operating temperature range, and power supply rejection ratio (Makinwa, 2010). Usually BJTs offer higher accuracy for the temperature range that falls between -55°C and 125°C. They also have low supply sensitivity of about 0.1°C/V; however, they usually require a supply voltage greater than 1.0 V to overcome the barrier voltage that is typically from 0.6 V to 0.8 V. This has the effect of restricting the use of BJTs for applications that are battery-operated, in addition to restricting the measurable temperature range for nanometer scale CMOS temperature sensors. For applications that require low voltage, typically below 1.0 V, MOSFET technology is recommended due to the very small value of the gate voltage, compared to the barrier voltage of BJTs.

Equation (3) illustrates the temperature dependence for BJTs:

$$I_C = I_s\left(e^{\frac{V_{BE}}{nV_T}} - 1\right) \quad , \quad V_T = \frac{kT}{q} \tag{3}$$

where I_C is the forward current, I_S is the saturation current, V_{BE} is the forward voltage, n is the ideality factor, V_T is the thermal voltage, k is Boltzmann's constant, and q is the charge of the electron. Many variants of Equation (3) exist to achieve better accuracy; the most famous one is to cancel the effect of I_S by obtaining two measurements for two different currents that are recorded at two different voltages and measuring the difference/ratio between them (Jones, 2012), as illustrated in Equation (4).

$$T = \frac{V_{BE1} - V_{BE2}}{\frac{nk}{q}\ln\left(\frac{I_{C1}}{I_{C2}}\right)} \tag{4}$$

BJTs can be easily calibrated and trimmed against any fluctuations in V_{BE}, process spread and/or packaging stress (Souri, 2016); however, this proved to be time consuming and significantly increases the cost during the production phase. Figure 7 illustrates the two most common structures of the vertical BJTs that are preferred over lateral structures due to their lower sensitivity to process spread and packaging stress. The vertical NPN transistors in modern CMOS technology supporting deep N-well, shown in (b), offers significantly larger current gain than PNPs design, shown in (a). In addition, NPNs offer more circuit design flexibility because of their accessible collector terminals.

CMOS technology is also employed in fabricating PTC RTDs. Most CMOS-compatible resistors offer linear PTC in the range of 1.0 – 4.0 mΩ/Ω/°C, which is a reasonable sensitivity. Although, these temperature sensors require low voltage supply, they suffer from the wide spread of resistances, in the range of 15 – 20%,

Figure 7. Vertical BJT structures (Souri, 2016)

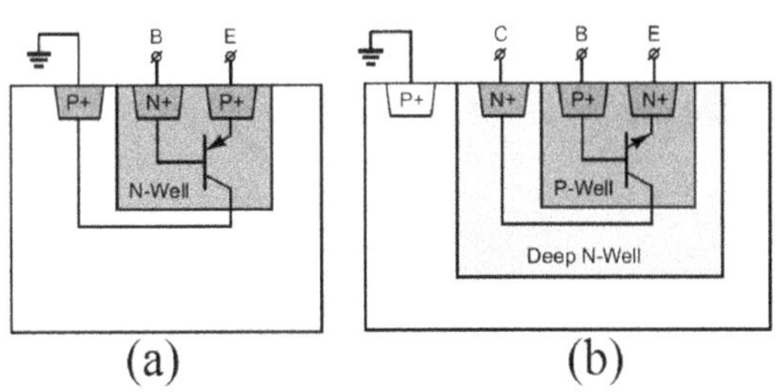

across the process corners; thus, calibration is required to achieve a good accuracy. Electro-thermal filters (ETFs) is another type of sensors that utilizes the temperature-dependence of thermal diffusivity of Silicon. The principle of operation is based on generating heat pulses at one end of a Silicon substrate (Kashmiri & Makinwa, 2009), and measuring the temperature changes at the other end, in terms of small voltage variations. Because the heat pulses generate a temperature time-varying gradient and the rate at which heat diffuses through the substrate is finite, the phase of the sensor's output will lag that of the power dissipation in the heater (Kashmiri & Makinwa, 2009). This measured phase shift, corresponding to low-pass filter, is almost a linear function in the temperature. Figure 8 shows an illustrative diagram for the ETF, in (a), and the linear relationship between the phase shift and the temperature, in (b), respectively.

The mathematical relationship, for the ETF, is given by Equation (5), where s is the distance between the heater and the temperature sensor, f is the frequency of the heat pulse, and ϕ is the phase shift.

$$\phi_{ETF} \propto s\sqrt{f_{heater}}\,T^{0.9} \tag{5}$$

Similar to the relationship, in BJTs, between the collector current, I_C, and the voltage across the base-emitter junction, V_{BE}, both the drain current, I_D, and the gate-source voltage, V_{GS}, in MOSFETs exhibit a temperature dependent performance. This suggests that MOSFETs can be used as temperature sensors, similar to BJTs, with the added advantage of employing less supply voltage, as V_{GS} is substantially smaller than V_{BE}, and can be easily controlled by adjusting the width and/or length of the sensor. On the other hand, MOSFETs are less accurate than BJTs and often require a time-consuming and costly calibration and trimming process (Revertera, Perpiñàb,

Figure 8. Structure of the EFT and its resulting signal (Souri, 2016)

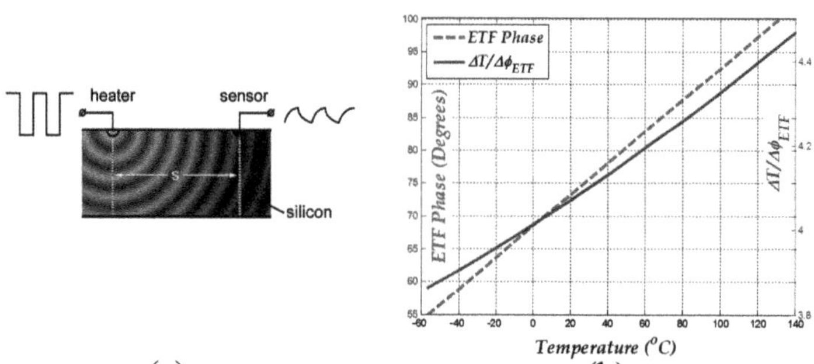

(a) (b)

Leónb, Vellvehib, Jordàb, & Altet, 2015). Moreover, MOSFET-based temperature sensors that use propagation delays in CMOS inverter chain are known to suffer from poor accuracy and too much dependence on the power supply; therefore, their usages are very limited to applications that require large power supplies.

Another type of MOSFETs, which has a dynamic threshold voltage, can be used to replace BJTs, through connecting the gate and the bulk. This special configuration is named DTMOST and is represented by Equation (6), which is very similar to Equation (3), where W and L represent the width and length of the device, respectively.

$$I_D \propto \frac{W}{L} e^{\frac{q}{kT}(V_{DS}-V_T)} \tag{6}$$

Figure 9 illustrates the configuration of the DTMOST, in (a), and a comparison between its performance with that of conventional diodes, in (b), respectively (Mansour, Haneef, Akhtar, DeLuca, & Udrea, 2015).

Deciding on which sensor to use depends mainly on the nature of the application and the working environment. Temperature range, accuracy, sensitivity, linearity, and noise level, are some important factors when taking such decision. Other factors that affect the choice of a particular sensor include cost, calibration and trimming efforts, and whether an analog or a digital output is required. Whether the sensor is used for monitoring, alarming, or control dictates the required circuitry that need to accompany the senor in order to build the complete measurement and/or control system.

Figure 9. Sturcture and performance of DTMOST

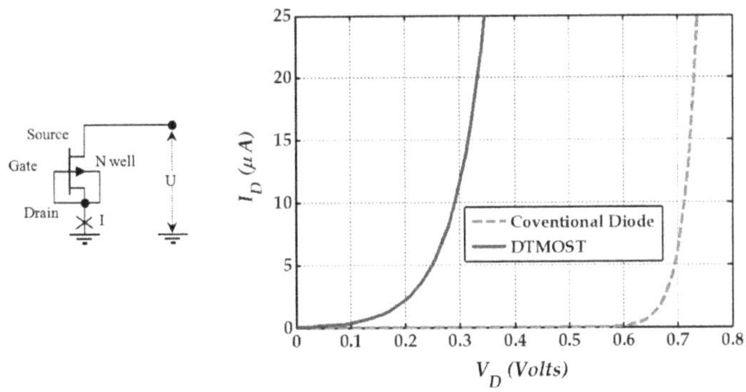

(a) (b)

MEASUREMENT SYSTEMS AND SMART SENSORS

It is a common practice to combine sensors, such as RTDs or semiconductor sensors, with signal processing and conditioning elements to build complete measurement systems. These systems can produce either an analog or a digital output signal of the temperature and are pre-calibrated and pre-scaled in units of Celsius or Fahrenheit. They also should provide means of calibration and noise rejection, as well as the ability of integration with SCADA systems and other intelligent networking or distributed system. Both RTDs and thermistors provide indirect measurement of temperature that requires additional circuitry to convert it into a readable signal. For analog measurements, a minimum of two more stages to condition and amplify the signal will be required. Analog to digital conversion, noise rejection, and data acquisition equipment will be required to convert the analog measurement into an equivalent digital one, when computers or microcontrollers are used.

Wheatstone bridge is the most famous analog device that is used to convert temperature into voltage, via employing the change in the electric resistance. In a balance condition, the ratio of the resistors, of the adjacent legs, is the same, resulting in zero voltage between its output terminals. If one of the resistors is replaced by a resistance-based temperature sensor, the balance condition no longer holds at different temperatures. Consequently, the output voltage is an indication of the temperature change (Fraden, 2010). Usually the voltage change is too small, and an additional amplification stage will be required, as shown in Figure 10. Equation (7) illustrates the dependence of the output voltage on the bridge resistors, where R_1, R_2, R_3, and R_4, represent the resistances of the individual arms of the bridge circuit.

$$V_o = \left(\frac{R_1}{R_1 + R_2} - \frac{R_3}{R_3 + R_4} \right) v_{ref} \tag{7}$$

Figure 10. Wheatstone bridge configuration for analog temperature measurement

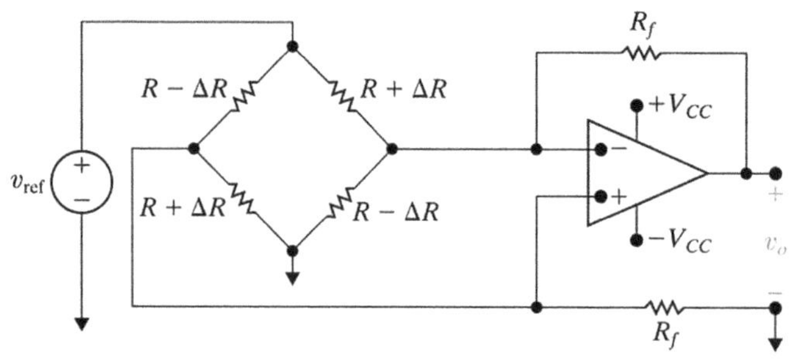

To improve both the linearity and the sensitivity, two sensors can be used, instead of one with opposite temperature coefficients. Using four sensors, as illustrated in Figure 10, will further improve the sensitivity of the output. Equation (8) shows the approximated output voltage, where *n* (number of sensors) can be one, two, or four.

$$V_o \cong \frac{n\Delta R}{4R} v_{ref} \tag{8}$$

Thermocouples have the ability to produce a direct voltage measurement that is almost a linear function of the temperature; however, they require additional compensation for the cold junction, prior to amplifying the voltage reading. This is depicted in the general circuit of Figure 11-a. There exist many linear technology ICs that can provide the required cold junction compensation, e.g. LT1025A. Figure 11-b shows a typical application circuit for a *K*-type thermocouple with a zero-drift Op-Amp that has a small current consumption, which makes it suitable for battery-operated applications. The output of the Op-Amp can be fed directly to an analog-to-digital-converter (ADC) of microcontrollers, e.g. Arduino, for further processing or connection with a computer.

Other types of temperature sensors that rely of semiconductors are usually available in compact IC configurations that integrate the necessary signal conditioning and the ADC in one package. This makes them more suitable for interfacing with microcontrollers and PCs for both wired and wireless applications. IC-based temperature sensors offer linear output with high accuracy, sensitivity and

Figure 11. Generic and practical connections of thermocouples (ourpcb.com, 2016)

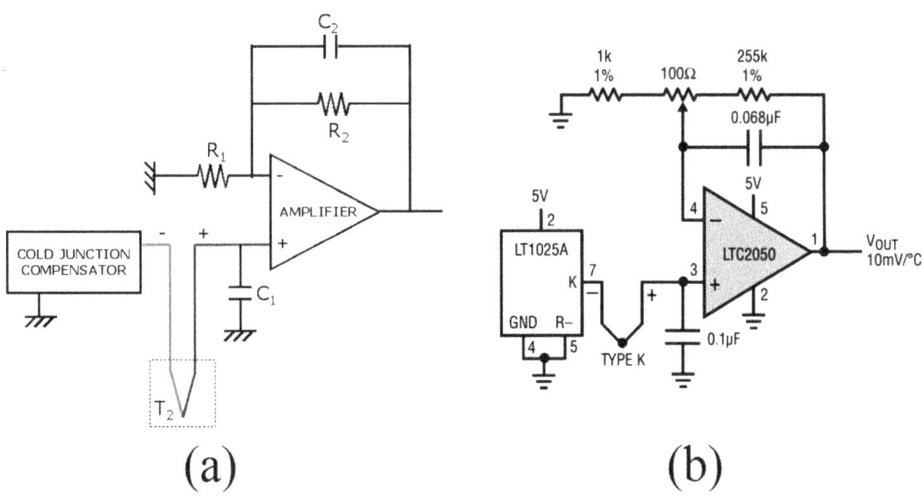

repeatability. They simplify the design of measurement and control systems. AD590, LM335, and TC1046, produce analog outputs, while ADT7410, LM76, and TC74 produce digital outputs, as shown in Figure 12 in (a)-(f), respectively.

APPLICATIONS FOR SMART TEMPERATURE SENSORS

Smart temperature sensors can be used to build a variety of applications in many different fields. Depending on the required tasks, used hardware, the programming language, and the available budget, many methodologies and architectures exist. Microcontrollers, Phidgets, and data acquisition kits are among the hardware components that are needed to interface the smart sensor to PCs and other digital circuitry. Visual Basic, C, and JAVA are usually the software languages to choose from. MATLAB, LabVIEW, and other IDEs for FPGAs that are based on HDL programming are among the developing environments for implementing these applications. Figure 13 illustrates two different connections, for two different smart temperature sensors; the first one in (a), uses Arduino microcontroller that is very cheap and is available in many different configurations with different number of pins and accessories, while in (b), a Phidget is used that is easily interfaced to the PC using a USB port. Arduino microcontrollers usually use C, as a programming language, while Phidgets can be easily programmed using Visual Basic. Both

Figure 12. IC-based temperature sensors that produce analog and digital outputs

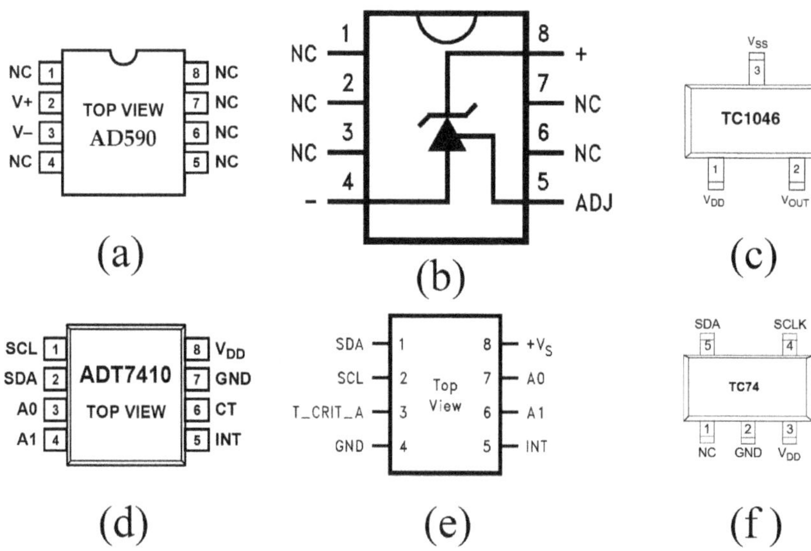

Figure 13. Using Arduino and Phidgets to interface smart temperature sensors

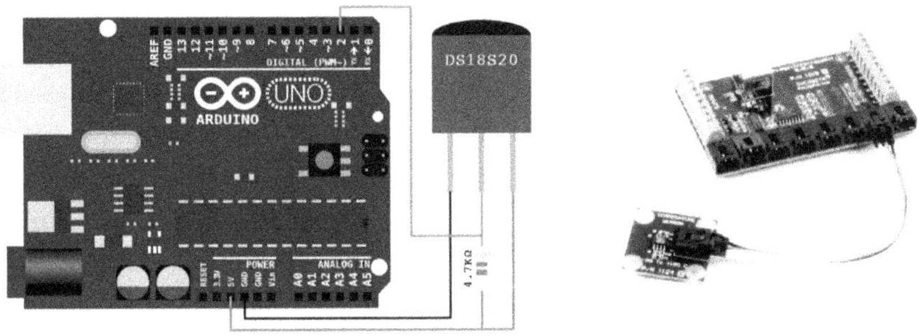

(a) (b)

Arduino and Phidgets have abundant resources and open-source codes that can greatly simplify their programming.

Smart Temperature Sensors in Automotive Applications

Temperature measurement in the automotive industry has many applications; checking the temperature of the engine oil and coolant liquid, turbo charger, brake system, and inside the car to control the temperature of the air conditioning system. These various applications have different ranges for the temperature, different connectivity constraints, and different functionality requirements. Engine oil temperature needs to be measured continuously in order to make sure that it is within the specified range, as too cold or too hot temperatures can damage the engine due to improper lubrication. The required sensor should be flexible, and not rigid, to be able to reach the oil. In addition, it should not interact with the oil and be able to be submerged all the time (Schwaller, 2005). Hermetically sealed RTDs, thermistors, and thermocouples are best suited for this task. Since analog measurements are employed, an additional signal conditioning stage will be required to display the temperature on the dashboard, while providing some alarming functions.

Measuring the temperature of the turbo charger is a challenging task due to the very high temperature of the exhaust stream that contains a considerable amount of contaminating gases. Usually a mineral-insulated thermocouple cable is used for such purpose that should provide efficient oxidation, carburization and chlorination resistance, in addition to the capability of withstanding high temperatures up to approximately 1200°C. Both K/N type thermocouples could be used for such purpose. Continuous monitoring of the brake temperature is usually required for high performance cars, such as high-speed sports cars. A temperature sensor is screwed into the caliber of the brake disk to provide continuous temperature measurement.

An RTD is usually used for such purpose that can withstand a temperature up to approximately 700°C. Both the turbocharger and the brake system use analog measurements, which is a characteristic shared with measuring the temperature of the engine oil and coolant liquid. In addition, they provide simple alarming functions and other basic control operations, such as switching the fan facing the radiator to cool down the engine. It is obvious that all the discussed temperature sensors rely on using direct contact with the heat source; however, for quick check-ups at the motor shops, remote sensors that rely on measuring the emitted infrared energy are usually used (Gilles, 2012). Figure 14 illustrates typical examples for three different sensors, used for measuring the temperatures of the engine oil, turbo charger, and the brake system, in (a), (b), and (c), respectively.

Climate control is now an available feature in most modern cars, where the temperature inside the car compartment is kept nearly fixed. This requires measuring the temperature in both inside and outside the car, in addition to measuring the sun load. A microcontroller is used to adjust the cooling load and the fan speed, according to the required temperature and the environmental conditions, as illustrated in Figure 15.

Figure 14. Analog sensors for measuring temperature of different car parts

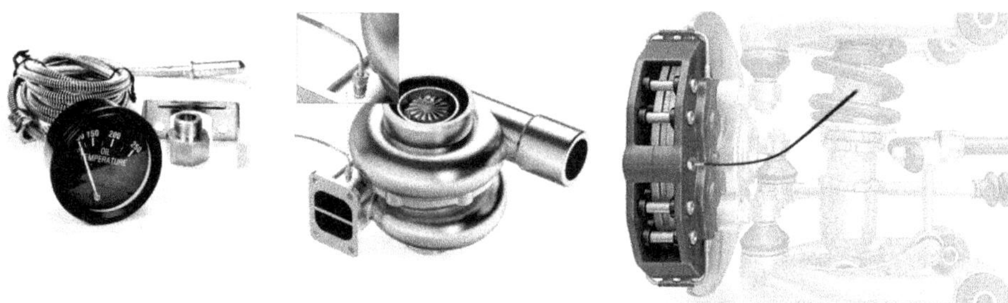

(a) (b) (c)

Figure 15. Simple closed-loop control system for regulating the car inside-temperature

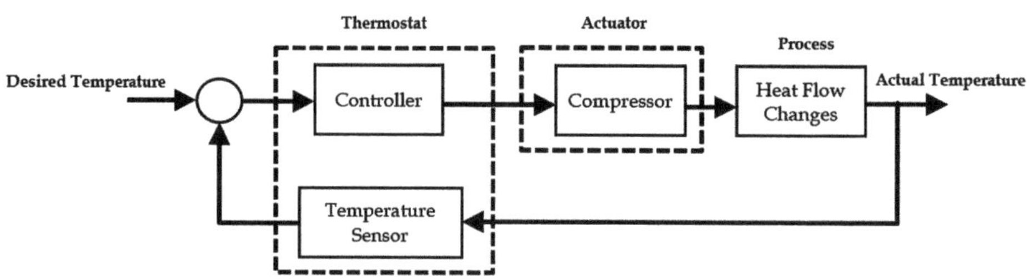

Smart Temperature Sensors in Weather Applications

All weather stations should provide accurate measurements of temperature. Many digital sensors that rely on IC-packages can be used for such purpose. The DHT11 temperature and humidity sensor could be interfaced to an Arduino microcontroller to perform such function using very low-cost components (Catelani, Ciani, Scotto, & Zanobini, 2016). Figure 16 shows two alternatives to using temperature sensors for reporting ambient temperature; in (a), the IC DS1621 is interfaced directly to the computer using standard serial communication; while, in (b) the IC TC1047A produces an analog output that can be further interfaced to a microcontroller, using an ADC. Both smart sensors are capable of providing linear and accurate measurements of temperature in the range from -40°C to 125°C.

Smart Temperature Sensors in Industrial Applications

Many industrial processes, especially in the field of petroleum and chemical industries rely on measuring and controlling temperature. The measurement of the temperature needs to be transmitted to the control element in order to operate the actuator. The instrument that interfaces a temperature sensor (e.g. thermocouple, RTD, or thermistor) to a measurement or control device (e.g. PLC, DCS, PC, loop controller, data logger, display, recorder, etc.) is called a temperature transmitter (TT). Figure 17 shows a schematic diagram for a chemical reactor, and typical temperature transmitters in (a), and (b), respectively.

Temperature transmitters perform several smart functions, as they isolate, amplify, filter the noise, linearize, and convert the signal from the sensor to a standardized range, before sending it to the control device. Typical electrical output signals, used in process industry, are 4-20 mA or 0-10V DC ranges. The signal range is scaled between these two extremes to represent 0% to 100% of the operating value,

Figure 16. Two different temperature sensing connections for weather forecast

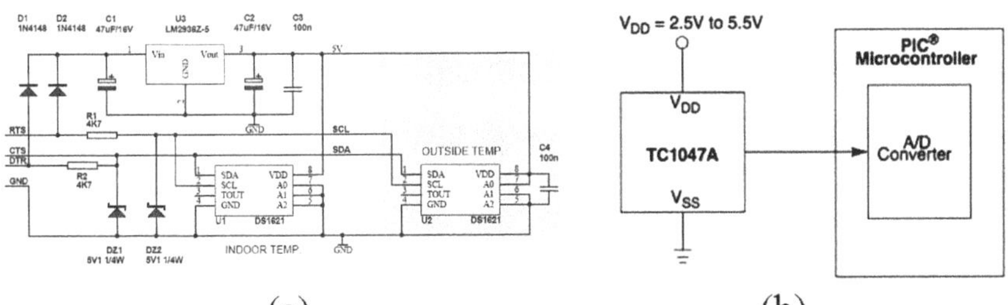

(a) (b)

Figure 17. Structure of the chemical reactor and its TTs

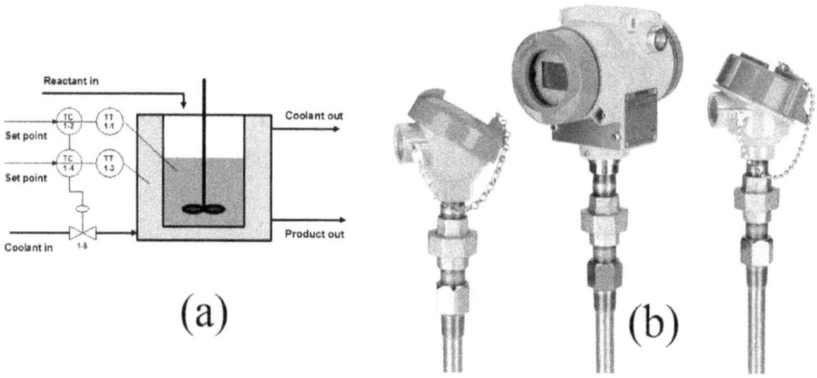

respectively. Precise temperature control of chemical reactors is required to have the best results for the quality of the product and speed of the required chemical reaction. Compromising the accuracy of the temperature control results in significant deteriorations in both the selectivity and the reaction speed of the chemical product, as illustrated in Figure 18a. Usually, the temperature of the exothermic chemical reactor is indirectly controlled using an external shell, called the reactor jacket. Thus, the temperature of the coolant, inside the jacket is controlled first, followed by the actual temperature of the chemical product, after some delay; this is illustrated in Figure 18b.

Figure 18. Optimization and control of the reactor temperature (huber-online.com, 2017)

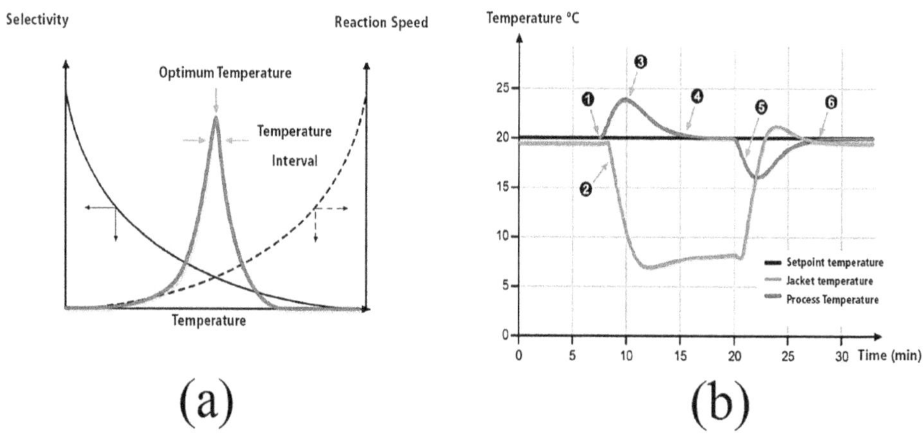

Smart Temperature Sensors in Health Care

Rapid advances in the technology led to the emergence of a new generation of wearable sensors that have many useful applications in providing physical and physiological measurements for the purpose of health care (Kenry, Yeo, & Lim, 2016). Inspired by the human skin, ultra-flexible, stretchable and biocompatible temperature sensors (BCTS), were successfully fabricated by integrating temperature sensitive materials with semipermeable polyurethane film through transfer printing in solution, which is targeting at human body temperature measuring in a 24/7 way (Chen, Lu, Chen, & Feng, 2015). The new generation of these smart temperature sensors provides an excellent PTC of 2.778 mΩ/Ω/°C, which is comparable to RTDs, with high accuracy and repeatability that was clinically proven very safe. Figure 19a shows the structure of a soft, stretchable, and breathable skin-inspired biocompatible temperature sensor based on the integration of a porous semipermeable PU film, as the substrate, and a temperature-sensitive patterned Au film, as the sensing component. It has five different layers of semipermeable film: the encapsulating layer; bonding layer; functional sensing layer; semipermeable film substrate; and adhesive layer stacked together to form a functional unit. This smart sensor is waterproof with outstanding air and water vapor permeability, while being impermeable to water and bacteria.

Figure 19b shows another smart wearable temperature sensor that is commercially available for an affordable price. It is also capable of wirelessly issuing alarming signals for the use with smart phones. More research is going on in this field and it is expected that wearable biomedical sensors will dominate conventional sensors in the near future.

Figure 19. Two different wearable smart temperature sensors

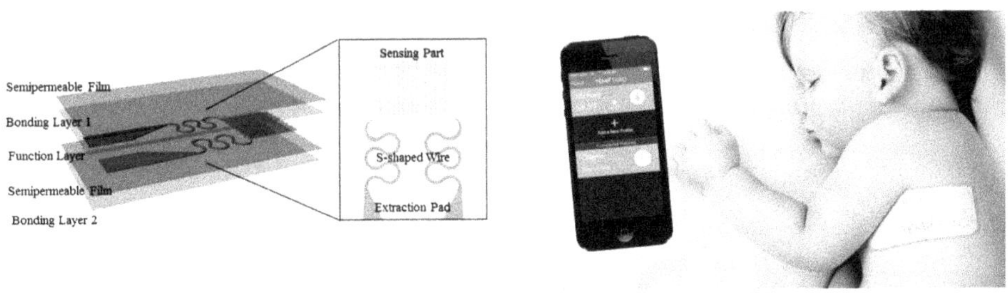

(a) (b)

Smart Temperature Sensors in Fire Fighting

Detecting fires in remote and isolated area, such as forests and buildings under construction is becoming a necessity nowadays, due to the need to prevent them from happening and to minimize the losses caused by them. Smoke detectors are usually not reliable and can produce false alarms; consequently, a more efficient way to detect fires is detecting heat (Agarwal, 2015). There are generally two types of heat detectors – simple fixed temperature detectors that alarm at a factory pre-set temperature, and rate-of-rise detectors that alarm when a rapid rise in temperature is detected. Many heat detectors have both capabilities built into them. When heat detectors are combined with smoke detectors, a more reliable alarming signal is obtained. Figure 20 shows three different choices to design smart fire detectors; a weather shield is shown in (a) that uses The Si7021 I2C humidity and temperature sensors, which is a monolithic CMOS IC, an analog-to-digital converter, signal processing, calibration data, and an I2C Interface. Figure 20b shows a typical heat detector, while Figure 20c shows a super-sensitive infrared sensor for non-contact thermal measurement that makes full use of proprietary Omron MEMS sensing technology.

Smart Temperature Sensors in Consumer Electronics

Many home appliances require measurement of temperature for monitoring, alarming, or control. Washing machines, refrigerators, and air conditioning systems are typical examples of these appliances. Control of the temperature is done using simple on-off relays. Another range of consumer appliances that include computers, both PCs and laptops, in addition to mobile phones and other electronics that use a microprocessor or a microcontroller, in the heart of it, require intelligent reporting of temperature. PCs usually have mechanical hard drives, in addition to multi-core processors,

Figure 20. Three different smart sensors for detecting fires

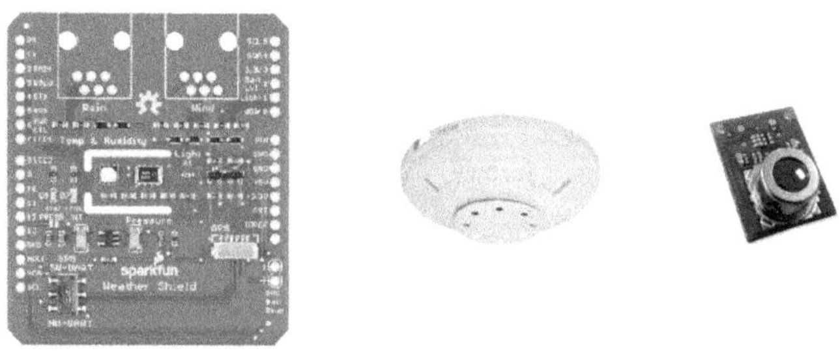

(a) (b) (c)

for which working at high performance might result in significant increase in the temperature. Hard drive manufacturers, both mechanical and solid-state (SS) use a self-monitoring, analysis, and reporting technology (SMART) to detect and report various measurements for the reliability of the drive. These measurements are sent to the operating system, and through special software package the temperature can be estimated, with very high accuracy. The same methodology is applied to microprocessors of computers and laptops. For mechanical hard drives, the temperature sensor is typically a thermistor integrated into the casing of the drive, as shown in Figure 21a, while for SSD, it is embedded into the Silicon gap itself. Figure 21b illustrates how to connect an RTD to the microprocessor. The smart functions of these types of temperature sensors are generated using the software package that extracts, from the operating system, the useful information and statistics to report to the user.

CONNECTIVITY CONSIDERATIONS

Most applications of temperature measurement require local sensing and wired connections to a display unit, a controller, or a data logger. Other applications use wireless connections that depend on measuring the IR energy; however, they are mostly hand-held and require being in a very close proximity to the heat source. Recently, more and more applications are in need to employ remote sensing, smart grids, distributed networking, and the use of modern communication techniques, such as the Internet of Things (IoT). During the last few decades, distributed temperature sensors (DTS) emerged as a new technology that measures temperatures by means of optical fibers. The measuring principles are based on detecting the back-scattering of light, e.g., using the Rayleigh, Raman, and Brillouin principles. Those optoelectronic devices provide a continuous profile of the temperature distribution along the fiber

Figure 21. Location of the temperature sensor in hard drives and microprocessors

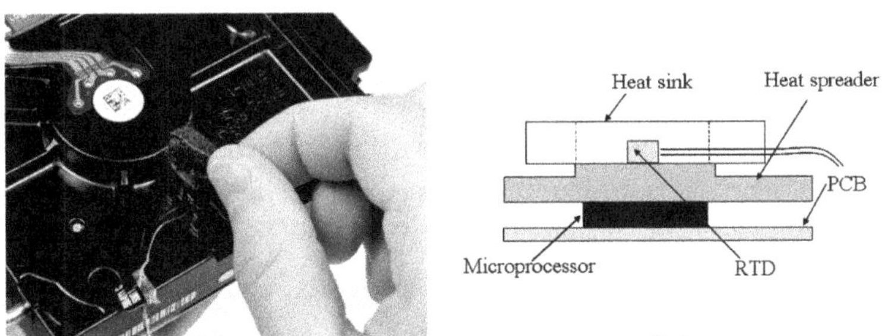

(a) (b)

cable with very high accuracy over significant distances (Ukil, Braendle, & Krippner, 2012), in the range of ±1.0°C at a resolution of 0.01°C to a spatial resolution of 1.0 m, over measurement distances as long as 30 km. This new technology is immune to electromagnetic interference (EMI), which makes it suitable for applications in power systems that require the transmission of medium to high voltage via long distance cables. Other applications include temperature monitoring in traction transformers in locomotives, switchgears, rotor winding of synchronous condenser of gas and electric substation, overhead transmission lines, just to name a few. Applications that are subject to geometry or space constraints also use temperature measurement via optical fibers to indirectly sense other physical variables that are hard to measure (Samiec, 2012). This includes using Brillouin scattering to measure failures in concrete structures and leakage of water in dams, in addition to Raman scattering to inspect fires. Other hazardous area like mines and oil wells make use of DTS to measure temperature.

IoT for Smart Temperature Measurement

As broadband Internet is becoming more widely available, and the cost of smart phones is dramatically decreasing, more and more sensors are being created with built-in connectivity capabilities to hook to the internet. This created the IoT that is becoming a huge network for connecting people-to-people, people-to-things, and things-to-things (Bahga & Madisetti, 2014). Thus, IoT is becoming the gateway for wirelessly connecting sensors for the purpose of display, monitoring, control, and data logging. The IoT allows for virtually endless opportunities and connections to take place, many of which have applications in manufacturing, smart homes/ cities, transportation, retail business, and healthcare. Many vendors are currently providing means of simple connectivity between different temperature sensors and the Internet, while providing open-source codes for programming these sensors to perform many smart tasks. Figure 22-a shows the layout of NodeMCU ESP 12-E, which is one of the most famous ICs to establish communication through the IoT, while Figure 22-b shows a complete layout of a typical IoT-based smart temperature sensing system that consists of a minimum of four components; the sensor, the communication link, the cloud, and the application. Many other IoT applications exist, where other wireless connection technology could be used, e.g. Wi-Fi, NFC, RFID, ZigBee for short range, HaLow and LTE for medium range, and LPWAN and VSAT for long range, communication, respectively. For wired communication, the Ethernet could be used, instead.

Figure 22. A simple IoT-based smart temperature sensor

Case Study: Integration Into SCADA Systems

Supervisory control and data acquisition (SCADA) systems provide a standard automation protocol to perform different actions for control, alarming, monitoring, and data logging. SCADA systems have special architecture that uses a smart layout of computers, networked data transmission, programmable logic devices, remote terminal units, PID controllers, and a wide range of sensors and actuators (Boyer, 2009). The architecture of SCADA is divided into levels; sensors belong to the first level, where different physical quantities are acquired and sent to the next level that contains the control units. MATLAB and LabVIEW are perhaps the most two important software packages that are considered as benchmarks in both industry and academia, when it comes to interfacing and data acquisition. Both have built-in libraries that support many hardware components, provided by different vendors. Fault finding, and noise rejection are two important goals, when using the signal coming from the sensor. Depending on an array of sensors provide useful redundancy, which results in better accuracy and robust sensing. Figure 23 shows the necessary components to build the first level of the SCADA system that consists of sensing and data acquisition DAQ. To ensure the required redundancy, two EI-1022 digital temperature sensors are connected to the wireless LabJack DAQ, which interfaces to the digital computer using a special built-in library inside LabVIEW. The hardware components are shown in (a) and (b), while the sensor layout and the LabVIEW program schematic are shown in (c) and (d), respectively.

The equations for the temperature measurement are given by:

$$T_{avgerage} = \frac{T_1 + T_2}{2} \quad , \quad T = \left(100v - 273.15\right)^{\circ} C \tag{9}$$

Figure 23. A simple system to exemplify SCADA, first level connection

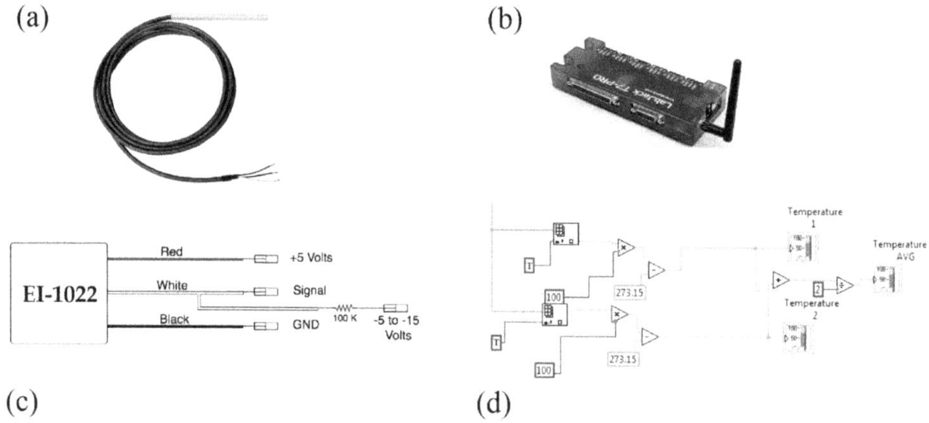

where v is the sensor reading and T is the recorded temperature (labjack.com, 2016).

CONCLUSION

In this chapter, the topic of smart sensors with emphasis on temperature sensing, was introduced. Traditional analog sensors that dominated the market were surveyed and their characteristics were carefully studied. In particular, RTDs, thermistors, and thermocouples that represent the first generation of temperature sensors were carefully investigated and contrasted against each other to highlight their advantages and limitations. In addition, noncontact IR-based sensors were covered as well. The study of these four different types emphasized the need for additional circuitry to convert the change of physical variables, either resistance or voltage, to a useful signal that is linear, noise-free, accurate, sensitive, and repeatable. This called for the next generation of temperature sensors that relied on CMOS technology. BJTs, DTMOSTs, MOSFETs, and ETFs were introduced, using different techniques that correlate thermal effects to variables, other than the resistance. This included biasing voltage, saturation current, introduced delays because of the propagation of heat waves, and the dimensions and the way of connecting the terminals of these semiconductor devices. These two different generations of the temperature sensors, although have different operating principles, motivated the introduction of smart temperature sensors, where the final product of the measurement device is useful for direct interfacing with modern digital devices that include, display units, controllers, and data loggers.

Many smart temperature sensors, in a variety of IC-packages, were introduced and a wide range of applications was explored to exemplify their performance; particularly, automotive industry, weather stations, process industry, wearable healthcare devices, fire fighting, and consumer electronics. Techniques for interfacing, signal conditioning, and calibration were discussed, while highlighting current practices and future trends. A combination of wired and wireless connections was investigated with examples from localized and distributed applications, including the IoT and SCADA systems. With the rapid advances in technology, the wide spread of wireless connectivity, and the decreasing cost of semiconductor devices, it is believed that smart sensors will prevail the market very soon; however, classical analog sensors will still exist, especially for applications with very high temperature ranges.

REFERENCES

Agarwal, T. (2015). *Heat Sensor Circuit and Working Operation*. Retrieved from: https://www.elprocus.com/heat-sensor-circuit-and-working-operation/

azom.com. (2010). *RTDs: Principle of Operation, Materials, Configuration and Benefits*. Retrieved from: https://www.azom.com/article.aspx?ArticleID=5573

Bahga, A., & Madisetti, V. (2014). *Internet of Things (A Hands-on-Approach)*. VPT.

Boyer, S. (2009). SCADA: Supervisory Control and Data Acquisition (4th ed.). ISA: The Instrumentation, Systems, and Automation Society.

Catelani, M., Ciani, L., Scotto, R., & Zanobini, A. (2016). Measurements and characterization of air temperature sensors for weather stations. In *Proceedings of the IEEE International Conference on Instrumentation and Measurement Technology* (pp. 269-273). Taipei, Taiwan: IEEE.

Chen, Y., Lu, B., Chen, Y., & Feng, X. (2015). Breathable and Stretchable Temperature Sensors Inspired by Skin. *Scientific Reports*, *5*(11505), 1–11. PMID:26095941

Dames, C. (2008). RTDs. In D. Li (Ed.), *Encyclopedia of Microfluidics and Nanofludics* (pp. 1782–1790). Springer. doi:10.1007/978-0-387-48998-8_1354

Fraden, J. (2010). Temperature Sensors. In J. Fraden (Ed.), *Handbook of Modern Sensors Physics, Designs, and Applications* (pp. 519–567). Springer. doi:10.1007/978-1-4419-6466-3_16

Gilles, T. (2012). *Automotive Science: Inspection, Maintenance, Repair*. Cengage Learning.

huber-online.com. (2017). *Temperature Control for Chemical Research and Production*. Retrieved from: http://www.huber-online.com/download/whitepapers_bulletins/Huber_TechDoc8.pdf

Huynh, T. (2015). Fundamentals of Thermal Sensors. In C. Jha (Ed.), *Thermal Sensors: Principles and Applications for Semiconductor Industries* (pp. 5–54). Springer.

Janata, J. (2009). Thermal Sensors. In J. Janata (Ed.); *Principles of Chemical Sensors* (pp. 51–62). Springer. doi:10.1007/b136378_3

Jones, M. (2012). Accurate Temperature Sensing with an External P-N Junction. *Linear technology Corporation – Application Notes, 137*, 1-12.

Kashmiri, S., & Makinwa, K. (2009). Measuring the thermal diffusivity of CMOS chips. In *Proceedings of IEEE Sensors Conference* (pp. 45-48). Christchurch, New Zealand: IEEE.

Kashmiri, S., Xia, S., & Makinwa, K. (2009). A Temperature-to-Digital Converter Based on an Optimized Electrothermal Filter. *IEEE Journal of Solid-State Circuits, 44*(7), 2026–2035. doi:10.1109/JSSC.2009.2020248

Kenry, J., Yeo, J., & Lim, C. (2016). Emerging flexible and wearable physical sensing platforms for healthcare and biomedical applications. *Microsystems & Nanoengineering*, *2*(16043), 1–19.

labjack.com. (2016). *EI-1022 sensor.* Retrieved from: https://labjack.com/support/datasheets/accessories/ei-1022

Love, J. (2007). Temperature Measurement. In J. Love (Ed.), *Process Automation Handbook – A Guide to Theory and Practice* (pp. 99–106). Springer.

Makinwa, K. (2010). Smart temperature sensors in standard CMOS. *Procedia Engineering*, *5*, 930–939. doi:10.1016/j.proeng.2010.09.262

Mansoor, M., Haneef, I., Akhtar, S., DeLuca, A., & Udrea, F. (2015). Silicon diode temperature sensors – A review of applications. *Sensors and Actuators. A, Physical*, *232*, 63–74. doi:10.1016/j.sna.2015.04.022

omega.com. (2016). *Thermocouples: Using Thermocouples in Temperature Measurement.* Retrieved from: https://www.omega.com/prodinfo/thermocouples.html

omega.com. (2017). *What are RTD Sensors? Why Use Them? How Do They Work?* Retrieved from: https://www.omega.com/temperature/pdf/rtd_gen_specs_ref.pdf

omega.com. (2018). *Technical Learning: Theory and application of the infrared temperature sensor.* Retrieved from: https://www.omega.co.uk/technical-learning/infrared-temperature-measurement-theory-application.html

ourpcb.com. (2016). *Temperature Sensor Thermocouple design considerations using LT ICs.* Retrieved from: https://www.ourpcb.com/thermocouple.html

resistorguide.com. (2018). *Thermistors.* Retrieved from: http://www.resistorguide.com/thermistor/

Revertera, F., Perpiñàb, X., Leónb, J., Vellvehib, M., Jordàb, X., & Altet, J. (2015). Characterization of MOSFET temperature sensors for on-chip dynamic thermal measurements. *Procedia Engineering*, *120*, 836–839. doi:10.1016/j.proeng.2015.08.699

Samiec, D. (2012). Distributed fiber-optic temperature and strain measurement with extremely high spatial resolution. *Optical Metrology*, *6*, 10–13.

Schwaller, A. (2005). *Total Automotive Technology*. Thomas Delmar Learning.

Souri, K. (2016). *Energy-Efficient Smart Temperature Sensors in CMOS Technology* (Unpublished doctoral dissertation). Electronic Instrumentation Laboratory of Delft University of Technology.

Ukil, A., Braendle, H., & Krippner, P. (2012). Distributed Temperature Sensing: Review of Technology and Applications. *IEEE Sensors Journal*, *12*(5), 885–892. doi:10.1109/JSEN.2011.2162060

ADDITIONAL READING

An, Y., Jung, D., Ryu, K., Woo, S., & Jung, S. (2015). An energy-efficient all-digital time-domain-based CMOS temperature sensor for SoC thermal management, *IEEE Transactions on Very Large Scale Integration (VLSI). Systems*, *23*(8), 1508–1517.

Bakker, A., & Huijsing, J. (2001). *High-Accuracy CMOS Smart Temperature Sensors*. Springer.

Bao, Y., & Li, W. (2018). A High-Speed Temperature Sensor with Low Supply Sensitivity for SoC Thermal Monitoring. *Journal of Circuits, Systems, and Computers*, *27*(7), 1850116. doi:10.1142/S0218126618501165

Chouhan, S., & Halonen, K. (2015). A low power temperature to frequency converter for the on-chip temperature measurement. *IEEE Sensors Journal*, *15*(8), 4234–4240. doi:10.1109/JSEN.2015.2414667

De Venuto, D., & Stikvoort, E. (2012). Low Power High-Resolution Smart Temperature Sensor for Autonomous Multi-Sensor System. *IEEE Sensors Journal*, *12*(12), 3384–3391. doi:10.1109/JSEN.2012.2198915

Najafi, K. (1991). Smart Sensors. *Journal of Micromechanics and Microengineering*, *1*(2), 86–102. doi:10.1088/0960-1317/1/2/002

Pertijs, M., Aita, A., Makinwa, K., & Huijsing, J. (2010). Low-Cost Calibration Techniques for Smart Temperature Sensors. *IEEE Sensors Journal*, *10*(6), 1098–1105. doi:10.1109/JSEN.2010.2040730

Pertijs, M., & Huijsing, J. (2006). *Precision Temperature Sensors in CMOS Technology*. Springer.

Shor, J., & Luria, K. (2013). Miniaturized BJT-based thermal sensor for microprocessors in 32-nm and 22-nm technologies. *IEEE Journal of Solid-State Circuits*, *48*(11), 2860–2867. doi:10.1109/JSSC.2013.2280039

Talataisong, W., Ismaeel R. & Brambilla G. (2018). A Review of Microfiber-Based Temperature Sensors, *Sensors, 18*(2), 461:1-26.

KEY TERMS AND DEFINITIONS

ADC: Analog-to-digital conversion, a system that converts analog continuous-time signals to equivalent digital signals that are binary based, with different accuracies and resolutions.

BCTS: Bio-compatible temperature sensors, low cost and high-performance range of temperature sensor that can be used in biological applications

BJT: Bipolar-junction transistor, a special type of transistors that use both electron and hole charge carriers. They are usually manufactured in PNP and NPN types, and can be used as switches or amplifiers, via controlling the current at the base terminal.

CMOS: Complementary metal-oxide semiconductor, a technology that is used in manufacturing microprocessors, microcontrollers, static RAM, and other digital logic circuits. It provides high noise immunity and low static power consumption.

CSTR: Continuous stirred-tank reactor, a continuous agitated-tank reactor, in chemical engineering that assumes perfect mixing, so that the output composition is identical to the composition of the material inside the reactor, which is a function of residence time, temperature, pressure, and rate of reaction.

DAQ: Data acquisition, the process of sampling signals that measure real world physical conditions and converting them into digital signals that can be manipulated by a computer. It usually includes a collection of sensors, signal conditioning units, and ADCs. It is mostly software-controlled.

DCS: Distributed control systems, a computerized control system, with a large number of decentralized control loops, in which autonomous controllers are distributed throughout the system. It provides higher reliability and reduced installation cost.

DTMOST: Dynamic threshold MOS transistor, a special type of MOSFET transistors that have a controlled threshold voltage. It can be used to replace traditional BJTs for better accuracy, when measuring temperature.

EMI: Electro-magnetic interference, a disturbance generated by an external source that affects an electrical circuit by electromagnetic induction, electrostatic coupling, and/or conduction.

HDL: Hardware description language, is a specialized computer language used to describe the structure and behavior of electronic circuits, and most commonly, digital logic circuits. It has many variants, such as VHDL and Verilog. It is widely used in mixed analog/digital codesigns.

IoT: Internet of things, a huge network for connecting people and things. It is becoming the gateway for wirelessly connecting sensors for the purpose of display, monitoring, control, and data logging.

MEMS: Micro-electro-mechanical system, a miniature machine that has both mechanical and electronic components, which are made using the techniques of microfabrication.

MOSFET: Metal-oxide semiconductor-field-effect transistor, a type of field-effect transistors (FETs), most commonly fabricated by the controlled oxidation of silicon. It has an insulated gate; whose voltage determines the conductivity of the device.

Phidgets: Physical widgets, an easy to use set of building blocks for low cost sensing and control from your PC, usually using the USB port.

PLC: Programmable logic controller, a special type of digital computers, which has been adapted for the control of manufacturing processes in industry, such as assembly lines, robotic devices, or any activity that requires high reliability control and ease of programming and process fault diagnosis.

RTD: Resistance-temperature detector, a special type of temperature sensors that relay on the effect of the increase in the electrical resistance, due to the increase in temperature. It has high accuracy and repeatability and can be used for different ranges of the temperature, depending on its material.

SCADA: Supervisory control and data acquisition, a special type of DCS, with an architecture that uses computers, networks, data communications, and graphical user interfaces for high-level process supervisory management. It also uses other peripheral devices such as PLCs and PID controllers.

TCR: Temperature coefficient of resistance, a measure of the sensitivity to changes in the electrical resistance, due to changes in temperature. It is positive in metals and negative in semiconductors. Usually, it provides a simple linear mathematical relationship for easy temperature measurements.

Chapter 9
Heat and Mass Transfer Characteristics of Evaporating Falling Films:
Application to Thermal Protection and Desalination

Monssif Najim
University of Ibn Zohr, Morocco

M'barek Feddaoui
University of Ibn Zohr, Morocco

Abderrahman Nait Alla
University of Ibn Zohr, Morocco

Adil Charef
University of Ibn Zohr, Morocco

ABSTRACT

This chapter presents a numerical investigation of heat and mass transfer characteristics during the evaporation of liquid films in vertical geometries. A two-phase model is developed to simulate laminar film evaporation into laminar gas flow. The liquid film evaporation is evaluated under adiabatic and heated wall conditions for both pure and binary liquid film. The model is based on a finite difference method to solve the governing equations of the two phases. The obtained results concerns two industrial processes. The first part of the chapter is devoted to the analysis of the thermal protection of vertical channel wall, while the second part is devoted to the desalination process by falling liquid film. The simulations results allowed the determination of the optimal operating conditions for both processes.

DOI: 10.4018/978-1-5225-7138-4.ch009

INTRODUCTION

A liquid film is thin layer of a liquid flowing on a wall. The mechanisms for forming liquid films are diverse. In fact, they can simply appear under the effect of gravity, for example, falling films can be observed during a rainfall day on cars windows. Most of industrial processes today include the phenomenon of evaporation that generally occurs from falling liquid films. In particular, falling liquid film evaporation process is for cooling and desalination purposes.

For cooling purposes, the most common applications are the evaporative cooling, wall thermal protection and electronic devices cooling. To evaporate a liquid film, it is primordial to provide an equivalent energy to the latent heat of vaporization. Thus, this energy may be provided by two modes: direct contact evaporation or heterogeneous evaporation. For heated surfaces (microchips), the heat flux is firstly exchanged by conduction then by convection through the flowing liquid film. When the liquid film interface reaches the saturation conditions, evaporation begins, which consists on the heterogeneous evaporation process. In contrast, when the liquid film flows over an adiabatic wall (cooling tower for example), the hot gas flow exchanges thermal heat flux with the flowing liquid film directly throughout the interface. However, the heat flux absorbed from the hot gas is not sufficient to ensure the liquid film evaporation. Consequently, the liquid film contributes to the evaporation process by its own internal energy. The liquid film experiences a temperature drop. The previous phenomenon are largely highlighted and investigated in the literature. Early, Yan et al. (Yan, Lin, & Tsay, 1991) conducted an experimental study to examine the heat and mass transfer of the evaporative cooling of a falling liquid film. Water and ethanol liquid film were investigated in the study. The results showed an important drop of the wall and interface temperature for both liquids. The highest temperature reduction was observed with high inlet temperature and low inlet mass flow rate of the liquid films. Yan et Lin (Yan & Lin, 1991) have numerically reproduced the previous experimental study. The authors took advantage of the numerical simulation to deeply investigate the heat and mass exchange and provided the interfacial latent and sensible heat flux along with the evaporation rate. He et al. (He, An, Li, & Jackson, 1998) investigated experimentally and numerically the liquid film evaporation inside a vertical tube considering counter current flows. The authors identified two cooling modes of evaporation: namely evaporation and direct film cooling. It is shown that when the cooling liquid film temperature is relatively high the system operates at evaporation mode, while for low temperatures of the system operates at the direct film-cooling mode. Feddaoui et al. (Feddaoui, Belahmidi, Mir, & Bendou, 2001; Feddaoui, Meftah, & Mir, 2006) investigated numerically the evaporative cooling of falling liquid film in laminar and turbulent mixed convection inside a vertical tube. A significant liquid film cooling is reported for high gas flow Reynolds number, high liquid film temperature and low mass flow rate. Narayanan et al. (Narayanan, Fedorov, & Joshi, 2013) examined the effect of liquid film flow rate on the cooling of a microscale device. It was found that for an equal evaporation flow rate and inlet flow rate, the heat dissipation is almost due to the evaporation, while with higher inlet flow rates, the advection contributes highly to the heat dissipation process, which is suitable for electronics cooling. Nait Alla et al. (Nait Alla, Feddaoui, & Meftah, 2015b) reported a numerical study of ethanol film evaporation in partially heated plate. The simulations showed that the number of zones does not affect the evaporation rate at both inlet and outlet of the channel. However, for partially heated wall, starting with the heating zone is favourable and enhances the evaporation. Nait Alla et al. (Nait Alla, Feddaoui, & Meftah, 2015a) examined in another study the heat and mass transfer during the evaporation of water and glycols inside a vertical channel. The authors observed that sensible heat transfer is higher than the latent heat transfer in the case of glycols contrarily to the case of water were latent heat transfer is dominant.

Falling liquid films are not exclusive to cooling processes only. Desalination units nowadays integrate vertical evaporators, especially in multi-effects distillation and Humidification/dehumidification desalination methods. Unlike cooling applications, an efficient desalination process must operate at moderate liquid film flow rate in order to enhance the evaporation. By decreasing the flow rate, the liquid film residence time increases allowing the absorption of high heat fluxes from the heat source and to elevate its temperature. In addition, it is favorable to operate at low pressure, so the diffusion coefficient increases and the thermal resistance of the gas flow decreases. However, the decrease of the flow rate to get more is not a general rule as reported by Orfi et al. (Orfi, Debbissi, Mohamed, & Nasrallah, 2004). Several attempts have been carried out to examine the falling film evaporation in desalination processes. Al-ansary and Owen (Al-Ansari & Owen, 1999) analysed the hydrodynamics and thermal performances of evaporation and condensation in horizontal tube. The results indicate that high brine mass flow rate causes high temperature but low evaporation. To analyse the air flow humidification by falling liquid film, Orfi et al.(2004) conducted a numerical study. The results show the high dependency of the interfacial heat and mass transfer on the liquid film inlet conditions. It was observed that the total evaporation rate might increase with the liquid film flow rate increasing for high liquid film temperatures while an opposite trend is observed for low liquid film temperature. Raach and Mitrovic (Raach & Mitrovic, 2005) attempted to minimize the thermal resistance of a falling liquid film over a heated plate. The authors highlighted the increase of the heat transfer by the immersion of turbulent wires in the liquid film. Yang and Shen (Dung & Yang, 2006) carried out an experimental study of falling film evaporation heat exchange outside horizontal tubes. It was found that the heat transfer coefficient increases with the increase of the flow density between 0.013 kg/m.s and 0.062 kg/m.s, with the increase of the heat flux and with the increase of tube height. However, there is a maximum height that must not be exceeded to avoid dry spots on the tube. An experimental study was carried out by Liu and Yi (Liu & Yi, 2002) to enhance the falling film evaporation of water-salt mixtures from roll-worked tubes and tube bundle. The results indicate a better evaporation heat transfer performance of the falling film by the use of low-cost roll-worked tubes instead of expensive commercial enhanced tubes. Ziqian et al. (Ziqian, Hongfei, Kaiyan, & Chaochen, 2007) developed a solar desalination unit using horizontal tube falling film evaporation. By several experiment tests, it was approved that the thermal performance of the system is improved by the evaporation of falling liquid film. The film evaporation enhancement in vertical annulus with preheated air flow was examined by Ben Radhia et al. (Ben Radhia, Harmand, Corriou, & Ben Jabrallah, 2014). The effect of the geometry dimensions and inlet conditions parameters was evaluated to determine the important factors for enhancing the film evaporation. Abraham et al. (Abraham & Mani, 2015) investigated numerically heat transfer characteristics of a falling liquid film evaporation outside a tube. Their results show that convective the evaporation heat transfer performance increases with feed rate, whereas it decreases with tube diameter. Recently, Najim et al. (Najim, Feddaoui, Charef, & Meftah, 2017) studied the effect of salinity on the heat and mass transfer during liquid film evaporation in a vertical tube. The authors investigated the variation of salinity along the tube and compared the effects of operating conditions on the efficiency of the desalination process.

This brief review of the literature shows that there is plenty of parameters to take into account when considering falling liquid film process for both cooling and desalination purposes. The appropriate determination of operating conditions is therefore necessary for an efficient application from economic and physic view. In the following, we will expose the impact of the inlet conditions and the operating conditions on the cooling and desalination systems.

Mathematical modelling

In most cases, falling liquid film flow in vertical channels takes place in a downward direction, thus, the fields of velocity, pressure, temperature and concentrations at each section depend only on the upstream solutions. This property gives the parabolic character to the governing equations. Moreover, the difference between the wall and the liquid film interface temperatures is small. In consequence, the vapour diffusion is done at very small velocities, which allows neglecting the interface shear stress. For Reynolds number less than 1500, the liquid film flow may be considered laminar as quoted by Ueda and Tanaka (Ueda & Tanaka, 1975).

Figure 1. Falling liquid film inside an insulated vertical channel

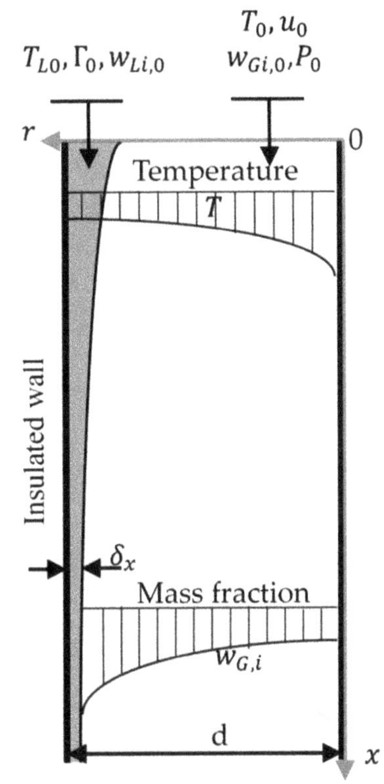

Proceeding with the above assumptions, the continuity equation for a laminar steady state flow is:

$$\nabla.\left(\rho_L \vec{u}_L\right) = 0 \tag{1}$$

The momentum conservation equation is:

$$\nabla.\left(\rho_L \vec{u}_L \bullet \vec{u}_L\right) = -\nabla P + \nabla.\left(\mu_L \nabla \vec{\mathrm{u}}_{\mathrm{L}}\right) + \rho_L \vec{g} \tag{2}$$

Since heat and mass transfers are mainly due to the evaporation, the species diffusion processes occur with very low concentration level, so that the diffusion-thermo (Dufour) and thermo-diffusion (Soret) effects can be neglected. Thus, for negligible Dufour and Soret effects and considering the conduction and convection through the liquid film, the energy and species diffusion equation for binary liquid film is:

$$\nabla.\left(\rho_L \vec{u}_L C_{pL} T_L\right) = \nabla.\left(\lambda_L \nabla \mathrm{T}_L\right) + \nabla.\left(\rho_L D_{L,12}\left(C_{pL,1} - C_{pL,2}\right)\nabla T_L \ \nabla w_{L,1}\right) \qquad i = 1,2 \tag{3}$$

$$\nabla.\left(\rho_L \overrightarrow{u_L} w_{L,i}\right) = \nabla.\left(\rho D_{L,im} \nabla w_{L,i}\right) \qquad i = 1,2 \tag{4}$$

The gas flow governing equation are similar to the liquid film, thus, with the presence of the air, the gas flow is ternary and the equations may be written as:

$$\nabla.\left(\rho_G \vec{u}_G\right) = 0 \tag{5}$$

$$\nabla.\left(\rho_G \vec{u}_G \bullet \vec{u}_G\right) = -\nabla P + \nabla.\left(\mu_G \nabla \vec{\mathrm{u}}_G\right) + \rho_G \vec{g} \tag{6}$$

$$\nabla.\left(\rho_G \vec{u}_G C_{pG} T_G\right) = \nabla.\left(\lambda_G \nabla \mathrm{T}_G\right) + \nabla.\left(\sum_{i=1}^{2} \rho_G C_{pG,i} D_{G,im} \Delta T_G \nabla w_{G,i}\right) \qquad i = 1,2,a \tag{7}$$

$$\nabla.\left(\rho_G \vec{u}_G w_{G,i}\right) = \nabla.\left(\rho D_{G,im} \nabla w_{G,i}\right) \qquad i = 1,2,\ a \tag{8}$$

BOUNDARY CONDITIONS

Boundary conditions are associated with the previous equations to solve the mathematical problem. Uniform velocity, temperature and mass fractions are specified at the channel inlet as follows:

In the liquid flow:

$$\Gamma = \Gamma_0; T_L = T_{L0}; w_{L,i} = w_{L0,i} \qquad i = 1,2 \tag{9}$$

In the gaseous phase:

$$u = u_0; T_G = T_{G0}; P_G = P_0;\ w_{G,i} = w_{G0,i} \qquad i = 1, 2, a \tag{10}$$

The channel walls are adiabatic and impermeable, thus the temperature and concentration gradients at the walls vicinity are null. In addition, the non-slip condition at both walls implies that the longitudinal velocity is null. In consequence:

$$u_L = 0; \frac{\partial T_L}{\partial r} = 0; \frac{\partial w_{L,i}}{\partial r} = 0 \qquad i = 1, 2 \tag{11}$$

$$u_G = 0; \frac{\partial T_G}{\partial r} = 0; \frac{\partial w_{G,i}}{\partial r} = 0 \qquad i = 1, 2, a \tag{12}$$

For the case of vertical tube, the problem is axisymmetric and the tube wall is subjected to a uniform heat flux:

$$q_W = -\lambda_L \left. \frac{\partial T_L}{\partial r} \right|_W \tag{13}$$

The boundary conditions at the centre line are:

$$\frac{\partial T_G}{\partial r} = 0; \ \frac{\partial u_G}{\partial r} = 0; \ \frac{\partial w_{G,i}}{\partial r} = 0 \qquad i = 1, 2, a \tag{14}$$

At the interface, the mass and heat balance should be satisfied between the two flows. With an evaporating liquid film, the total evaporation rate at the interface is defined as:

$$\dot{m} = -\frac{\rho_G \sum_{i=1}^{2} D_{G,im} \left. \frac{\partial w_{G,i}}{\partial r} \right|_I}{1 - \sum_{i=1}^{2} w_{G,i}} \tag{15}$$

Accordingly, the evaporation rate of each component can be written as:

$$\dot{m}_i = \dot{m}w_{L,i} - \rho_L D_{L,12}\frac{\partial w_{L,i}}{\partial r} = \dot{m}w_{G,i} - \rho_G D_{G,im}\frac{\partial w_{G,i}}{\partial r} \tag{16}$$

Considering the non-slip condition at the interface, the shear stress at the interface is equal between the gas and liquid flow:

$$\tau_I = \left[\mu\frac{\partial u}{\partial r}\right]_{L,I} = \left[\mu\frac{\partial u}{\partial r}\right]_{G,I} \tag{17}$$

The heat at the interface is released through latent and sensible heat fluxes. The energy balance translates this as follows:

$$\left[\lambda\frac{\partial T}{\partial r}\right]_{L,I} = \left[\lambda\frac{\partial T}{\partial r}\right]_{G,I} + \dot{m}h_{fg} \tag{18}$$

With the assumption of liquid-vapor interface at thermodynamic equilibrium state, the interface vapor mass fraction of each component is:

$$w_{Gi,I} = \frac{M_i P_{vi,I}}{\sum_{i=1}^{n} M_i P_{vi,I}} \tag{19}$$

Heat and Mass Transfer Parameters

After the possession of velocities, temperatures and species concentration fields, the evaluation of heat and mass transfer is proceeded by plotting different parameters such as sensible and latent heat flux, accumulated evaporation rate etc. the Definition of those parameters is giving below.

The main advantage of numerical studies is the ability to express the energy budget. Let us consider a unit length element with surface coinciding with the wall and the liquid film interface. The energy balance in the considered element could be written as follows:

$$q_W = \int_{d-\delta_x}^{d}\left(\rho C_p u\frac{\partial T}{\partial x}\right)_L dr + \dot{m}h_{fg} - \lambda_G\left.\frac{\partial T}{\partial y}\right|_{G,I} = q_{fc} + q_L - q_s \tag{20}$$

- The local latent heat flux at the interface, which reflects the quantity of energy needed to change the phase, is:

$$q_L = \dot{m} h_{fg} \tag{21}$$

- The local sensible heat flux at the interface expresses the temperature variation of a heated fluid:

$$q_s = \lambda_G \left. \frac{\partial T}{\partial r} \right|_{G,I} \tag{22}$$

- The convective heat of the liquid film is:

$$q_{fc} = \int_{d-\delta_x}^{d} \left(\rho C_p u \frac{\partial T}{\partial x} \right)_L dr \tag{23}$$

- The accumulated evaporation rate Mr is:

$$Mr = \frac{1}{\Gamma_0} \int_0^x \left| \dot{m}_I \right| dx \tag{24}$$

The bulk quantities are evaluated by the following calculation:

$$T_b = \frac{\int_0^{d-\delta_x} \rho_G Cp_G u_G T_G dr}{\int_0^{d-\delta_x} \rho_G Cp_G u_G dr} \tag{25}$$

Numerical Approach

The resolution of the coupled governing equations is possible by many numerical models and many of them are proposed in the literature. A very important number of studies uses finite difference schemes. In fact, the parabolic form of those equations gives advantage to finite difference method, for its simplicity and for computing time saving. To solve the parabolic equations described above along with their boundary conditions, a fully implicit scheme is employed. The numerical solution is realized using finite difference method. Each finite-difference equation's system forms a tridiagonal matrix, which can be solved using the TDMA method (Patankar, 1980). Moreover, to satisfy the global mass flow constraint a correction of the pressure gradient and axial velocity profile at each axial step, according to Raithby and Schneider method (Raithby & Schneider, 1979) is applied.

It is also important to use a dynamic non-uniform mesh. In fact, the liquid film interface does not conserve the same position along the channel. Therefore, due to the evaporation the interface location changes and decreases along the channel. Consequently, the mesh have to deal with the liquid film thickness. This is established by first locating the interface at a given location and then dividing the transversal domain to a liquid film domain and to a gas flow domain. In this manner, the interface is rigorously tracked. Moreover, to take into account the strong variation of velocity, temperature, and concentrations near the interface the grid nodes are non-uniformly positioned with higher density near the interface.

With respect to the mentioned model, the following steps are involved in the computation code:

For I=1, the liquid film longitudinal velocity is obtained by solving the momentum equation neglecting the inertia and convection terms. The initial liquid film thickness is then calculated by:

$$\delta_0 = \left(\frac{3\mu_L \Gamma_0}{\rho_L^2 g} \right)^{1/3} \tag{26}$$

For I=2 to N_I:

Step 1: In the first iteration, an arbitrary pressure gradient *dp/dx* is assumed.

Step 2: An arbitrary value of the liquid film thickness δ_x^{it} is assumed.

Step 3: Calculate the longitudinal velocities and species mass fractions in the liquid and vapor phases.

Step 4: Integrate continuity equation to obtain the transversal velocities in both phases.

Step 5: The temperatures in the liquid film and gas flow are then calculated.

Step 6: Calculate the relative error E_L^{it} of the liquid film mass balance using equation.

Step 7: A better approximation to the liquid film thickness is then obtained using the secant method. Thus:

$$\delta_x^{it+1} = \delta_x^{it} - \frac{\delta_x^{it} - \delta_x^{it-1}}{E_L^{it} - E_L^{it-1}} E_L^{it} \tag{27}$$

The convergence criteria used is $E_L^{it} = 10^{-6}$. Usually five to seven iterations suffice to obtain converged solution.

Step 8: The mass flow in vapor phase is then calculated and for a given inlet vapor velocity the error in the mass flow E_G^{it} is determined.

Step 9: A better approximation to the pressure gradient is then obtained using the secant method. Thus:

$$\frac{dp}{dx}^{it+1} = \frac{dp}{dx}^{it} - \frac{\frac{dp}{dx}^{it} - \frac{dp}{dx}^{it-1}}{E_G^{it} - E_G^{it-1}} E_G^{it} \tag{28}$$

Five to six iterations are usually sufficient to satisfy the convergence criteria for the error in the mass balance

Step 10: Check the convergence of all calculated variables: longitudinal velocities, species mass fractions and temperatures. If the convergence criteria ($Err \leq 10^{-6}$) is satisfied then the new thermophysical properties are calculated, if not, the steps 1to 9 are repeated for a new iteration (it+1).

The calculations are then advanced in the longitudinal direction *x*.

Validation of the Model

Before dealing with the simulation results, we first compared our results with evaporative cooling experiment in an adiabatic vertical channel, cooled by a pure ethanol liquid film. The comparison shows the effect of various inlet mass flows on the variation of wall temperature along the channel. A very satisfying agreement between our simulations and the experiment curves is found with a maximum relative error of 4.7% as illustrated in Figure 2a. The difference observed for the lowest inlet mass flow rate is due to the difficulty to maintain a pure inlet air flow (without ethanol vapour trace) during the experiment as stated by the authors. A second comparison is done with the study of Ali Cherif and Daif (Cherif & Daïf, 1999) for the case of binary liquid film evaporation under an applied heat flux. The plot shows a very good agreement between our simulation and those of the numerical study for different wall heat fluxes. The maximum relative error is less than 1.9%. Therefore, the present numerical model is suitable for describing the studied problem.

THERMAL PROTECTION OF WALLS BY THE EVAPORATIVE COOLING

Effect of the Film Mixture on the Exchanged Heat Fluxes and Mass Transfer

The examination of the mixture effect on the heat transfer is done by evaluating the latent, sensible and film convective heat fluxes variation along a channel with 1 m of height and 1 cm of width. To run the simulation, a liquid film is feed at $T_{L0} = 300\text{K}$ and at $\Gamma_0 = 0.01\text{kg}/(\text{m.s})$, while a dry air flows enters with $Re = 1000$ at $T_{G0} = 400\text{K}$ and at atmospheric pressure. The water mass fraction in the liquid mixture is varied from 0 to 100% by 50% step. Two mixtures are considered: water-ethanol and water-methanol.

Figure 3a shows the variation of the local sensible heat flux as a function of the channel length for various water mass fractions. The first analyses of the curves reveals that the heat transfer by the sensible mode is negative for all cases. This indicates that the liquid film absorbs heat flux from the gas flow at the interface. This is obvious because the gas flow is much hotter than the liquid film. The pure water presents lower sensible heat flux compared to the ethanol and methanol mixtures. Ethanol and methanol specific heats are much lower than the water one.

Figure 2. Comparison of our simulations with previous experimental (a) and numerical (b) studies

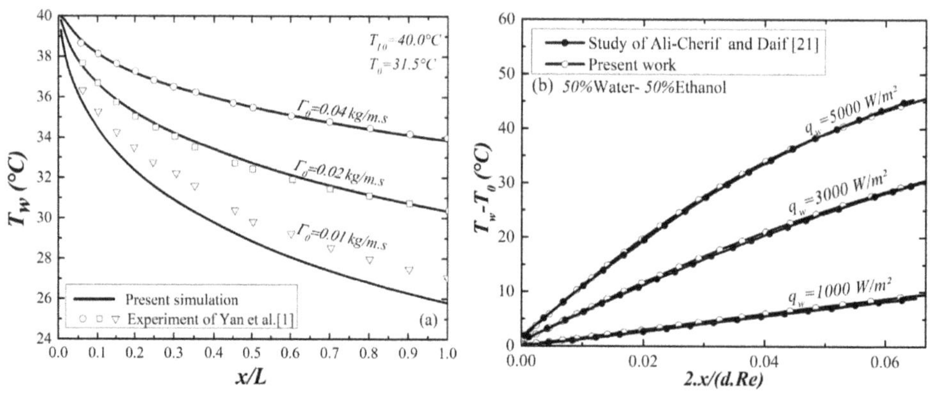

The addition of one of the two alcohols to the water decreases the specific heat of the obtained mixture. Consequently, decreasing the amount of water in the mixtures leads to the decreases of the mixtures capacity to carry out the absorbed heat flux without heating effect (due to the decrease of the mixture thermal capacity). The water-methanol liquid mixture presents the highest sensible heat flux compared to the water-ethanol mixture. This is due to the higher conductivity of methanol. The ratio $q_S / q_{S,w}$ of mixture sensible heat flux to pure water sensible heat flux plotted in Figure 3b, represent the enhancement of the sensible heat flux by reducing the water mass fraction. Figure 3b shows that this ratio varies considerably for low mass fractions of water in the both mixtures. The heat exchange by sensible mode is enhanced by 6 to 21% for the water-methanol mixture, while it is enhanced by 3% to 9% for water-ethanol mixture, by reducing the water mass fraction to 75% and to 0%. It is also observed that for distances higher than 0.6m from the inlet, the enhancement ratio is almost stable and does not vary. This is explained by the dissipation of energy by evaporation instead of the temperature rising of the liquid film.

The local latent heat flux evolution is illustrated in Figure 4a. As the evaporation takes place against *y* direction, the latent heat flux is negative as can be seen in Figure 4a. The lowest value of latent heat flux corresponds to the pure water liquid film. Since the highest saturation pressure corresponds to methanol followed by ethanol, the addition of alcohols to the water changes the saturation pressure of the mixture. Thereby, the decrease of water mass fraction increases the latent heat flux especially for water-methanol mixture. Basing on the last finding, we can conclude that the highest evaporation will occur for high concentration of alcohols in the studied mixtures and for water-methanol mixture specifically. In fact, by eliminating water

Figure 3. Evolution of local sensible heat flux (a) and mixture to pure water sensible heat flux ratio (b) along the channel

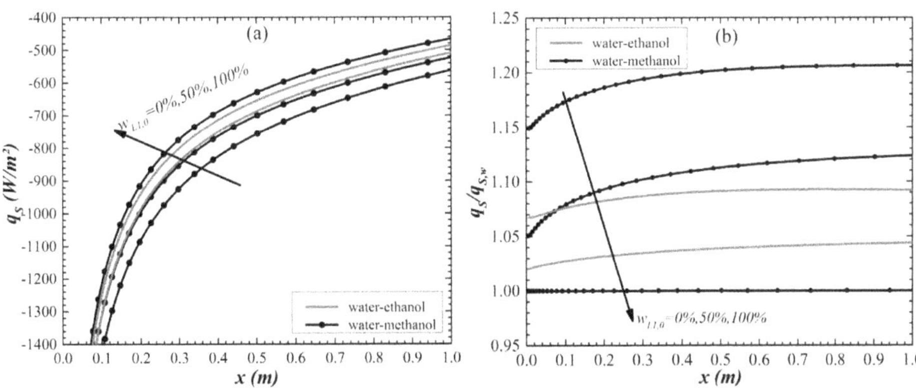

from the mixtures, the latent heat flux enhancement reaches 45% for methanol and 25% for ethanol at the channel outlet (Figure 4b). It can be also deduced that the evaporative cooling of the wall is more efficient with methanol presence in the mixture. This ascertainment could be verified by plotting the accumulated evaporation rate.

To evaluate the mass transfer along the channel, the accumulated evaporation rate is presented in figure 5. As it was expected, the accumulated evaporation rate is very low for pure water. It can be seen from figure 5.a that as the quantity of alcohols increases in the mixture, the evaporation rate increases due to the increase of mixture volatility. The water-methanol mixtures presents always higher evaporation rates along the channel due to the higher saturation pressure as already explained. However, pure methanol have higher evaporation rate compared to pure ethanol but as the liquid films approached the channel outlet, the difference decreases. This may result from the decrease of the heat exchange between the gas and liquid film flows due to the high reduction of methanol liquid film temperature. This assumption would be verified afterward. The total accumulated evaporation rate is enhanced by about 300% and 280% by using pure methanol and ethanol respectively.

Figure 6 presents the evolution of the local liquid film convective heat flux. The convective heat flux of pure water and water-ethanol mixture is positive along the channel, which reflects the heating of the liquid film. However, for pure ethanol, water-methanol mixture and pure methanol, the convective heat flux of the liquid film is negative. This means that as the liquid film advances in the channel, its temperature decreases for pure ethanol and water-methanol mixtures, while it decreases for pure water and water-ethanol mixtures. This implies again that the process of evaporative cooling of the wall is more important with the presence of methanol.

Figure 4. Evolution of local latent heat flux (a) and mixture to pure water latent heat flux ratio (b) along the channel

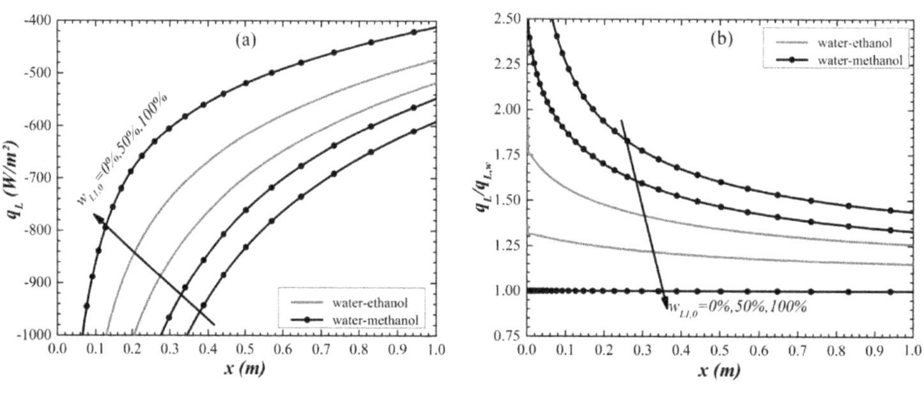

Figure 5. Distribution of local accumulated evaporation rate along the channel

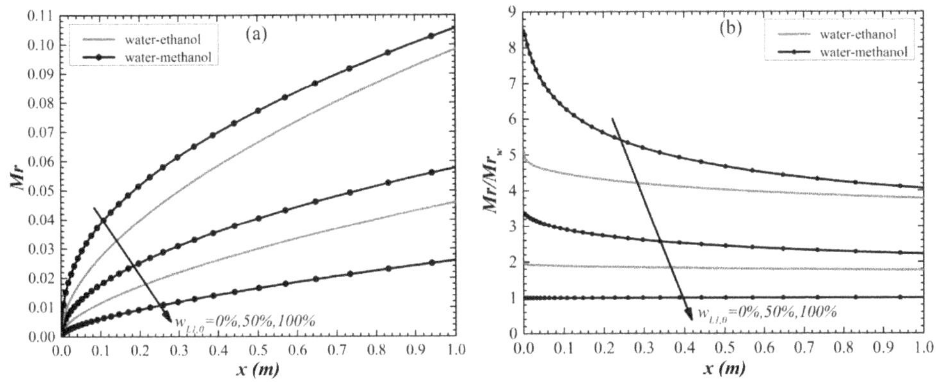

Figure 6. Variation of local convective heat flux of the liquid film along the channel

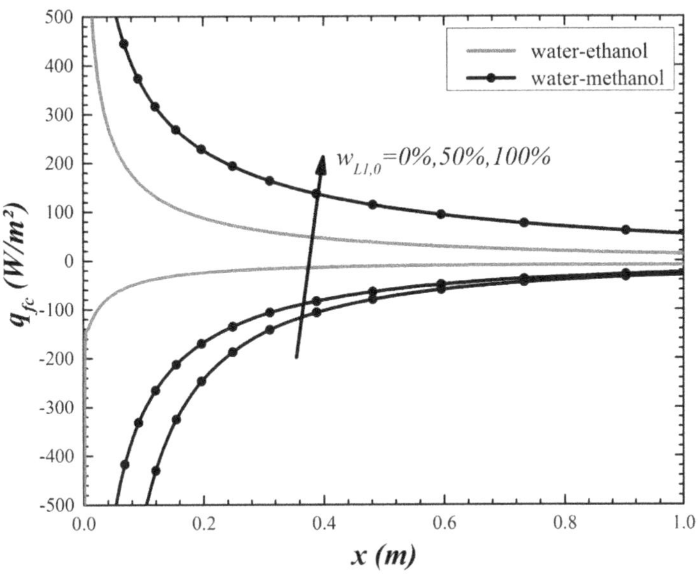

Effect of the Film Mixture on the Wall and Gas Temperature

To highlight all the previous ascertainment, the axial evolution of gas bulk temperature and wall temperature is plotted in Figures 7a and 7b. The bulk temperature presented in Figure 7a decreases from the channel inlet to its outlet. The addition of alcohols to the mixtures decreases considerably the bulk temperature, especially for methanol. Therefore, pure ethanol and methanol cases have the lowest gas bulk temperature. The reasons are, as explained before, the high volatility of alcohols. In fact, in the absence of a heat source (adiabatic wall), the gas flow plays this role and releases a quantity of its heat flux to the colder liquid film. The absorbed amount of the heat flux may contribute in one hand to the increase of the liquid film temperature and in the other hand to its evaporation. The Figure 7b shows the distribution of the wall temperature. The wall temperature decreases along the channel for water-methanol mixture, pure methanol and pure ethanol. In contrast, it increases for pure water and water-ethanol mixture. This could be explained by the evaporative cooling process. In fact, when the liquid film is highly volatile, an important amount of the absorbed heat flux is converted to latent heat flux to evaporate the liquid film. However, the liquid film contributes also to the evaporation process. Consequently, its temperature decreases also leading to the decrease of the wall temperature. It is important to mention that pure ethanol and water-methanol mixture with $w_{L1,0} = 75\%$, allows to keep relatively constant temperature (around 298K and 300K) along the channel. From the economic point of view, the water-methanol mixture is a better alternative to the pure ethanol liquid film since only 25% of methanol concentration is necessary to obtain approximatively the same wall temperature.

CONCLUSION

In this part, the heat and mass transfer in wall protection process is analyzed. Comparisons have been made between pure water and water-alcohols mixtures. The main outcomes of this study are:

1. The presence of ethanol or methanol with high concentration in the mixtures allowed a noticeable wall temperature reduction,

Figure 7. Variation of gas bulk temperature and wall temperature along the channel

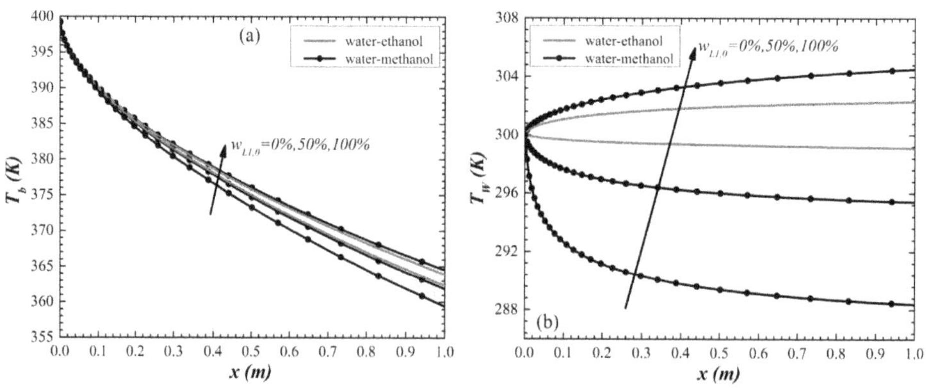

2. For a pure water liquid film or a mixture of water and ethanol, the temperature rises along the wall by 6k.

A mixture of only 25% of methanol with water is sufficient to maintain almost the same wall temperature obtained with pure ethanol.

Saline Water Film Evaporation for Desalination

In the introduction, we have discussed the important role of falling liquid films in desalination units. A conventional humidification/dehumidification (HDH) desalination process is presented in Figure 8. A humidification-dehumidification system may take different configurations (open or closed cycle, air heated or water heated cycle ...). However, the basic parts of the HDH process are the humidifier, the dehumidifier and the preheater. The seawater is initially fed to the condenser inlet and enters in an indirect contact with humid air stream. This permits the condensation of the water vapour carried by the air stream and the heating of the seawater at the same time. A part of the seawater leaving the dehumidifier is then transported to the preheater while the other part is rejected. The preheater rises the seawater temperature to a well-designed value. The preheated seawater flows then to the humidifier with an inlet temperature T_{L0}. Under the effect of wall heat flux, a part of the seawater is evaporated to the flowing air stream and the remaining flow exit the evaporator as brine. Therefore, the air stream leaves the humidifier with a higher humidity and temperature to the condenser where it is condensed. The condensed fresh water is collected and the cold air leaves the condenser and maybe reused in a new cycle depending on the nature of the HDH configuration (open or close cycle).

As the quantity of the purified water (fresh water) depends highly on the amount of evaporated water at the humidifier and thus the air humidity at the humidifier outlet, it is very important to understand the impact of seawater temperature and mass flow rate at the humidifier inlet on the evaporation process.

In this section, we will numerically examine the impact of operating conditions on the evaporation of falling liquid film in vertical tube with the presence of salt. A similar model to that described in the previous section is used to solve the governing equations. Since the geometry under consideration is a vertical tube, the governing equations are considered in the cylindrical coordinates. With one component falling liquid film (saline water film), the species equations in the liquid phase are not solved. Thus, the governing equations are the same presented in the study of Najim et al. (Najim, Feddaoui, Nait Alla, Charef, & Labsi, 2017), and the results are presented in a conference paper (Najim, Feddaoui, Nait Alla, Charef, & Labsi, 2017) .

Figure 8. Schematic of humidification/dehumidification desalination process

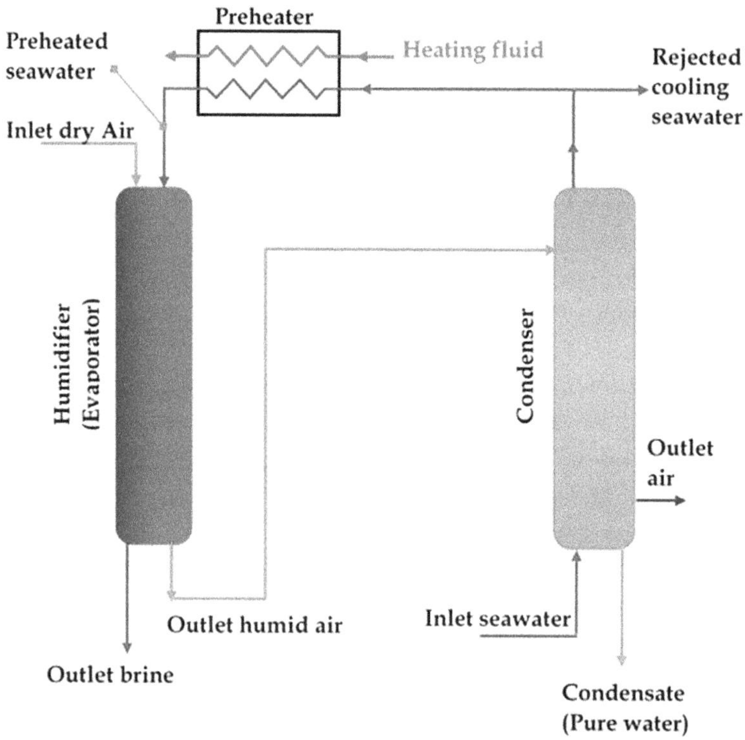

To analyze the effect of salinity and inlet conditions on the evaporation process, several runs were carried out using water (pure fluid) and seawater (S=39g/kg) in evaporation inside a vertical tube of 3m of height and 2cm of diameter. All simulations were carried out at 0.25 atm. The tube is subjected to a uniform heat flux of 5000W/m^2 at the wall. A dry gas flow enters the channel at $T_0 = 20°C$ and at $Re = 2000$.

A first study is carried out in order to determinate the best conditions for evaporation. The total evaporating rate $\overline{\dot{m}_{ev}}$ is presented in the figure 9. When increasing the inlet mass flow, three different behaviours of total evaporating rate are revealed. The first behaviour is where $\overline{\dot{m}_{ev}}$ decreases by increasing the inlet mass flow and the difference between the seawater and pure water is negligible. This behaviour is identified for the lowest liquid film temperature ($T_{L0} = 20$ °C). In the second behaviour $\overline{\dot{m}_{ev}}$ increases with increasing the inlet mass flow ($T_{L0} = 60$ °C). While for the third behaviour, almost no variation of $\overline{\dot{m}_{ev}}$ occurs. In the last two cases, the seawater presents higher total evaporating rate comparing to pure water.

Figure 9. Total evaporating rate in function of inlet mass flow in different conditions

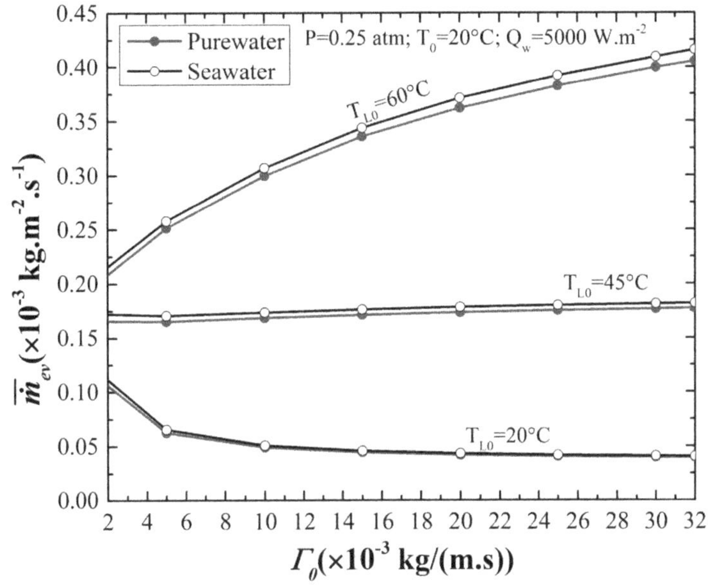

To establish the following results we have chosen the case where evaporation is important: $T_L = 60°C$. As plotted in figure 10, latent heat flux decreases from the inlet to a defined distance x_i (inversion distance) where it begins to increase. Moreover, when increasing the inlet mass flow of the liquid, the distance x_i increases, and the latent heat flux becomes smaller. The distance x_i increases from 0.5 to 1.5 m. The examination of the figure shows also that salinity enhances latent heat transfer, especially for high inlet mass flows. The reason behind those different trends is the thermodynamic state of the liquid film. In fact, for lower temperature (20°C), the heat must migrate through the liquid film to the interface. The heat migration is completely done by conduction for instance. Therefore, the liquid film requires a minimum distance to reach the saturation state, so that evaporation begins. In this case, the increase of the liquid film increases the minimum of this distance. In contrast, when the film enters at a warm state that is close enough to the saturation state, the evaporation begins immediately at the channel inlet. Consequently, the elevation of the liquid film flow rate increases the evaporation rate. In other words, when the liquid film temperature is high enough to ensure that evaporation begins already near the channel entrance, a higher evaporating liquid film flow rate produces higher evaporation rate.

The Figure 11 shows the variation of the sensible heat flux along the tube. It is clear that pure water presents slightly higher sensible heat fluxes comparing to seawater. The difference increases as the fluid goes to the outlet. This is due to the thermal conductivity that becomes smaller with salinity increasing (Figure 12). In fact, as the fluid flow downstream the tube, the amount of the evaporated water increases, so the ratio of salt in the liquid flow increases and in consequence, the salinity increases too. It is worth noting that latent heat flux is dominating this process of evaporation since it is much larger then sensible heat flux.

Attention is turned now to mass transfer by evaluating the evaporating flux presented in Figure 13. When increasing the salinity (seawater) or the inlet mass flux, the evaporation mass flux increases too. Notice that the salinity impact is only significant after a distance x almost equal to 0.75m in this study conditions. The evaporation mass flux increases continuously along the tube. Consequently, air stream humidification increases.

CONCLUSION

The behavior of seawater in the case of the evaporation in vertical tube has been investigated computationally. The effects of inlet mass flow and temperature of liquid film on heat and mass transfer are reported. The following conclusions are demonstrated:

1. The latent heat flux is bigger whereas the sensible heat flux is smaller in the case of seawater.
2. Liquid film evaporation is enhanced by the use of seawater.
3. The liquid film evaporation when increasing the inlet mass flow rate depends on its inlet temperature: for inlet liquid film temperature less than 45°C it decreases, otherwise evaporation increases.

The findings of this study are expected to benefit the design and the optimization of desalination units based on the evaporation of liquid films. The study revealed that the increase of the inlet mass flow rate does not necessarily conduct to the decrease of the evaporation. In this context, it is necessary to design the preheating temperature value. Thus, a higher temperature at the preheater outlet is recommended, to have an increasing evaporation with the increase of the liquid film flow rate at the evaporator (humidifier).

Figure 10. Axial evolution of latent heat flux for different inlet mass flow of liquid

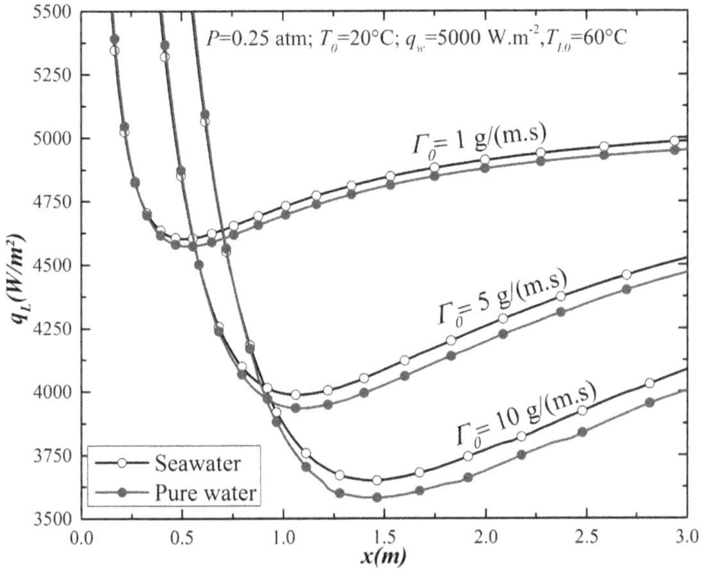

Figure 11. Axial evolution of sensible heat flux for different film flow rates

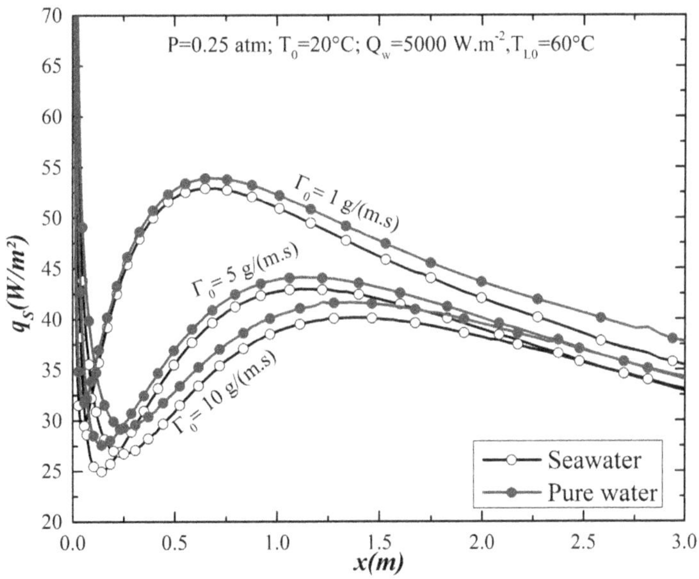

Figure 12. Variation of thermal conductivity of water and seawater with temperature

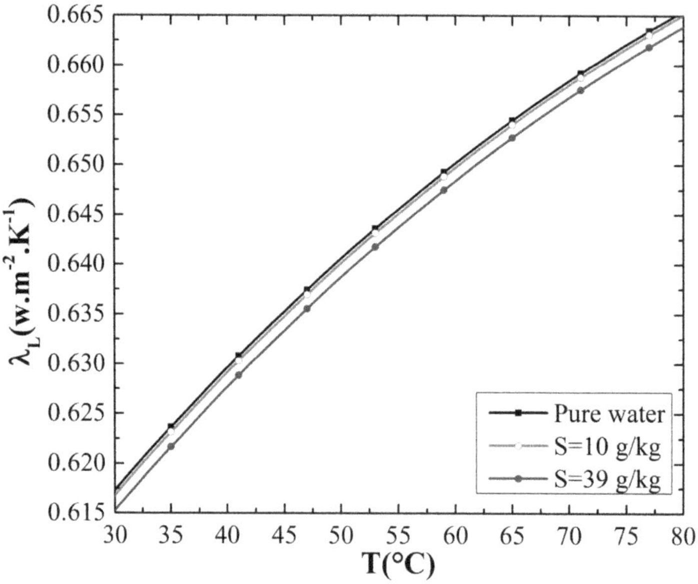

Figure 13. Local evaporating mass flux at the interface along the channel

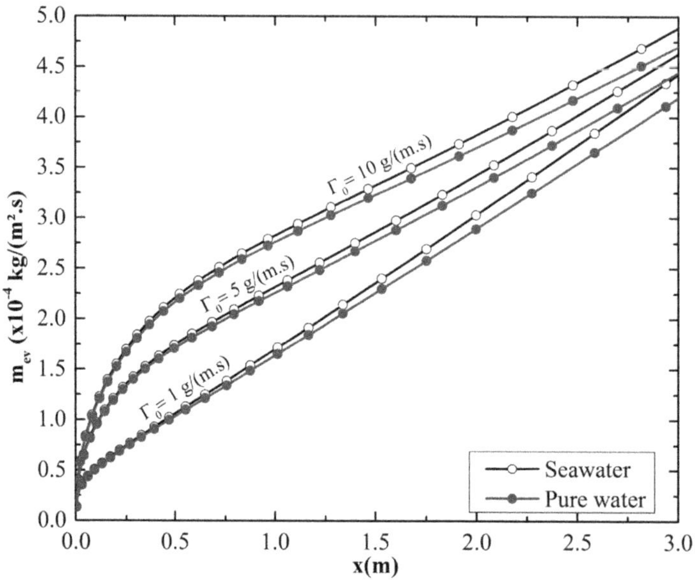

NOTE

The authors declare that there is no conflict of interests regarding the publication of this paper.

REFERENCES

Abraham, R., & Mani, A. (2015). Heat transfer characteristics in horizontal tube bundles for falling film evaporation in multi-effect desalination system. *Desalination*, *375*, 129–137. doi:10.1016/j.desal.2015.06.018

Al-Ansari, D., & Owen, I. (1999). Thermal and hydrodynamic analysis of the condensation and evaporation processes in horizontal tube desalination plant. *International Journal of Heat and Mass Transfer*, *42*(9), 1633–1644. doi:10.1016/S0017-9310(98)00240-3

Ben Radhia, R., Harmand, S., Corriou, J. P., & Ben Jabrallah, S. (2014). Liquid film evaporation enhancement in a vertical annulus with preheated air flow. *Desalination and Water Treatment*, *52*(7–9), 1654–1661. doi:10.1080/19443994.2013.806999

Cherif, A. A., & Daïf, A. (1999). Etude numérique du transfert dechaleur et de masse entre deux plaques planes verticalesen présence dun film de liquide binaireruisselant sur lune des plaques chauffée. *International Journal of Heat and Mass Transfer*, *42*(13), 2399–2418. doi:10.1016/S0017-9310(98)00339-1

Dung, S., & Yang, S. (2006). *Second law based optimization of free convection film-wise condensation on a horizontal tube*. Academic Press. doi:10.1016/j.icheatmasstransfer.2006.01.013

Feddaoui, M., Belahmidi, E. M., Mir, A., & Bendou, A. (2001). Numerical study of the evaporative cooling of liquid film in laminar mixed convection tube flows. *International Journal of Thermal Sciences*, *40*(11), 1011–1020. doi:10.1016/S1290-0729(01)01286-8

Feddaoui, M., Meftah, H., & Mir, A. (2006). The numerical computation of the evaporative cooling of falling water film in turbulent mixed convection inside a vertical tube. *International Communications in Heat and Mass Transfer*, *33*(7), 917–927. doi:10.1016/j.icheatmasstransfer.2006.04.004

He, S., An, P., Li, J., & Jackson, J. D. (1998). Combined heat and mass transfer in a uniformly heated vertical tube with water film cooling. *International Journal of Heat and Fluid Flow*, *19*(5), 401–417. doi:10.1016/S0142-727X(98)10021-8

Liu, Z.-H., & Yi, J. (2002). Falling film evaporation heat transfer of water/salt mixtures from roll-worked enhanced tubes and tube bundle. *Applied Thermal Engineering*, *22*(1), 83–95. doi:10.1016/S1359-4311(01)00061-8

Nait Alla, A., Feddaoui, M., & Meftah, H. (2015a). Numerical Simulation of Heat and Mass Transfer in the Evaporation of Glycols Liquid Film along an Insulated Vertical Channel. *American Journal of Heat and Mass Transfer*, *2*(2), 59–75. doi:10.7726/ajhmt.2015.1005

Nait Alla, A., Feddaoui, M., & Meftah, H. (2015b). Numerical study of the evaporation by mixed convection of ethanol in partially heated plate along a vertical channel. *International Journal of Heat and Mass Transfer*, *89*, 206–215. doi:10.1016/j.ijheatmasstransfer.2015.04.076

Najim, M., Feddaoui, M., Charef, A., & Meftah, H. (2017). Computational study of saline water film evaporation in a vertical tube. *Desalination*, *408*, 81–91. doi:10.1016/j.desal.2017.01.008

Najim, M., Feddaoui, M., Nait Alla, A., Charef, A., & Labsi, N. (2017). Gas flow humidification enhancement by evaporating seawater liquid film. *23ème Congrès Français de Mécanique*.

Narayanan, S., Fedorov, A. G., & Joshi, Y. K. (2013). Heat and mass transfer during evaporation of thin liquid films confined by nanoporous membranes subjected to air jet impingement. *International Journal of Heat and Mass Transfer*, *58*(1–2), 300–311. doi:10.1016/j.ijheatmasstransfer.2012.11.015

Orfi, J., Debbissi, C., Mohamed, A. B., & Nasrallah, S. B. (2004). Air humidification by free convection in a vertical channel. *Desalination*, *168*, 161–168. doi:10.1016/j.desal.2004.06.182

Patankar, S. V. (1980). *Numerical heat transfer and fluid flow*. New York: Hemisphere/McGraw Hill. doi:10.1201/9781482234213

Raach, H., & Mitrovic, J. (2005). Seawater falling film evaporation on vertical plates with turbulence wires. *Desalination*, *183*(1–3), 307–316. doi:10.1016/j.desal.2005.02.053

Raithby, G. D., & Schneider, G. E. (1979). Numerical solution of problems in incompressible fluid flow: Treatment of the velocity-pressure coupling. *Numerical Heat Transfer*, *2*(4), 417–440. doi:10.1080/10407787908913423

Ueda, T., & Tanaka, H. (1975). Measurements of velocity, temperature and velocity fluctuation distributions in falling liquid films. *International Journal of Multiphase Flow*, *2*(3), 261–272. doi:10.1016/0301-9322(75)90014-2

Yan, W. M., & Lin, T. F. (1991). Evaporative cooling of liquid film through interfacial heat and mass transfer in a vertical channel—II. Numerical study. *International Journal of Heat and Mass Transfer*, *34*(4–5), 1113–1124. doi:10.1016/0017-9310(91)90021-6

Yan, W. M., Lin, T. F., & Tsay, Y. L. (1991). Evaporative cooling of liquid film through interfacial heat and mass transfer in a vertical channel—I. Experimental study. *International Journal of Heat and Mass Transfer*, *34*(4–5), 1105–1111. doi:10.1016/0017-9310(91)90020-F

Ziqian, C., Hongfei, Z., Kaiyan, H., & Chaochen, M. (2007). Steady-state experimental studies on a multi-effect thermal regeneration solar desalination unit with horizontal tube falling film evaporation. *Desalination*, *207*(1–3), 59–70. doi:10.1016/j.desal.2006.08.003

ADDITIONAL READING

Al-Sahali, M., & Ettouney, H. M. (2008). Humidification dehumidification desalination process: Design and performance evaluation. *Chemical Engineering Journal*, *143*(1-3), 257–264. doi:10.1016/j.cej.2008.04.030

Belessiotis, V., Kalogirou, S., & Delyannis, E. (2016). *Thermal Solar Desalination: Methods and Systems*. Elsevier.

Charef, A., Feddaoui, M., Alla, A. N., & Najim, M. (2018). Computational Study of Liquid Film Condensation with the Presence of Non-Condensable Gas in a Vertical Tube. In M. Eyvaz & E. Yüksel (Eds.), *Desalination and Water Treatment*. Rijeka: IntechOpen. doi:10.5772/intechopen.76753

Charef, A., Feddaoui, M., Najim, M., & Meftah, H. (2017). Liquid film condensation from water vapour flowing downward along a vertical tube. *Desalination*, *409*, 21–31. doi:10.1016/j.desal.2017.01.018

Feddaoui, M., Belahmidi, E. M., Mir, A., & Bendou, A. (2001). Numerical study of the evaporative cooling of liquid film in laminar mixed convection tube flows. *International Journal of Thermal Sciences*, *40*(11), 1011–1020. doi:10.1016/S1290-0729(01)01286-8

Meftah, H., & Feddaoui, M. (2016). Numerical study of wall protection by the evaporation of liquid film of ethylene glycol. *Heat Transfer Research*, *47*(5), 433–447. doi:10.1615/HeatTransRes.2016007192

Najim, M., Feddaoui, M., Charef, A., & Meftah, H. (2017). Computational study of saline water film evaporation in a vertical tube. *Desalination*, *408*, 81–91. doi:10.1016/j.desal.2017.01.008

Najim, M., Feddaoui, M., Nait Alla, A., Charef, A., & Kabeel, A. E. (2018). New cooling approach using successive evaporation and condensation of a liquid film inside a vertical mini-channel. *International Journal of Heat and Mass Transfer*, *122*, 895–912. doi:10.1016/j.ijheatmasstransfer.2018.02.034

Reay, D., McGlen, R., & Kew, P. (2013). *Heat pipes: theory, design and applications*. Butterworth-Heinemann.

KEY TERMS AND DEFINITIONS

Boiling: Is a phase change phenomenon from liquid state to vapor state that occurs only at a well-defined temperature called boiling point.

Desalination: Is a process of salt removal from seawater or brackish water in order to produce suitable water for drinking or irrigation.

Evaporation: Is a progressive transformation from liquid state to vapor state that occurs from liquid surfaces without boiling.

Evaporative Cooling: A heat and mass transfer process for cooling purpose based on evaporation phenomenon. A liquid absorbs sensible heat from an airflow and dissipates this later by the latent heat transfer necessary to the phase change (evaporation).

Finite-Difference Method: A numerical method to solve partial differential equations by discretizing the derivatives into approximated finite differences.

Laminar Flow: A flow in which a viscous fluid particles move in parallel layers with a constant velocity at each layer.

Liquid Film: A very thin liquid layer flowing over a wall.

Chapter 10
Computer Simulation of the Aerodisperse Systems Hydrodynamics in Granulation and Drying Apparatus

Artem Artyukhov
Sumy State University, Ukraine

Nadiia Artyukhova
Sumy State University, Ukraine

Jan Krmela
Alexander Dubcek University of Trencin, Slovakia

ABSTRACT

The chapter presents validation of the selection criteria for the construction of the granulation and drying equipment. New forms of the organization of the flows motion in a fluidized bed regime are offered. The steering mechanisms for the motion and the residence time of the disperse phase in a workspace of the apparatus are described. The results of the simulation of the hydrodynamic conditions for the motion of the directional gas flow and the two-phase aerodisperse system are presented. Based on the simulation results, various constructions of the apparatus components are offered. The well-minded construction of the workspace of the apparatus helps to achieve optimized consumption of the heating agent and the minimal residence time of the disperse phase in the apparatus. The computer simulation allowed the authors to offer an algorithm of the optimization calculations of the granulation and drying equipment. New constructive solutions to perform the granulation and the drying processes in the devices with a directional fluidized bed are offered.

DOI: 10.4018/978-1-5225-7138-4.ch010

INTRODUCTION

The flows of the single-phase and multiphase systems play a key role in the operating process of many modern engineering devices. A design of these apparatuses according to given parameters requires a reliable forecast of the properties of these flows (Crowe, 2006; Khalatov, 2010). Since the production of many modern engineering devices is expensive and labor-intensive, physical simulation with experimental determination of the operating parameters in different regimes induces much time and financial expenses (Solanki, Basuri, Thakkar, & Patel, 2010). Therewith, limited capacities of the modern experimental detecting and measuring devices do not give a full picture of the phenomenon under study. The nature of these environments itself causes complicated flows with the formation of the nonstationary effects, stagnant zones and vortex structures. The methods of the Computational Fluid Dynamics, executed as the software products, proved to be effective for the design of new equipment, and improvement of the existing devices (Li & Li, 2016; Tahvildari, Tavakoli, Mashayekhi, & Golmohammadzadeh, 2016; Zimmermann & Taghipour, 2005). Although calculations of the flow studies can not replace experimental methods, the use of calculations helps to reduce the volume of the necessary experimental studies. Apart from that, after the calculations are tested, they can be used to evaluate the existing successful elements (or failures) of the flow channel and to explain the reasons for that (Liu, Gu, Ba, Li, Du, Yue, & Yang, 2012). Implementation of these methods in a form of the user-friendly computer programs allows us to enter the data in a few hours, to solve tasks on fluid flows calculations, to analyze and to select the most relevant configuration of the flow chamber in the installation under study (Chen, Zhai, & Wang, 2007; Trojan, 2015).

Calculations of the devices with a fluidized bed are performed with the use of the well-known techniques. However, these techniques are effective only for the classical fluidized bed, in which there is no controlled directional motion of the dispersed particles due to the installation of various types of guided (accelerating) elements and swirling inserts of a variety of configurations. The processes of the workspace sectioning by the shelf contacts in the dryer, selection of the construction for each shelf and their quantity must be based on the reliable mathematical model, the relevance of which is proved experimentally.

An additional area of interest is the study of the dispersed phase motion in the mode of the dense (constraint) flow, when it is necessary to take into account the mutual influence of the dispersed phase particles (the ensemble of particles) and the elements of the construction on the velocity of the particles and their residence time in the apparatus. Nowadays, scientists use the predictive models, which give an approximate result, to describe this process. It will be possible to determine the characteristic features of the particles motion in a fluidized bed with a high degree of accuracy by providing a directional motion of the dispersed phase in various configurations of the fluidized bed and the mathematical description of this process, developed within the framework of this work.

BACKGROUND

The devices with active hydrodynamic regimes are widely used in the chemical engineering to process solid disperse materials in the gas flow. Classification, separation, thermal treatment, drying and granulation can be the target processes in such devices (Dvornikov & Belousov, 2011; Kaewklum, Kuprianov, & Douglas, 2009; Volchkov, Nikolaev, Kustod, Nikolaev, & Tarovatyu, 2013; Voynov, Nikolaev, Kustod, Nikolaev, & Tarovatyu, 2008).

One of the most effective methods for performing heat and mass exchange in the direct contact of the phases is to apply the fluidized and semi-fluidized bed (Parikh, 2009). The operational time for the target process may be reduced through the formation of the active hydrodynamic regime, the increased relative velocity of the flow movement and a high specific surface area of the phases contact (Boyadjiev & Babak, 2000; Goldschmidt, 2001; Patel, Telange, & Sharma, 2011).

Despite numerous advantages, the apparatuses with the fluidized bed have a lot of disadvantages: differential residence time of the disperse phase and a low stability of the bed in a wide range of the load change in the liquid, solid and gas phases (Solanki, Basuri, Thakkar, & Patel, 2010; Saikh, 2013).

In these devices, intensification of the processing due to the turbulization of the flows is limited, since it is accompanied by an increase in the physical impact on the disperse phase. Herewith, an increased operating velocity of the gas phase leads not only to the intensified work of the fluidized bed, but to the higher energy costs of the system as well. The increase in the intensity of the fluidized bed operation is not proportional to the increase in the energy necessary to fluidize the disperse phase. It should be noted that the above-described target processes are followed by the problems of the accurate fractioning of the disperse phase, the differential regime for the processing of the separated fractions in the polydisperse system, and the control of the residence time of this system in a workspace of the apparatus.

Implementation of the new forms of organization of mutual flows motion (while keeping the same principles of the fluidized phases contact), which would intensify processing of the dispersed materials without considerable increase in the energy costs, is a promising direction for the development of heat and mass exchange processes in the heterogenous systems (Artyukhov, Obodiak, Boiko, & Rossi, 2015).

At present, the authors of this work (based on the hypothesis that it is possible to find a joint solution of the basic equations of the hydrodynamics of flow and the kinetics of change in the temperature-humidity characteristics of the interacting flows) are launching the development of the scientifically validated methods for calculating the granulation and drying units. These units will contain multifunctional devices with the implemented technologies for the multiuse of the drying agent potential, a possibility of harmful substances utilization and recovery of heat and energy. The results of the presented work solve the first part of the task - simulation of the hydrodynamic conditions for the directional motion of single-phase and two-phase flows.

The novelty of the approach lies in the application of the original methods for determining the hydrodynamic characteristics of the equipment, a complex evaluation of the results of theoretical descriptions and experimental studies, as well as the additional implementation of the computer simulation as a kind of a multifactorial experiment to determine an optimal construction of installations at the stage of their design.

The work will result in the development of one of the blocks of the scientific methodology for calculating granulation and drying installations with the use of the multistage apparatuses, based on the original results of the theoretical and experimental research. At the same time, the methodology will optimize the constructive and technological parameters of both the main process – granulation (drying) – and related processes (classification and separation of dispersed particles, waste gas utilization). The results will not contradict the well-known theories of hydrodynamics and heat mass exchange (in particular, a theory of the dispersed particles motion based on the Navier-Stokes and continuity equations, the k-ε turbulence model, the trajectory method for calculating the dispersed phase motion, the vortex dynamics, and the Lagrangian force analysis of the dispersed phase motion). Moreover, the results will have a specifying nature for certain classes of apparatuses with the multistage contact of phases, a variable cross-section area and a directional fluidized-bed motion.

MAIN FOCUS OF THE CHAPTER

To provide a directional motion of the fluidized bed (which is a precise way to ensure the necessary "hydrodynamic" residence time of the solid phase in the device) the following methods are used:

- The use of the gas-distributing devices with directional elements;
- The installation of various inserts, which change the direction of the solid phase motion by force, inside the apparatus;

- A local change of the cross-sectional area of the workspace;
- A simultaneous inlet of the fluidizing agent to the different parts of the device.

When selecting the method of how to control the motion of the fluidized bed in the granulation and drying apparatuses, the following factors should be considered:

- The necessity to ensure the operating range for the fluidized bed (the range of the sustainable operation of the apparatus is between the first and the second critical velocities of the gas flow);
- The possibility of dehydration and thermal treatment of the polydisperse solid phase;
- The constraint motion of the solid phase;
- The probability of the solid phase destruction due to intensive collision of the particles;
- Possible agglomeration of the damp particles.

Under these conditions, the use of the last method is unduly energy-consuming and inefficient. For this reason, to create the directional motion of a fluidized bed the first three methods will be used separately and in combination.

For instance, in most cases the polydisperse materials are subjected to hot treatment by the heat-transfer agent during the convection drying. Thus, the necessary average final moisture is reached throughout the system, but the overheating and the dehydration of the fines (and heat-labile materials in particular) lead to their destruction. When the porous ammonium nitrate is produced through bloating of the ordinary ammonium nitrate, thermal treatment of the polydisperse system of the granules results in the formation of pores of various configurations and in the loss of the special properties of granules (like absorption and retaining as related to the diesel distillate). Thus, creation of different time and space conditions for the dispersed phase motion within the device volume and the development of the steering mechanism for the dispersed phase motion (ensuring the necessary residence time) are the problems to be solved.

The presented constructive solutions are used in the author's constructions of the vortex granulators (Pat. No. 112393 UA, 2016; Pat. No. 112394 UA, 2016; Pat. No. 112622 UA, 2016; Pat. No. 113141 UA, 2017) and the multistage gravitational shelf dryers (Pat. No. 74070 UA, 2012; Pat. No. 81720 UA, 2013).

The vortex granulator (Figure 1):

- A type of the fluidized bed – vortex;
- A method of formation of various hydrodynamic conditions for the motion of the disperse phase in the apparatus – the height-variable cross-section of the

workspace (the body form is an opening or closing cone, or a combination of the opening and closing cones with a cylinder stabilizing area between them), gas-distributing devices with guide blades.

The multistage gravitational shelf dryer (Figure 2):

- A type of the fluidized bed – gravitational falling or cross-moving;
- A method of formation of various hydrodynamic conditions for the motion of the disperse phase in the device – sectioning of the workspace by the perforated shelf contacts with a varied angle of slope towards the horizon, a varied length and a varied level of perforation.

The vortex granulator operates as follows.

The heat-transfer agent is supplied through the pipe in the lower cylindrical part of the granulator. Then it is transferred to the vortex gas-distributing hub 4. The axisymmetric vortex flow of the heat-transfer agent is moving upside the inner cone 3 towards the material. Simultaneously, the liquid material (solution or melt) is supplied to the swirling flow of the heat-transfer agent through the pipe in the upper part of the external body 1 to the distribution hub 6. The melt stream moves through the distribution hub and becomes granulated in spherical particles. Depending on the size, granules are classified into fine and large fractions due to the change of the circular and axial velocity of the axisymmetric vortex flows of the heating agent within the height of the workspace. Getting into the vortex flow of the heat-transfer agent, the fines are transferred to the upper section of the inner cone and got removed through the annular space 3. The gravity force makes the fines move down the annular space to the lower section of the external body.

These granules reach the exhaust area in the lower part of the outer case. This area is formed around the gas stream, which is supplied through the pipe in the central part of the granulator (not shown in the figure). The granules are drawn by this stream and ejected to the core of the vortex fluidized bed in the central part of the workspace of the inner cone. When contacting the surface of the fines, the melt becomes crystalized. The size of the fines increases. The heavy fraction remains in the workspace, and while the size is growing, the fraction circulates in the inner cone and then moves down the section. When the target size is reached, the granules fall down the inner cone, move through the gas-distributing hub and get removed from the apparatus. The exhaust heat-transfer agent is removed from the outer case through the relevant pipe.

Advantages of the vortex granulators:

Figure 1. Suspended-layer vortex granulator with internal circulation of seeding agent: 1 – outer case; 2 – inner cone; 3 – interannular space; 4 – swirler; 5 – atomizer

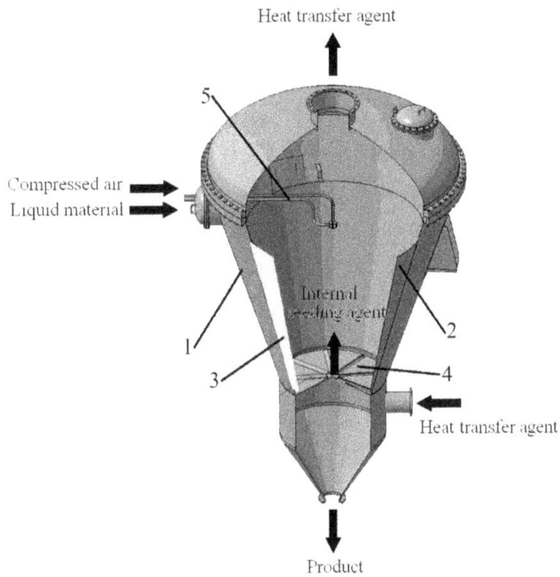

- A possibility to reduce significantly the dimensions (the height in particular) of the workspace;
- Increased residence time of the granule in the workspace of the vortex apparatus;
- A possibility to control the motion of the granule in the workspace of the granulator;
- Capacities for the intensive turbulence in the workspace of the granulator;
- Multipurposeness (granulation and drying are conducted within one apparatus);
- Constructability and ease-of-production;
- A possibility to adjust and to change the constructional and technological parameters if necessary.

The gravitational shelf dryer operates as follows.

The drying agent is supplied to the lower shell of the unit 1. It is uniformly distributed within the body section, and after that it is lifted. The drying agent goes through the sections of the lower, middle, upper and cone shells and then it is removed from the apparatus. The dispersed material is supplied to the upper shelf 2 for drying. Under the counterflow impact of the drying agent on the upper shelf, the dispersed material is heated, and the fines are removed. The length of the shelf ensures the residence time of the dispersed material, which is necessary for the full heating of the material, separation of the dispersed fines and initial removal of the unbound moisture from the dispersed material surface at the permanent velocity of the heating. While drying by the dying agent streams, the dispersed material gradually moves on the upper shelf to the middle one (further intense removal of the unbound moisture from the dispersed material surface at the permanent velocity of the heating). Then it moves to the lower shelf (the removal of the bound moisture from the material depth at the reducing velocity of the heating). After moving through all the shelves, the dispersed material is removed from the lower part of the dryer.

Advantages of the multistage shelf dryer:

- Formation of the counterflow mode by the interacting streams, which is the most effective and with the largest value of the average moving force (an average difference of temperatures and humidity concentration);
- A possibility to conduct drying and classification of the material in the same apparatus simultaneously;
- The reduction of the apparatus dimensions by means of making separated contact stages for flows in the one operating space;
- A possibility to control the residence time of the disperse phase in the workspace of the apparatus due to the change of technological characteristics of drying and constructional parameters of the dryer;
- Reuse of the drying agent, which enables us to reduce the energy costs and to enhance the environment safety of thc production.

The shelf contact elements can be shifted within the unit and demounted for a changeover. It is easy to change their location, constructional parameters and the quantity. They have no special load-transfer components, and no blocking tendency. The impact on the intensity of the phases contact and the formation of various hydrodynamic regimes of the fluidized bed (on the shelves and within the height of the apparatus) are achieved by means of the change of the cross-section area of the shelf (a change of the quality of holes and their diameter), an angle of its slope, a distance from the edge of the shelf to the wall of the unit (the width of the discharging space), and a quantity of shelves.

The variation of the length, an angle of the installation, and a level of the perforation of the upper shelf (a shelf is conventional here, and the number of shelves depends on the operational conditions) ensure the minimal required contact time for the disperse material with the given physicochemical properties, functional components and parameters of the drying agent. Here, if the shelf (or the upper shelf in a cascade of shelves) is a separator, the fine fractions are removed. The medium contact shelf ensures an increase in the residence time of the disperse material and its contact

Figure 2. Multistage shelf dryer: 1 – case; 2 – shelf

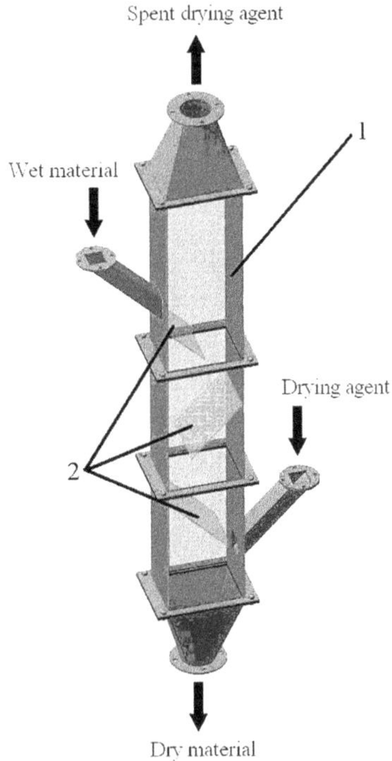

with the drying agent, which intensifies the dehydration. The contact of the disperse material with the drying agent on the lower shelf provides a long residence time of the particles, which ensures the dehydration in the depth of the material.

Theoretical Basics

The calculation starts with the given initial approximation. The fields of velocities and pressure, obtained upon the expiry of the time integration step, are calculated at each global iteration. The stationary solution, if it exists, is obtained upon expiration of many global iterations corresponding to a long period of time.

During the calculations, the programme automatically divides the calculation area into the sub-areas, the size of which is proportional to the relevant processing speed.

The algorithm of the calculations supposes that each processor calculates the flows in its sub-area only, and the obtained result is summed up after each global iteration. Used in the modern software, the algorithms of parallel computing allow users to achieve the speed-up with mostly 100% performance capability. Put it otherwise, the calculation time is inversely proportional to the total computing capacity. As a rule, the boundary conditions are the condition of sticking on all the solid walls (the zero velocity), distribution of all the velocity components in the input section, and the zero first derivatives (in the downstream direction) of the velocity components in the output section. In practice, if the velocity of the flow in the inlet is uniformly distributed across the section, a user specifies the average velocity (or the consumption) at the input, and no values for the output. Herewith, a user is supposed to select the output section far enough from the areas of the intensive flow reorganization.

The relevance of the developed calculation methods to the flows under study is connected with the mathematical model, based on the general hydrodynamic equations, the common approach to the numerical solution of these equations, and the systematized comparison of the calculation results with the known experimental results. The use of common research methods and a fair accuracy of the values validate the obtained experimental results.

Analytical and numerical calculations are used to describe the hydrodynamic characteristics of the flows motion in the workspace of the granulation and drying equipment.

The analytical description of the vortex flow is based on one of the following approaches (Artyukhov & Sklabinskyi, 2015):

1. The flow is presented as a plain stream imposed by the potential rotation. The calculations, based on the empirical coefficients, consider the determination of the hydraulic resistance only.
2. Application of Bernoulli equation, a point of extremum in the function of one of the components of the vortex flow velocity under the conservation of the angular momentum along the radius of the processing chamber. The disadvantage of the approach is a rough mapping of the flows in the near-axial area of the processing chamber.
3. Application of Bernoulli equation to describe the liquid motion in the chamber of a spiral type. This method requires preliminary determination of the characteristics that depends on the geometry of the processing chamber. It does not allow a user to determine all the components of the total velocity.

The abovementioned disadvantages show that the field, where these methods are applied to describe and to calculate the directional flows, is considerably narrow.

There is also a method to describe the axisymmetric motion of the swirling gas flow, based on a system of Navier-Stokes equations (1) and the Continuity Equation (2):

$$\frac{\partial}{\partial t}\left(\rho_g V_i\right) + \frac{\partial}{\partial q_j}\left(\rho_g V_i V_j\right) = -\frac{\partial p}{\partial q_i} + \frac{\partial}{\partial q_j}\left[\varepsilon\left(\frac{\partial V_i}{\partial q_j} + \frac{\partial V_j}{\partial q_i}\right)\right] + f_i, \tag{1}$$

$$\frac{\partial \rho_g}{\partial t} + \frac{\partial}{\partial q_j}\left(\rho V_j\right) = 0 \tag{2}$$

where ε is a coefficient of turbulent viscosity; t – time; V_i – instantaneous component of the velocity on the given axis; $\acute{A}_g$ – gas density; p – pressure; f_i – component, characterizing the action of the bulk forces; q_j – axes of coordinates (it is better to use the curvilinear coordinates to simulate the hydrodynamics in the workspace of a vortex granulator and a gravitational shelf dryer), i, j – 1…3; for a cylindrical system of coordinates (Figure 3) index «1» is an axial direction (z), index «2» is a radial direction (r), index «3» is a circumferential direction (φ).

This method allows a user to determine the hydrodynamic characteristics of the flows motion in more details. However, it is most applicable for the laminar flows.

A system of Navier-Stokes equations and the Continuity Equations are mostly nonlinear and there is no analytical solution without simplifications and suppositions.

The direct simulation of the turbulent flows through a quantitative solution of Navier-Stokes equations, written for the instantaneous velocities, is a task to be solved. However, for the solution of individual tasks it becomes possible to apply this fundamental equation of hydrodynamics under certain conditions:

Figure 3. A scheme of the workspace of the apparatus (an example of a vortex granulator) and a system of coordinates

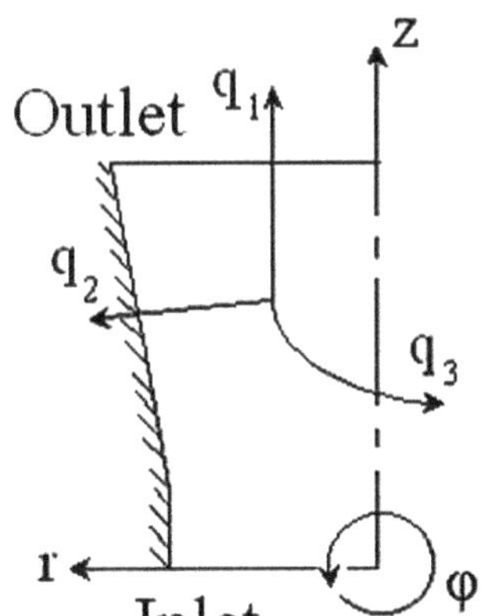

- To work with the time-averaged values of velocities when the turbulent flows are simulated;
- To select a system of coordinates, which meets the demands of a certain case and simplifies recording of the equations.

The numerical methods (DEM - Discrete Element Method in particular) should be used to solve the mentioned equations during the simulation (Artyukhov, 2016). Herewith, instead of a continuous solution, a discrete set of target values is determined in the particular place (a cell or a grid node) of the area (given the stationary movement of flows). To achieve the maximum precision of the solution, one should select a way of the presentation of the discrete values, which corresponds to the analogues of the algebraic equations for such a discrete sampling. Thus, a mathematical solution of the system of the differential and integral equations resolves into a solution of the system of algebraic equations.

In practice, we use various models of the numerical solution of the classical equation of hydrodynamics for turbulent flows. These models are used in certain cases, and they have their advantages and disadvantages (Polascik, Cottet, & Huberson, 2013): Direct numerical simulation (DNS), Reynolds Averaged Navier-Stokes (RANS), Large Eddy Simulation (LES).

Considering that not instantaneous but time-averaged velocities are of interest in most practical cases, the mathematical description of the turbulent vortex gas flow uses Reynolds equations as a modification of Navier-Stokes equations (1):

$$\frac{\partial}{\partial t}\left(\rho_g \overline{V_i}\right)+\frac{\partial}{\partial q_j}\left(\rho_g \overline{V_i V_j}\right)+\frac{\partial}{\partial q_j}\left(\rho_g \overline{V_i' V_j}\right)=-\frac{\partial p}{\partial q_i}+\frac{\partial}{\partial q_j}\left[\varepsilon\left(\frac{\partial \overline{V_i}}{\partial q_j}+\frac{\partial \overline{V_j}}{\partial q_i}\right)\right]+f_i \tag{3}$$

where $\overline{V}$ is a time-averaged value of velocities; $\overline{V'}$ – components of the velocity pulsation.

The main advantage of the method of describing and solving hydrodynamics tasks, based on the numerical solution of Reynolds equations, is its accuracy and multipurposeness.

The solution of Reynolds equations (3) requires a Boussinesq theory (Saqr, Aly, Wahid, & Sies, 2009). According to this theory, components with the velocity pulsations $\left(\rho \overline{V_i V_j'}\right)$ in the Equation (3) are connected with the averaged properties of the flow as shown in the formula:

$$\rho_g \overline{V_i' V_j'} = -\varepsilon \left(\frac{\partial \overline{V_i}}{\partial q_j} + \frac{\partial \overline{V_j}}{\partial q_i} \right) + \frac{2}{3} \rho_g \delta_{ij} k \tag{4}$$

where $k = 0,5\left(\overline{V_j' V_j'}\right)$ is a kinematic energy of turbulence, $\delta ij = 1$ if $i = j$, and $\delta i = 0$ if $i \neq$ j.

A system of Reynolds equations is elliptical. It is applied to calculate the flows in cases when the properties of the flow in a random point depend on the structure of the flow, both upstream and downstream, i.e. when the dominating direction of the flow is absent or weak. The elliptical system of equation means that its solution requires specification of the boundary conditions for all the variables and on all the boundaries of the calculation area.

The calculation is performed through determination: during the first approximation, a user specifies the values by the random but most probable distribution of values in all the calculation nodes; then the equations of the mathematical model are numerically solved in all the calculation area in order to receive the next, more precise approximation. This integration process lasts until the sustainable solution is obtained, that is, as long as the values of the required variables from iteration to iteration do not change by more than a certain fixed value.

When simulating the axisymmetric flows, the equations of motion (3) and continuity of flows (2) are considerably simplified. The curvilinear (cylindrical) system of coordinates shows the equations as follows (if the Reynolds number $Re = V_0 D/\nu$ is in the equation, where the characteristic parameters D is a hydraulic diameter of the input section of the calculation area; V_0 is an average supply velocity in the input section; ν is a kinematic viscosity):

- The Reynolds equation projected on the axial direction q_1:

$$\frac{V_1}{H_1}\frac{\partial V_1}{\partial q_1} + \frac{V_2}{H_2}\frac{\partial V_1}{\partial q_2} - \frac{V_2^2}{H_1 H_2}\frac{\partial H_2}{\partial q_1} - \frac{V_3^2}{H_1 H_3}\frac{\partial H_3}{\partial q_1} + \frac{1}{H_2}\frac{\partial\left(\overline{V_1' V_2'}\right)}{\partial q_2} + \frac{\left(\overline{V_1' V_2'}\right)}{H_1^2 H_2 H_3}\frac{\partial\left(H_1^2 H_3\right)}{\partial q_2} +$$

$$+\frac{1}{H_1}\frac{\partial\left(\overline{V_1' V_1'}\right)}{\partial q_1} + \frac{\left(\overline{V_1' V_1'}\right)}{H_1^2 H_2 H_3}\frac{\partial\left(H_1 H_2 H_3\right)}{\partial q_1} - \frac{\left(\overline{V_2' V_2'}\right)}{H_1 H_2}\frac{\partial H_2}{\partial q_1} - \frac{\left(\overline{V_3' V_3'}\right)}{H_1 H_3}\frac{\partial H_3}{\partial q_1} = -\frac{1}{H_1}\frac{\partial p}{\partial q_1} +$$

$$+\frac{1}{\mathrm{Re}}\left(\frac{1}{H_1^2}\frac{\partial^2 V_1}{\partial q_1^2}+\frac{1}{H_2^2}\frac{\partial^2 V_1}{\partial q_2^2}+\frac{1}{H_1H_2H_3}\frac{\partial V_1}{\partial q_1}\frac{\partial(H_2H_3/H_1)}{\partial q_1}+\frac{1}{H_1H_2H_3}\frac{\partial V_1}{\partial q_2}\frac{\partial(H_1H_3/H_2)}{\partial q_2}+\right.$$

$$-\frac{2}{H_1H_2^2}\frac{\partial H_2}{\partial q_1}\frac{\partial V_2}{\partial q_2}+\frac{V_1}{H_1}\frac{\partial}{\partial q_1}\left(\frac{1}{H_1H_2H_3}\frac{\partial\left(H_2H_3\right)}{\partial q_1}\right)+$$

$$\left.+\frac{V_2}{H_1}\frac{\partial}{\partial q_1}\left(\frac{1}{H_1H_2H_3}\frac{\partial\left(H_1H_3\right)}{\partial q_2}\right)-\frac{V_2}{H_2H_3}\frac{\partial}{\partial q_2}\left(\frac{H_3}{H_1H_2}\frac{\partial H_2}{\partial q_1}\right)\right) \tag{5}$$

- The Reynolds equation projected on the radial direction q_2:

$$\frac{V_1}{H_1}\frac{\partial V_2}{\partial q_1}+\frac{V_2}{H_2}\frac{\partial V_2}{\partial q_2}-\frac{V_1V_2}{H_1H_2}\frac{\partial H_2}{\partial q_1}-\frac{V_3^2}{H_2H_3}\frac{\partial H_3}{\partial q_2}+\frac{1}{H_1}\frac{\partial\left(\overline{V_1'V_2'}\right)}{\partial q_1}+\frac{\left(\overline{V_1'V_2'}\right)}{H_1H_2^2H_3}\frac{\partial\left(H_2^2H_3\right)}{\partial q_1}+$$

$$+\frac{1}{H_2}\frac{\partial\left(\overline{V_2'V_2'}\right)}{\partial q_2}+\frac{\left(\overline{V_2'V_2'}\right)}{H_1H_2^2H_3}\frac{\partial\left(H_1H_2H_3\right)}{\partial q_2}-\frac{\left(\overline{V_3'V_3'}\right)}{H_2H_3}\frac{\partial H_3}{\partial q_2}=-\frac{1}{H_2}\frac{\partial p}{\partial q_2}+$$

$$+\frac{1}{\mathrm{Re}}\left(\frac{1}{H_1^2}\frac{\partial^2 V_2}{\partial q_1^2}+\frac{1}{H_2^2}\frac{\partial^2 V_2}{\partial q_2^2}+\frac{1}{H_1H_2H_3}\frac{\partial V_2}{\partial q_1}\frac{\partial(H_2H_3/H_1)}{\partial q_1}+\frac{1}{H_1H_2H_3}\frac{\partial V_2}{\partial q_2}\frac{\partial(H_1H_3/H_2)}{\partial q_2}+\right.$$

$$\left.+\frac{2}{H_1H_2^2}\frac{\partial H_2}{\partial q_1}\frac{\partial V_1}{\partial q_2}+\frac{V_2}{H_2}\frac{\partial}{\partial q_2}\left(\frac{1}{H_1H_2H_3}\frac{\partial\left(H_1H_3\right)}{\partial q_2}\right)\right) \tag{6}$$

- The Reynolds equation projected on the circumferential direction q_3:

$$\frac{V_1}{H_1}\frac{\partial V_3}{\partial q_1}+\frac{V_2}{H_2}\frac{\partial V_3}{\partial q_2}+\frac{V_1V_3}{H_1H_3}\frac{\partial H_3}{\partial q_1}+\frac{V_2V_3}{H_2H_3}\frac{\partial H_3}{\partial q_2}+\frac{1}{H_2}\frac{\partial\left(\overline{V_2'V_3'}\right)}{\partial q_2}+\frac{\left(\overline{V_2'V_3'}\right)}{H_1H_2H_3^2}\frac{\partial\left(H_1H_3^2\right)}{\partial q_2}+$$

$$+\frac{1}{H_1}\frac{\partial\left(\overline{V_1'V_3'}\right)}{\partial q_1}+\frac{\left(\overline{V_1'V_3'}\right)}{H_1H_2H_3^2}\frac{\partial\left(H_2H_3^2\right)}{\partial q_1}=\frac{1}{\mathrm{Re}}\left(\frac{1}{H_1^2}\frac{\partial^2V_3}{\partial q_1^2}+\frac{1}{H_2^2}\frac{\partial^2V_3}{\partial q_2^2}+\right.$$

$$\left.+\frac{1}{H_1H_2H_3}\frac{\partial V_3}{\partial q_1}\frac{\partial(H_2H_3/H_1)}{\partial q_1}+\frac{1}{H_1H_2H_3}\frac{\partial V_3}{\partial q_2}\frac{\partial(H_1H_3/H_2)}{\partial q_2}+\frac{V_3}{H_1H_2}\frac{\partial}{\partial q_2}\left(\frac{H_1}{H_2H_3}\frac{\partial H_3}{\partial q_2}\right)\right) \tag{7}$$

- The Continuity Equation:

$$\frac{1}{H_1H_2H_3}\left(V_1\frac{\partial(H_2H_3)}{\partial q_1}+V_2\frac{\partial(H_3H_1)}{\partial q_2}\right)+\frac{1}{H_1}\frac{\partial V_1}{\partial q_1}+\frac{1}{H_2}\frac{\partial V_2}{\partial q_2}=0 \tag{8}$$

where H_1, H_2, H_3 are the Lamé coefficients.

When simulating the vortex gas phase in the workspace of a vortex granulator with a fluidized bed, it is possible to simplify the system of equations (5-8) by assuming that:

- There is a dominant direction of the flow, along which the axial component of the gas flow velocity is positive, and it exceeds the radial component;
- The component of the gas flow velocity changes in the axial direction much slowlier then in the radial direction;
- The values of the velocities and the pressure in each elementary volume of the gas flow depend on the downstream conditions exclusively, and there is no dependence on the upstream conditions.

These assumptions allow us to analyze the orders of the components in the Equations (5) – (8) and neglect those having no influence on the calculation results.

Considering the assumptions for the axisymmetric gas flow, the Equations (5) – (8) will be:

$$\frac{V_1}{H_1}\frac{\partial V_1}{\partial q_1}+\frac{V_2}{H_2}\frac{\partial V_1}{\partial q_2}-\frac{V_3^2}{H_1H_3}\frac{\partial H_3}{\partial q_1}+\frac{1}{H_2}\frac{\partial\left(\overline{V_1'V_2'}\right)}{\partial q_2}+\frac{\left(\overline{V_1'V_2'}\right)}{H_1^2H_2H_3}\frac{\partial\left(H_1^2H_3\right)}{\partial q_2}=$$

$$= -\frac{1}{H_1}\frac{\partial p}{\partial q_1} + \frac{1}{\mathrm{Re}}\left(\frac{1}{H_2^2}\frac{\partial^2 V_1}{\partial q_2^2} + \frac{1}{H_1 H_2 H_3}\frac{\partial V_1}{\partial q_2}\frac{\partial (H_1 H_3 / H_2)}{\partial q_2}\right) \tag{9}$$

$$\frac{\partial p_r}{\partial q_2} = \frac{V_3^2}{H_3}\frac{\partial H_3}{\partial q_2} \tag{10}$$

$$\frac{V_1}{H_1}\frac{\partial V_3}{\partial q_1} + \frac{V_2}{H_2}\frac{\partial V_3}{\partial q_2} + \frac{V_1 V_3}{H_1 H_3}\frac{\partial H_3}{\partial q_1} + \frac{V_2 V_3}{H_2 H_3}\frac{\partial H_3}{\partial q_2} + \frac{1}{H_2}\frac{\partial \overline{(V_2' V_3')}}{\partial q_2} + \frac{\overline{(V_2' V_3')}}{H_1 H_2 H_3^2}\frac{\partial (H_1 H_3^2)}{\partial q_2} =$$

$$= \frac{1}{\mathrm{Re}}\left(\frac{1}{H_2^2}\frac{\partial^2 V_3}{\partial q_2^2} + \frac{1}{H_1 H_2 H_3}\frac{\partial V_3}{\partial q_2}\frac{\partial (H_1 H_3 / H_2)}{\partial q_2} + \frac{V_3}{H_1 H_2}\frac{\partial}{\partial q_2}\left(\frac{H_1}{H_2 H_3}\frac{\partial H_3}{\partial q_2}\right)\right) \tag{11}$$

$$\frac{1}{H_1 H_2 H_3}\left(V_1\frac{\partial (H_2 H_3)}{\partial q_1} + V_2\frac{\partial (H_3 H_1)}{\partial q_2}\right) + \frac{1}{H_1}\frac{\partial V_1}{\partial q_1} + \frac{1}{H_2}\frac{\partial V_2}{\partial q_2} = 0 \tag{12}$$

This system of equations is rounded up with the flow continuity equation:

$$\int_0^{Q_2} V_1 \quad H_2 H_3 dq_2 = const \tag{13}$$

where Q_2 is a q_2 coordinate on the wall of the workspace of the apparatus.

The obtained system of equations (9) – (13) is paraboloidal, and its solution is based on the Patankar-Spalding method and processed with SIMPLE (Semi-Implicit Method for Pressure-Linked Equations) and its modifications (Patankar and Spalding, 1972).

The numerical solution of the equations of the mathematical model is performed in one march passage from the input section to the output section of the working volume. The finite volumes method with components of the finite-element approach is used in the solution. Calculations are preceded by the building of the calculation grid, where unknown values of the velocities are found in the grid nodes.

The main requirement to the calculation grid is that it should be sufficiently thick to allow physical effects that occur inside the calculation area. To achieve the uniformly accurate calculation, the grid nodes must be concentrated in the places of quick change of the parameters of flows, particularly on the walls. Herewith, the calculation grid should not contain unduly drawn or deformed cells with a shape different from a regular polyhedron. Such cells make obtaining of the converging result complicated.

The study of the marching algorithm sustainability was executed with the commonly used Neyman analysis, the pattern analysis, and a numerical experiment as well.

To create a structural grid, a user divides a calculation area into blocks according to the partition topology, specified by a user. Each block has a calculation grid with the nodes defined by the numbers in the 3D array. The application of such a grid allows a user to organize the most efficient calculation algorithms.

The motion of granules in the workspace was simulated on the basis of the system of differential equations of the particle motion (Artyukhov, Obodiak, Boiko, & Rossi, 2017):

Figure 4. Creation of the calculation grid (the case of a vortex granulator)

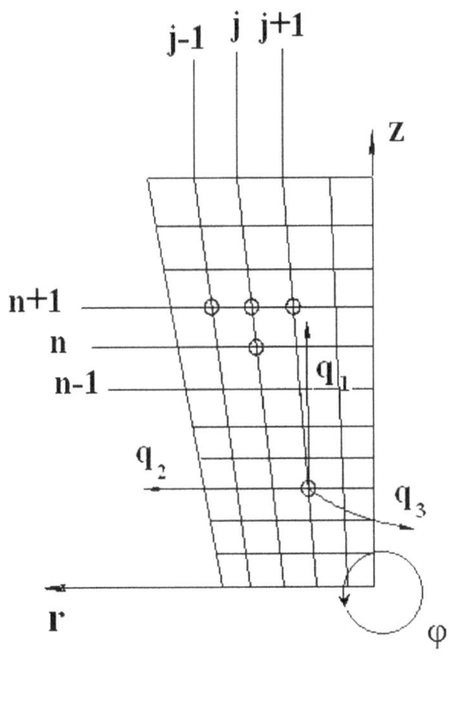

$$\left.\begin{aligned} m\frac{d^2q_1}{dt^2} &= -g + \psi\cdot\frac{\pi\cdot\mu_g\cdot d}{8\cdot m}\left(V_1 - W_1\right), \\ m\frac{d^2q_2}{dt^2} &= \frac{W_3^2}{r} + \psi\cdot\frac{\pi\cdot\mu_g\cdot d}{8\cdot m}\left(V_2 - W_2\right), \\ m\frac{d^2q_3}{dt^2} &= -\frac{W_2W_3}{r} + \psi\cdot\frac{\pi\cdot\mu_g\cdot d}{8\cdot m}\left(V_3 - W_3\right) \end{aligned}\right\}, \tag{14}$$

where m – granule's mass; r – a current radius of the workspace of a unit; W_1, W_2, W_3 – radial, circumferential and flow rate components of the granule's velocity correspondingly; g – gravitational acceleration; ψ – a linear coefficient of resistance of a granule to the gas flow; μ_g – gas flow viscosity; d – a granule's diameter.

The continuous phase flow, simulated by a system of the Navier-Stokes equations (1) and the Continuity Equation (2), gives a part of angular momentum to the discrete particles. If a liquid phase appears in a workspace of the apparatus, the vortex motion absorbs it by the gas flow energy. Upon that, an entry of the disperse phase into the continuous phase considerably changes the value of the circumferential component of the gas flow velocity V_3.

The angular momentum of the gas phase in the absence of the disperse phase in the operating volume is

$$M_g = 2\cdot\pi\cdot\rho_g\cdot V_3\cdot r^2 drdz. \tag{15}$$

The angular momentum of the gas phase after interaction with the disperse phase in the operating volume is

$$M_g' = 2\cdot\pi\cdot\rho_g\cdot V_3'\cdot r^2 drdz, \tag{16}$$

where V_3' is a circumferential velocity of the gas flow after interaction with the disperse phase.

The angular momentum of the disperse phase, obtained after interaction with the gas flow, is

$$M_d = m_d \cdot W_3 \cdot r \tag{17}$$

Due to the disperse phase and gas flows ratio Q_d/Q_g, an Equation (17) is:

$$M_d = 2 \cdot \pi \cdot \rho_d \cdot \left(\frac{Q_d}{Q_g} \right) W_3 \cdot r^2 drdz. \tag{18}$$

According to (Artyukhov, 2016), an equation is:

$$M_g = M_g^{'} + M_d, \tag{19}$$

or

$$\rho_g \cdot V_3 = \rho_g \cdot V_3^{'} + \rho_d \cdot \left(\frac{Q_d}{Q_g} \right) W_3. \tag{20}$$

In this case the circumferential velocity of the gas flow after interaction with the disperse phase is:

$$V_3^{'} = V_3 - \frac{\rho_d}{\rho_g} \cdot \left(\frac{Q_d}{Q_g} \right) W_3. \tag{21}$$

Considering that a granule takes the amount of time τ to pass the distance in a radial S_r, circumferential S_φ and vertical S_z directions, the system of Equations (14) is:

$$\left.\begin{aligned} m\frac{d^2 S_z}{d\tau^2} &= -g + \psi \cdot \frac{\pi \cdot \mu_g \cdot d}{8 \cdot m}\left(V_1 - W_1\right), \\ m\frac{d^2 S_r}{d\tau^2} &= \frac{W_2^2}{r} + \psi \cdot \frac{\pi \cdot \mu_g \cdot d}{8 \cdot m}\left(V_2 - W_2\right), \\ m\frac{d^2 S_\varphi}{d\tau^2} &= -\frac{W_2 W_3}{r} + \psi \cdot \frac{\pi \cdot \mu_g \cdot d}{8 \cdot m}\left(V_3 - W_3\right), \end{aligned}\right\} \tag{22}$$

The system of Equation (22) with respect to the variable S in each direction is solved for a certain amount of residence time of a particle in the workspace of the unit. The residence time of the disperse phase inside the apparatus is determined by the results of calculating the kinetics of the particle heating and dehydration.

The optimization criterion in this case is to provide the minimum required residence time of the granule in the working space of the granulator, which will allow the formation of a complete crystalline structure of the granule with the definite durability and mono-dispersion of the product fraction.

This is highly important for the designing of vortex granulators that produce granules with special properties, including porous ammonium nitrate. Apart from the above-mentioned quality indicators of the final product, porous ammonium nitrate granules must meet specific targets, like capacities of diesel oil constraining and absorption. In this case, the "hydrodynamic" residence time of the particle in the device should not be less than the "thermodynamic" time (this parameter is determined by the dehydration kinetics). Herewith, the particle's core remains intact unless the "hydrodynamic" time exceeds the "thermodynamic" time by more than 5-10%. The regulation of the hydrodynamic properties of flows is specifically intended to receive an optimal construction of a vortex granulator, which meets the optimization criteria.

A numerical solution of the differential equations is applied in the Equation (22).

To accomplish this, a cone body of the granulator was divided into several cylindrical areas as shown in Figure 5.

The calculation is done for the cylindrical areas, and the calculation results on one area are the input data for the calculations on further areas of motion.

Figure 5. A calculation scheme to determine the trajectory of the particle's motion

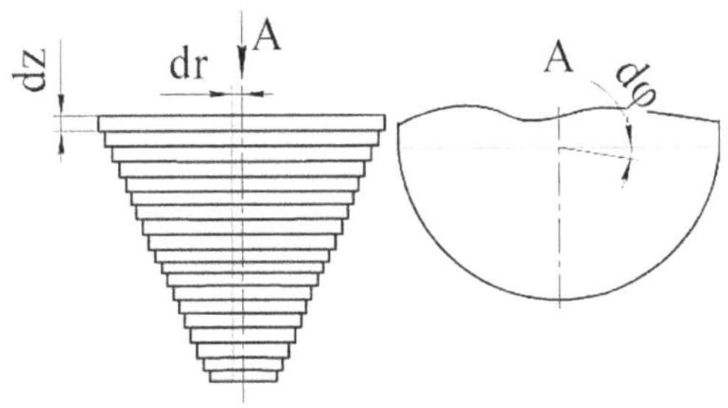

RESULTS, DISCUSSION, SOLUTIONS AND RECOMMENDATIONS

Vortex Granulator

The workspace of the vortex granulator should contain three separated zones for the gas flow and granules motion within the height and three separated zones within the radius of the unit (fig.6). The rate of the gas flow and granules motion in each zone is determined by the velocity components and the dominant direction of the total velocity.

Figure 6. The main zones of granules motion in the vortex granulator: a – within the height of the workspace; b – within the radius of the workspace; I – a zone of the dominant vortex motion of granules; II – a zone of the combined vortex and the upward motion of granules; III – a zone of the dominant upward motion of granules; IV – an operation zone of the upward gas flow of the fluidizing agent; V – a zone of formation of the vortex fluidized bed; VI – a zone of the reduced rate (retardation) of the vortex fluidized bed motion

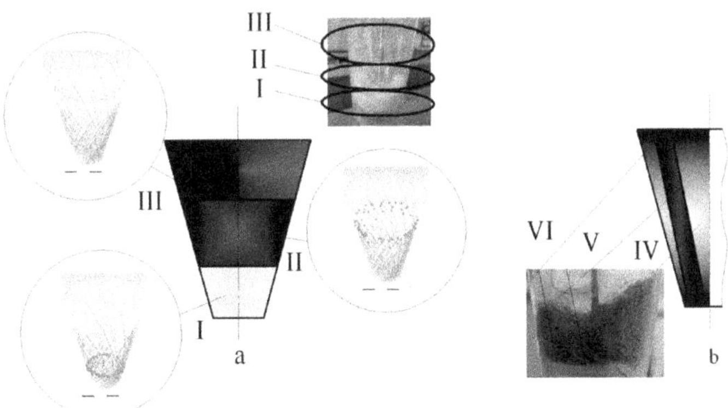

The free motion of a granule in the workspace of the granulator is determined by a differential equation of the granule's motion in the vertical direction. If the granule's motion is constraint, the constraint ratio and the residence time of the granule is determined for the zones I, II and III separately. The heights of the zones I, II and III and the widths of IV, V, VI are determined by the initial conditions of swirling of the gas flow and the volume content of the granules in the workspace of the vortex granulator.

The examination of the granules motion shows that an intensive initial swirling of the gas flow in the zone I results in the motion of granules in a form of the dense ring at the distance S from a wall of the granulator (Figure 7).

This phenomenon can be explained by a big difference of pressures in the center and at the periphery; this difference of pressures opposes the centrifugal force. This phenomenon may be further used to prevent a contact of granules with a wall of the apparatus and their distortion due to rubbing.

The peculiar feature of the granulators under study is a height-variable cross-section area of the workspace. Granulators with a permanent cross-section area do not ensure complete classification of granules and separation of the off-spec fractions in the workspace. This is explained by the fact that the workspace of the vortex granulator sustains the upward velocity of the gas flow, which corresponds to the operating velocity of the granule's motion (or a fraction of granules in a narrow

Figure 7. Granules' motion in the dense ring at the distance S from a wall of the granulator

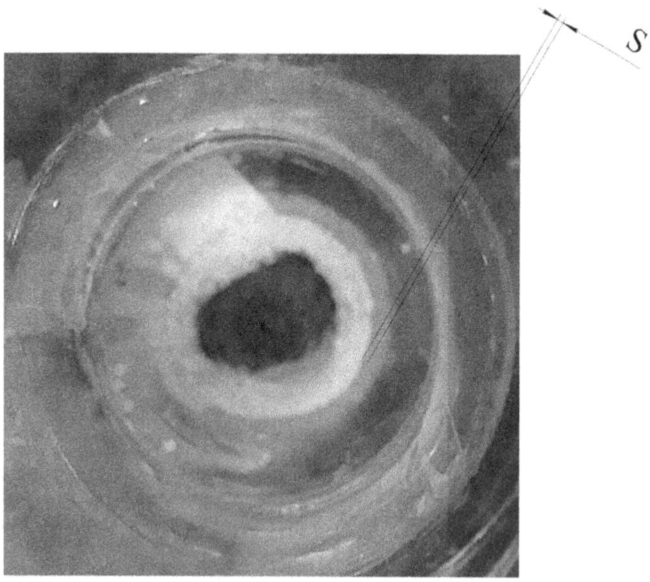

range). Classification of granules in the apparatuses with a permanent cross-section area becomes possible only if the gas is supplied in several flows with different locations of the takeoff points. Such a method of classification is energy-consuming and, thus, not popular.

To connect a mathematical model to a particular physical (engineering) task and to the area of the space in which it is solved (the so-called calculation domain), the initial and boundary conditions are given.

When solving the internal task, the calculation domain is filled with the fluid medium and limited by the walls of the model, for which the boundary conditions are set: the velocity profile is set on the inlet of the model, and a full pressure profile is set on the outlet.

Then, the boundary conditions are set inside the calculation domain. In addition to the initial and boundary conditions, we also specify the fluid medium (air) with certain properties, which are set in the programme (and changeable if necessary). After that, the programme builds a calculation grid that consists of the basic grid (obtained by partitioning the space of the grid construction on layers by means of the planes parallel to the coordinate planes of the Cartesian coordinate system model). Since the faces of the calculating cells do not approximate surfaces of solids tangential to the fluid medium, for the separation of the calculation grid in relation to small geometric peculiarities of these surfaces (highly curvilinear areas), the procedures of local splitting of the grid cells near these areas of the surfaces are used before the calculation. Then the calculation itself is carried out, which gives the visualization of the calculation results in the graphic window of the simulation system. The results of the visualization, their description and the analysis are presented below.

Figure 8 shows an example of building the calculation grid.

The use of the fluidized bed granulators with a variable cross-section area of the operational chamber and the internal circulation of the fines enables classifying of the polydisperse system with a wide range of fractions. Creation of different hydrodynamic conditions for the gas flow at different heights allows classification of granules into the necessary number of fractions and separation of the fines.

The analysis of the simulation results, presented on Figures 9 and 10, shows that:

- Different configuration of the gas flow with some dominant direction depends on the value of the velocity component of the gas flow;
- The dominant axial component of the velocity leads to the narrowing of the gas flow zone;
- The dominant circumferential component of the velocity leads to the growing height of the vortex gas flow zone;
- The dominant radial component of the velocity leads to the intensified transfer of the gas flow to a wall of the vortex granulator;

Figure 8. An example of building the calculation grid

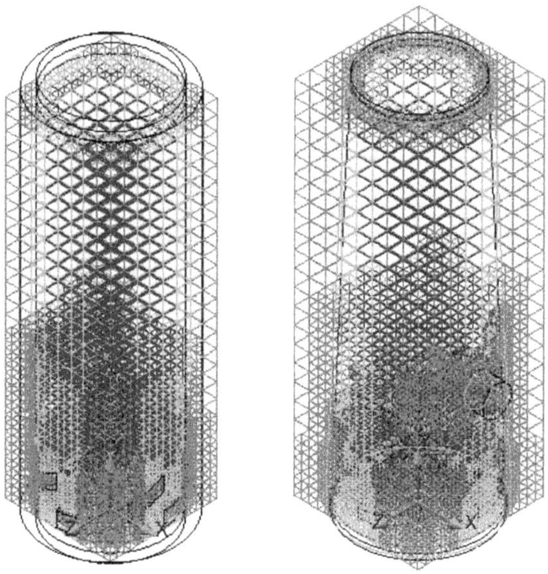

**For a more accurate representation see the electronic version.*

Figure 9. Influence of the construction of the workspace and components of the gas flow velocity on the total velocity of the gas flow and its trajectory (the workspace is a "diffuser" with a cylindrical insert in the upper part): $V_1=V_2=1$ m/s; $V_3=8$ m/s; d=0.3 m; l=0.2 m; L=1 m; 1 – pressure; 2 – velocity; 3 – velocity in the cross section at a height of 0.1 m; 4 – velocity in the cross section at a height of 0.5 m; 5 – trajectory

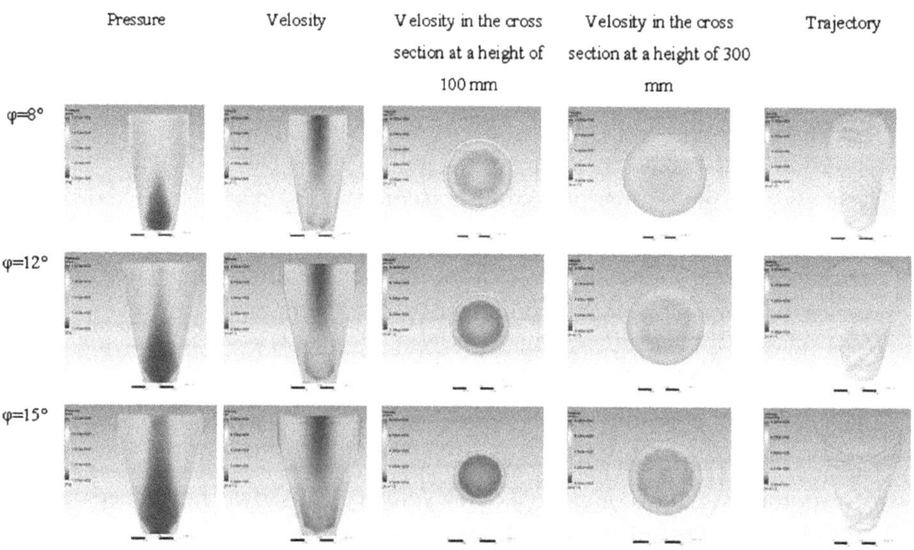

**For a more accurate representation see the electronic version.*

- Under certain conditions, the rate of the initial swirling of the gas flow provides no influence on the trajectory of its motion, but it influences the value of the resultant velocity of the gas flow;
- An axial component of the velocity decreases gradually when the gas flow moves within the height of the granulator (this is connected with the expanding section area of the granulator);
- The circumferential component of the velocity shows its maximum value in the outlet of the gas-distributing unit;
- The radial component of the velocity shows its maximum value on the axis of the apparatus;
- Pressure in the workspace of the vortex granulator grows with the height and within the radius on a certain height of the workspace.

The analysis of the results (Figures 11 and 12) reveals the laws of how the construction of a vortex granulator influences the distribution of the gas flow velocity.

The configuration of the swirler:

- A growing angle of slope of blades reduces the gas flow velocity on the cross-section of the workspace;
- A growing angle of slope of blades creates a chaotic swirling of the gas flow in a gas-distributing unit;
- A growing number of blades results in the non-uniform distribution of velocities – an inlet velocity of the gas flow is considerably higher than the outlet one;
- A growing angle of slope of blades creates a large dead area of granules.

The configuration of the workspace:

- A growing opening angle of the cone enhances formation of large dead areas in the center;
- The cylindrical configuration of the workspace provides non-uniform distribution of the circumferential velocity of the gas flow – the inlet velocity is higher than the outlet velocity, and there are zones with the reduced velocity in the center;
- A convergent configuration ensures a growing velocity of the gas flow during its motion through the workspace.

Due to different distribution character of the velocity fields in the workspace of a vortex granulator, it becomes possible to obtain granules from materials with different properties (strength, thermal resistance, porousness etc.).

Figure 10. Influence of the components of the gas flow velocity on the trajectory of the gas motion (the workspace is a "diffuser" with a cylindrical insert in the lower part): 1 – direction of gas flow movement (axonometry); 2 – direction of gas flow movement (plane); 3 – direction of gas flow movement in cross section in height 0.1 m; 4 – direction of gas flow movement in cross section in height 0.3 m

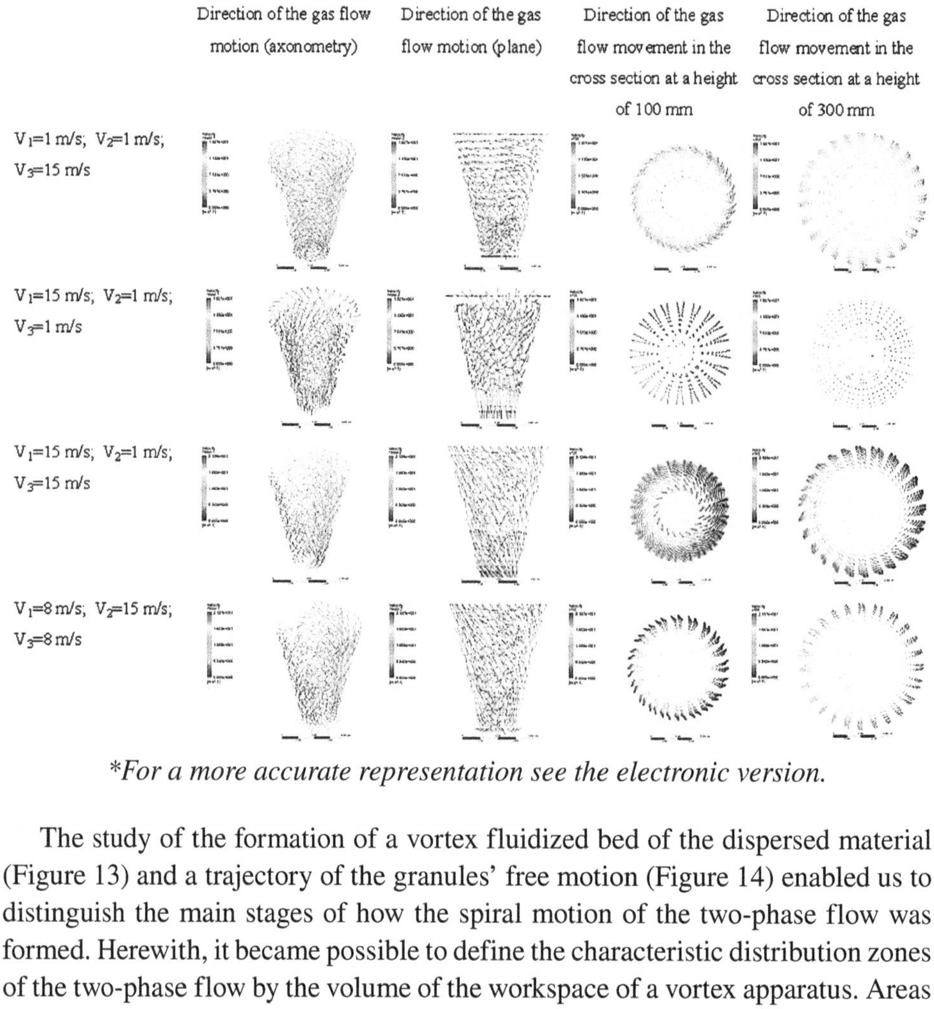

**For a more accurate representation see the electronic version.*

The study of the formation of a vortex fluidized bed of the dispersed material (Figure 13) and a trajectory of the granules' free motion (Figure 14) enabled us to distinguish the main stages of how the spiral motion of the two-phase flow was formed. Herewith, it became possible to define the characteristic distribution zones of the two-phase flow by the volume of the workspace of a vortex apparatus. Areas with boosting motion of the granular material and the dead areas were forecasted as well. The development of the vortex fluidized bed was studied step-by-step from its early formation till the operational stage.

A visual presentation of the simulation results of the polydisperse granule system in the conditions of changing the opening angle of the cone in the vortex granulator is presented in Figure 15 (a relative composition of granules in the workspace is φ=0.25).

Figure 11. A velocity field and a trajectory of gas flow motion in various constructions of the gas-distributing unit (a vortex), an angle of slope of blades 50 degr: a – an amount of blades n=3; b – n=5; c – n=6; d – n=8

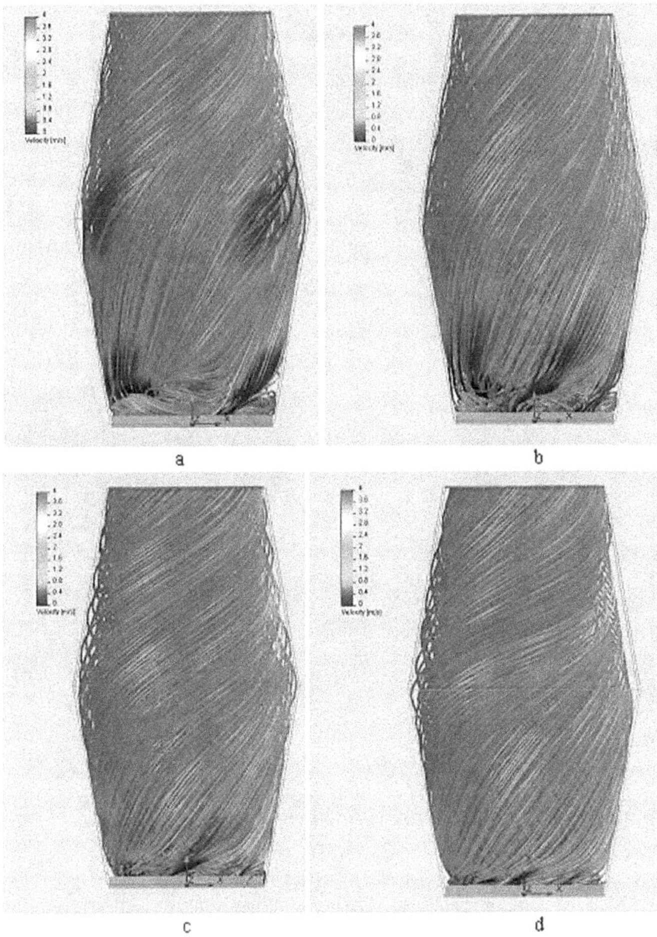

**For a more accurate representation see the electronic version.*

The results of the study have proved the fact that the trajectories of different fractions cross over. It is shown that even the small constraint of the particle motion leads to the deviations of the calculated trajectories. It has also been detected that a trajectory of the granules' constraint motion does not differ from the free motion of granules with the same physicochemical properties. The only exception occurs due to the detection of the polydisperse system on each height point of the workspace. This causes random changes in the trajectory of certain amount of the main fractions of granules (on the calculated height), if the main fractions collide with granules of other fractions.

Figure 12. A velocities field and a trajectory of the gas flow motion in various configurations of the workspace of the vortex granulator

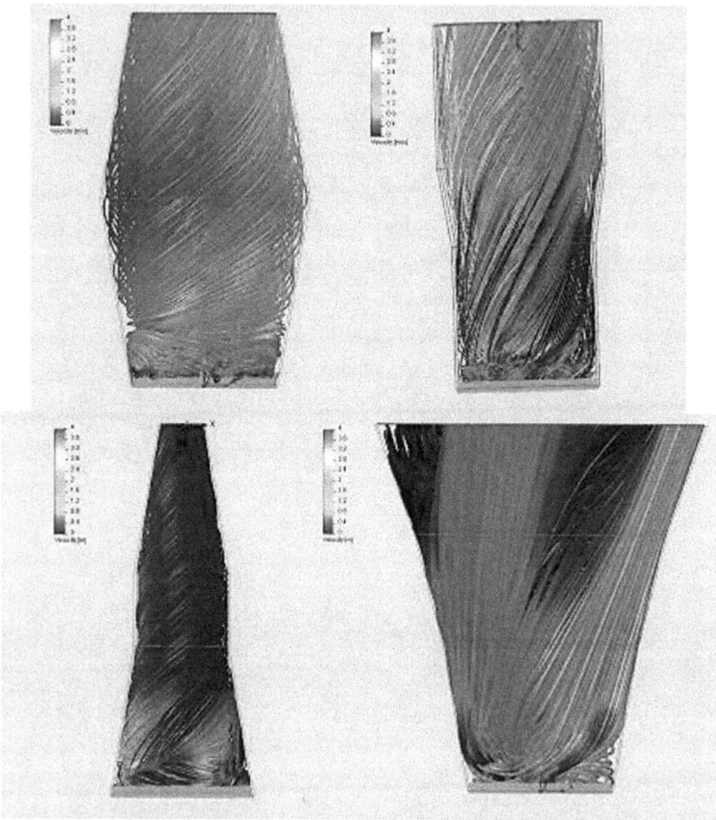

**For a more accurate representation see the electronic version.*

The above-mentioned fact requires an additional continual description of how granules of different sizes move and get their possible distribution within the workspace of the vortex granulator. An additional computer simulation with respect to the constraint degree of the granules' flow was performed to solve the task. The methods for this part of the investigation allows entering of the following input data: the calculated number of granules with the given size (an analogue of the fraction composition of the polydisperse system) and the total number of granules (an analogue of the constraint degree of the disperse flow). Herewith, different porousness of the vortex flow of the disperse phase within the height and the radius of the workspace is taken into consideration. The simulation results of the motion of granules in the range of the relative composition of granules in the workspace φ=0,2-0,4 is presented on Figure 16.

The analysis of the results brings the conclusions as follows:

Figure 13. Formation of a vortex fluidized bed

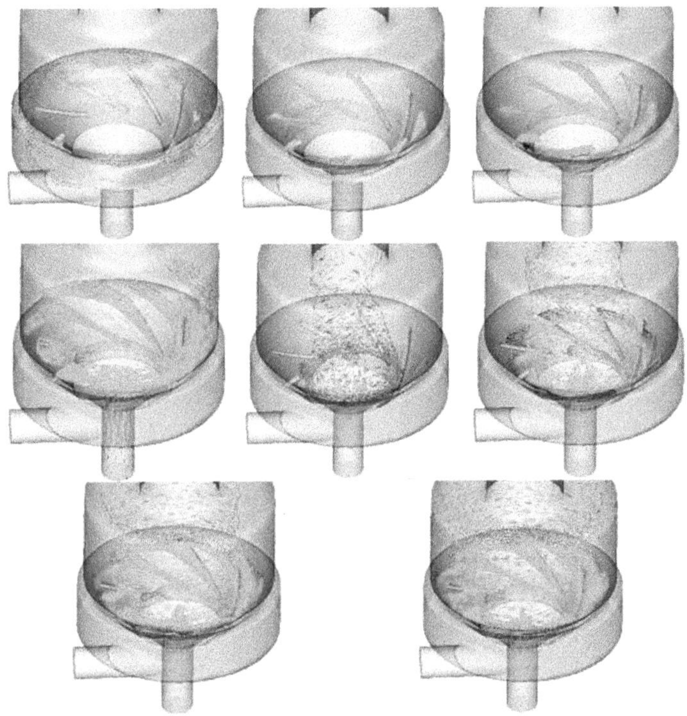

**For a more accurate representation see the electronic version.*

- A growing constraint degree of the flow shows a wider range of the fractions at the given height;
- The amount of the main fraction (relevant to this calculated height point) grows in response to the increasing constraint degree of the flow;
- The amount of the fraction of another range decreases within the height of the vortex granulator.

Gravitational Shelf Dryer

The presence of the tilted perforated shelves in the workspace of the shelf dryer reduces its free cross-section area. This constructive solution brings local increase in the velocity and the turbulence degree of the drying agent flow, and changes the distribution of velocities around the shelf.

At certain velocities of the drying agent (which is much lower than the first critical velocity), the material, which is permanently supplied to the dryer, moves in a form of the "fast" bed on the tilted perforated shelf. This type of motion is explained by the fact that there is no enough momentum for the particles on the outlet of the

Figure 14. A trajectory of the granules' motion under different components of the velocity

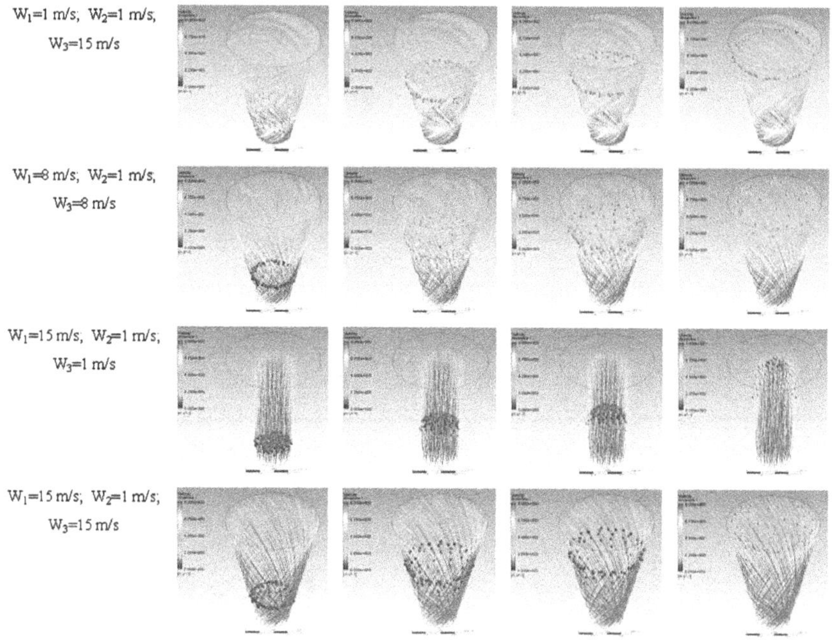

**For a more accurate representation see the electronic version.*

batcher. When the consumption of the drying agent grows, the porousness of the "fast" bed approaches its maximum, the structure of the bed becomes suspended and it removes the fines. After moving along the tilted perforated shelf, the particles reduce their velocity upon approaching the walls of the dryer in the discharging space. The surface of the walls accumulates the particles. The upward flow of the drying agent blows the formed dispersed phase; the volume of the bed is growing due to the intake of new particles from the dispersed phase. This process lasts until the balance between the gravity of the particle and the aerodynamic force of its interaction with the upward flow is broken. After the "critical" volume of the disperse phase accumulated and the balance of the forces broken, a part of the material is removed through the discharging space. Under this type of motion on the tilted shelf and in the discharging space, the aerodynamic forces of the upward flow constrain the motion of the solid particles to some extent only, and most particles in the thin bed move due to the gravity. This type of motion is called a "gravitational falling bed".

Thus, the "gravitational falling bed" is characterized by the minimal contact time of particles with the drying agent, induced by the high velocity of particles. That is the reason why drying on the surface of the shelf is limited by the lengths of the shelf, where particles' motion is fast, and the residence time is minimal, rather

Figure 15. A trajectory of the polydisperse granules under different opening angles of a cone (a constraint motion of granules): a – 10°; b – 13°; c – 16°

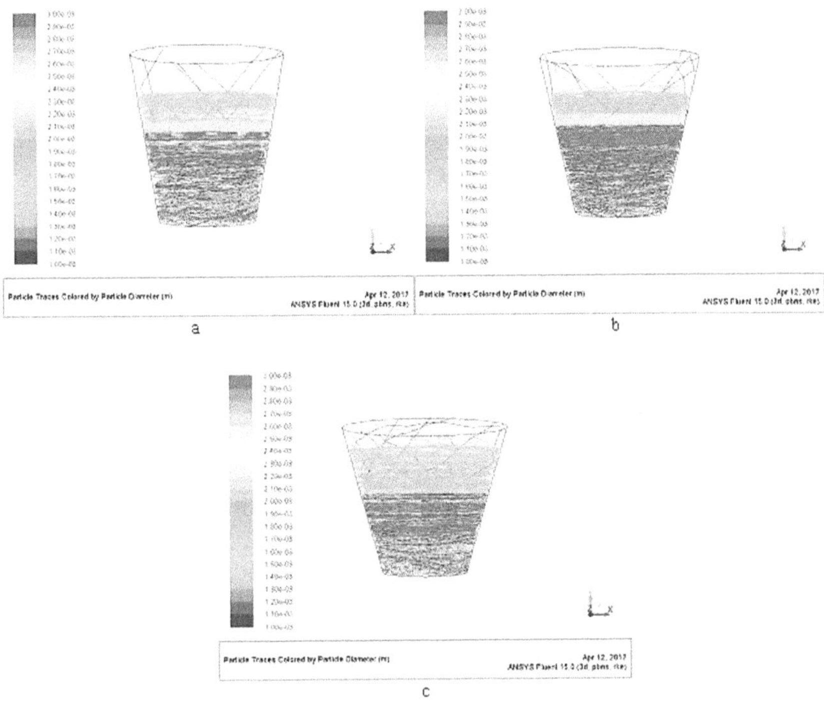

**For a more accurate representation see the electronic version.*

than by the hydrodynamic conditions. The residence time will be increased in the accumulated layer of the particles on the wall. This results in the more efficient contact of phases and, correspondingly, more intensive dehydration by the drying agent in this area. The gravitational falling regime, executed on the upper (according to the direction of the particles' motion) shelf, is complemented by the heating of the particles. Thus, the process on the upper shelf, executed within a short period of time, corresponds to the period of heating of the material with the small liberation of the surface moisture due to the growing velocity of drying. Herewith, the moisture moves from the internal layer to the surface of the particles. The period of heating is usually shorter than other periods of drying. And if the residence time of the particles on the one shelf is shorter than the kinetically necessary time of heating, the upper part of the dryer must contain more shelves, which operate in the "gravitational falling bed".

The growing velocity of the drying agent causes greater aerodynamic effects on the layer of the material and creates conditions for the fluidized bed on the surface of the perforated shelf and near the wall of the apparatus in the discharging gap.

Figure 16. The motion of granules in case the opening angle of a cone is 10 degr (a constraint motion): a – φ=0.2; b – φ =0.3; c – φ =0.4

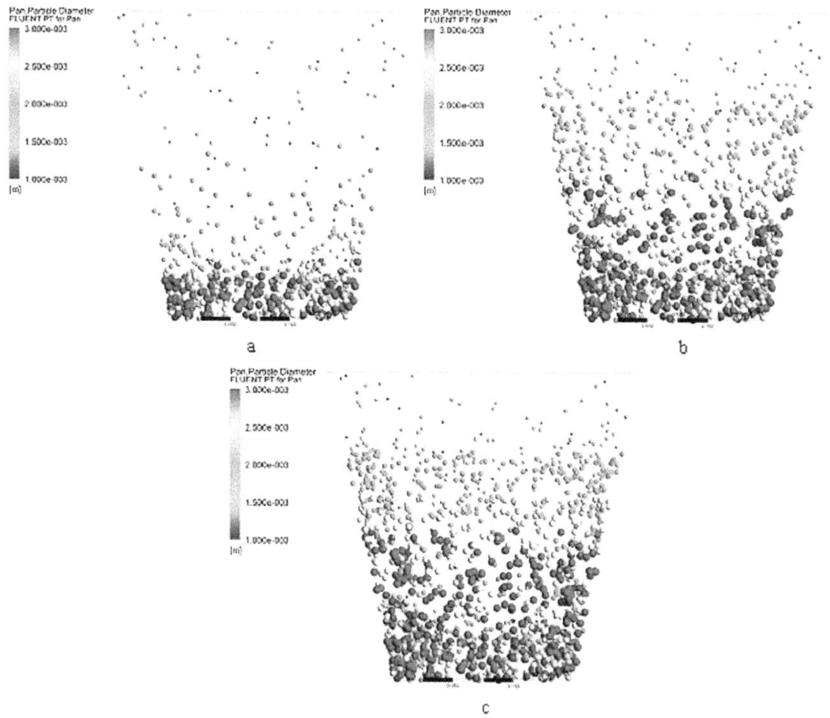

**For a more accurate representation see the electronic version.*

While moving from the discharging gap, the drying agent streams create a pressure zone. A part of the kinetic energy of the moving drying agent compensates the friction force, which provides deceleration of the drying agent. The disperse phase, that moves in this area and has a small amount of the kinetic energy, turns into the fluidized bed. While discharging from the upper shelf, the inertia of the fine phase is not enough to compensate the kinetic energy of the air flow. The fine phase iterates the trajectory of the drying agent stream, which leaves the discharging area. The phase is transmitted by the agent to the outlet tube for the drying agent in the separation zone. The polydisperse system in the separation zone is divided into two fractions: the larger particles, whose velocity is higher than the velocity of the air flow in this section of the apparatus, move down the surface of the shelf; and the smaller particles are removed from the apparatus by the air stream.

The fluidized bed shows more effective dehydration of the material due to the intensive contact of phases, given that the particles interact actively with the air streams both on the surface of the perforated shelf and in the discharging area. The reduced gap between the edge of the shelf and the wall of the apparatus speeds up the air stream, which enters the fluidized bed through the discharging area. And this is the main force to intensify the contact of phases and mass exchange correspondingly. The increased residence time of particles due to the mass circulation in the zone over the discharging area contributes to the enhanced dehydration of particles. Thus, taking into consideration the hydrodynamic features of the fluidized bed, the drying of the damp material at the permanent (the first period) and reducing (the second period) velocities becomes rather efficient.

The results of the investigations and the computer simulation (Figures 17-20) clarify the defining features of the disperse material motion in the workspace of the apparatus, and detect the areas with the reduced velocity, intensive swirling, separation etc.

The motion of the disperse phase is of the same quality on each cascade of the shelves. And yet, the single zones of its motion on the cross-section have different lengths depending on the altitude where the shelf is located.

While moving within the height of the dryer, the drying agent reduces its velocity due to the aerodynamic frictional resistance and the local resistance. The deceleration of the drying agent is not significant, but still creates different regimes of the motion of the disperse phase. The upper shelf of the cascade shows higher velocity of the particles and lower accumulation of the material on the walls. Reaching the inlet of the drying agent, the velocity of particles on the shelf begins to decrease. This picture is not much observable, but experimental studies prove this fact.

When determining the contact of the shelves with the discharging gap, the distribution of the gas flow velocity is characterized by the following:

- The distribution of the gas flow velocity is descending from the wall of the apparatus to the edge of the shelf contact. This is induced by the pressure jumps under and over the shelf contact;
- The shortening of the discharging gap from the middle of the shelf contact intensifies the deceleration of the gas flow, induced by the vortexes in the discharging gap. The shortening of the discharging gap intensifies the vortexes on the edge of the shelf contact, that disrupts the upward gas flow;
- The minimum value of the gas flow velocity on the shelf contact grows in response to the shortening of the discharging gap.

The analysis of the experiments, where the gas flow velocity is distributed within different stages of the dryer depending on the altitude of the stage and under the same construction of the stages, shows that:

- The distribution of the gas flow velocity becomes more uniform due to the enlarging of the discharging gap;

Figure 17. A colored plot of the gas flow velocity fields (a vector interpretation)

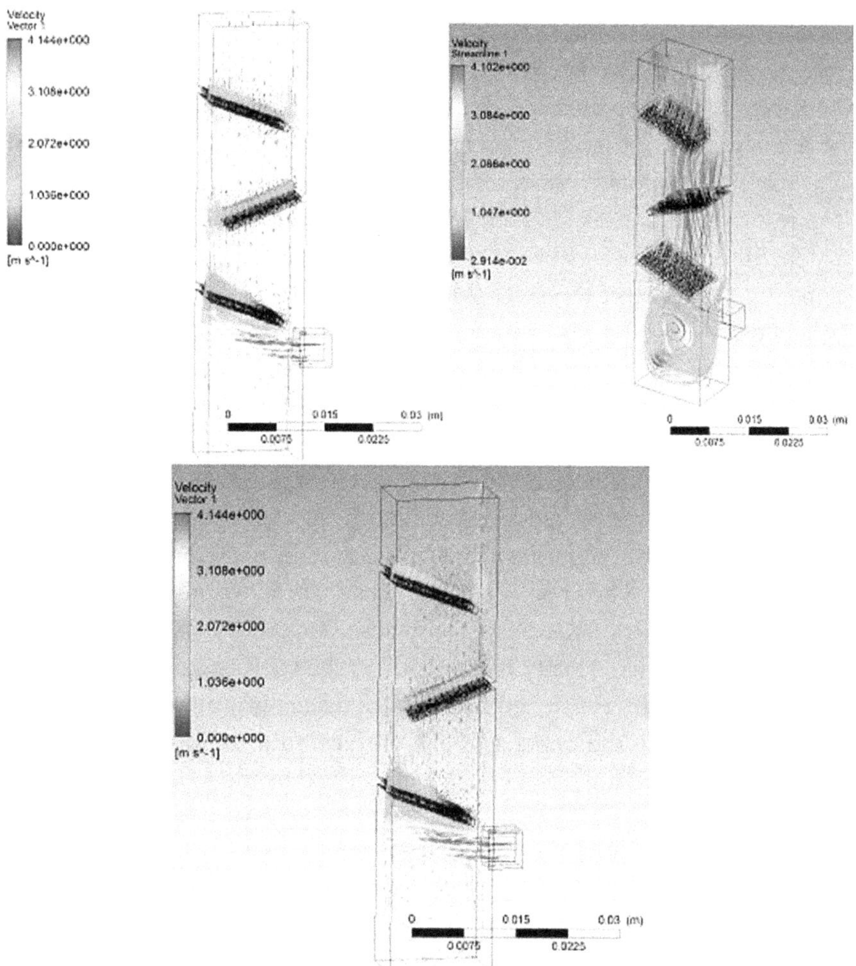

**For a more accurate representation see the electronic version.*

- The diagram of the gas flow velocity for the shelf contact of the one construction changes the quantitative feature of its profile. This is explained by the height redistribution of the gas flow within the cross-section of the dryer. Herewith, it becomes more uniform while moving upward.

An insert of the shelf contact with different value of the free cross-section (an increasing value) shows that:

- The profile of the gas flow velocity on the shelf contact flattens out partially;
- The peak velocity of the gas flow in the discharging gap is reduced;

Figure 18. A colored plot of the gas flow velocity fields (a location of the shelf contact)

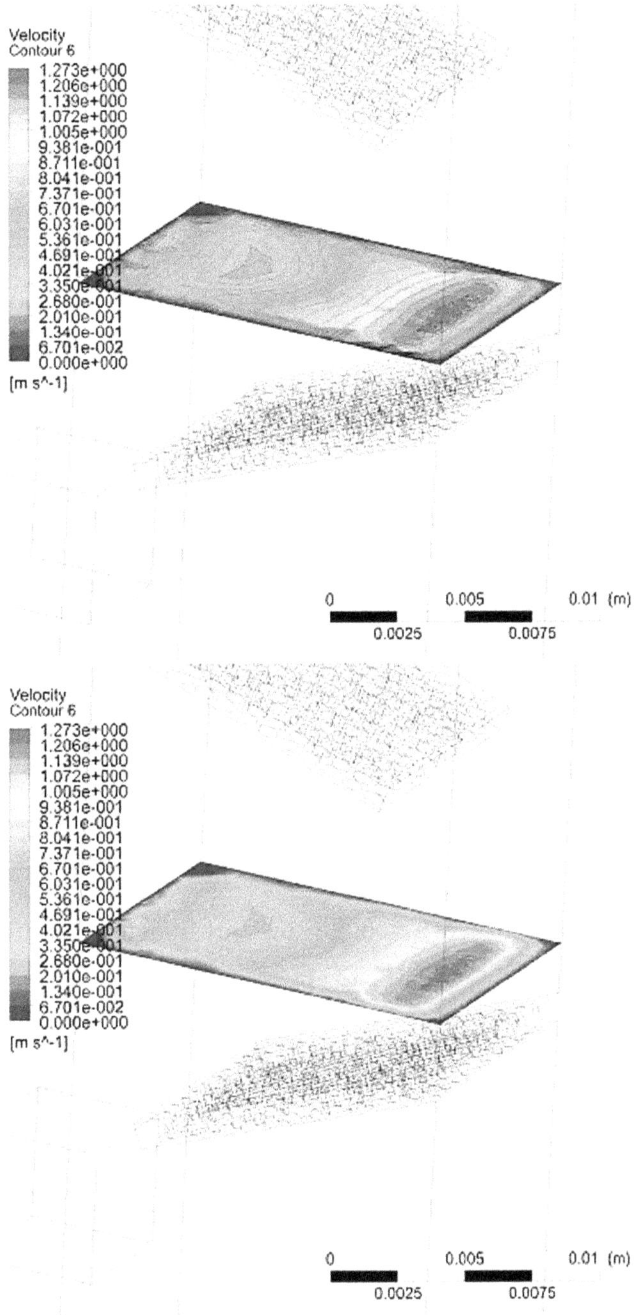

*For a more accurate representation see the electronic version.

Figure 19. A colored plot of the gas flow velocity fields (a cross-section of the dryer)

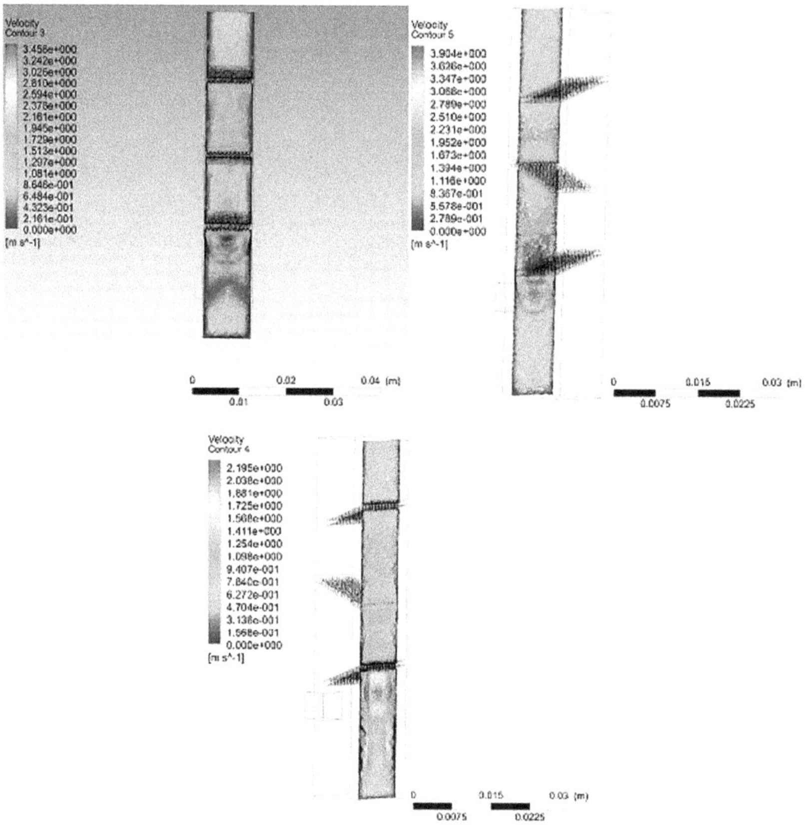

**For a more accurate representation see the electronic version.*

- The diagram of the gas flow velocity in the area from the shelf contact to the discharging gap is smoother.

If the angle of the shelf contact construction is reduced under the same value of the discharging gap, the distribution diagram of the gas flow velocity acquires the following features:

- The peak velocity of the gas flow on the shelf decreases, the diagram becomes smoother;
- The peak velocity of the gas flow in the discharging gap decreases;
- The area of the maximum velocity of the gas flow in the discharging gap expands, while the diagram becomes smoother.

Figure 20. A colored plot of the gas flow velocity fields (a longitudinal cross section of the dryer)

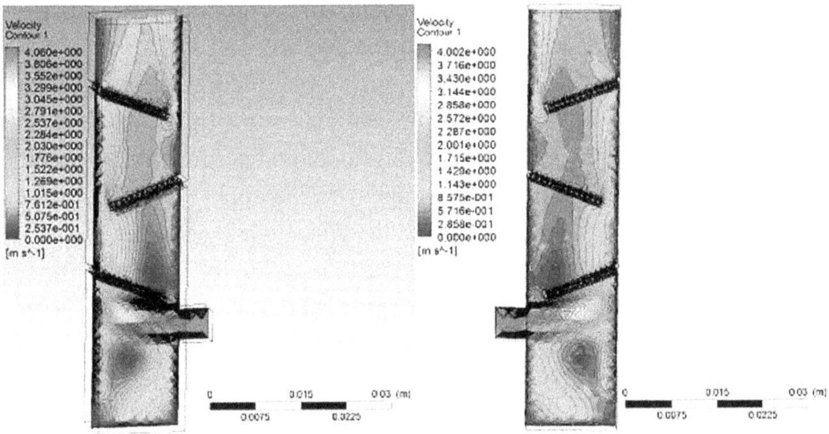

**For a more accurate representation see the electronic version.*

The distribution diagram of the gas flow velocity with the growing consumption of the gas flow has the same qualitative features. But the distinctive features are as follows:

- The peak is leveled up in the middle of the shelf contact;
- The velocity is equalized within the length of the shelf contact;
- The peak velocity of the gas flow is more evident in the discharging gap.

The same area of the cross-section of the shelf, but different diameters and a different amount of perforations influence the direction of the gas flow motion around the shelf:

- The gas flow moves around the shelf more evidently if a diameter is smaller and the number of perforations is bigger;
- A growing diameter and a smaller number of perforations make the motion of the gas flow around the shelf less evident, and the gas moves through the perforations rather than around the shelf.

The necessary hydrodynamic conditions for the motion of the material on each shelf in the cascade are formed due to the changes in the construction parameters of the shelf contacts. It should be noted that the construction of the gravitational shelf dryer must ensure the uniform contact of the drying agent with the dispersed material on each shelf of the dryer. This uniformity makes it possible to control the residence time of the particles on the shelves in respect to their physicochemical properties. Any non-uniform contact of the drying agent with the dispersed material may lead to underheating (and incomplete drying) or overheating with the distortion of particles and lowering of the consumer grade.

Thus, due to the variations of the length, a construction angle or a level of perforation it is possible to ensure the minimal necessary time for the contact of the dispersed material with the specified physicochemical properties, granulation composition and parameters of the drying agent on the upper shelf. The same shelf operates also as a separator and it removes the fines. Changes in the construction parameters of the middle shelf make it possible to increase the residence time of the dispersed material and its contact with the drying agent, that intensifies the removal of the moisture. The contact of the dispersed material with the drying agent on the lower shelf is complemented by the removal of the moisture from the depth of the material.

FUTURE RESEARCH DIRECTIONS

Further research aims to develop an algorithm for exporting the constructive calculation results, obtained for the main technological facilities of the granulation unit, into the solid modeling programmes with further modeling of the devices and preparation of the engineering documentation of their production.

Thus, the scientific calculation methodology for the vortex granulation devices, that produce porous ammonium nitrate granules, will be developed on the basis of the original results of the theoretic and experimental researches. Furthermore, the methodology will describe not only the target granulation, but the complementary processes (like separation, classification, or exhaust gas disposal) as well.

For the first time, the scientific methodology, based on the theoretical and experimental researches and used for calculations of the vortex granulators, will enable obtaining of the complex algorithm for the projecting of the multistage and multitask vortex granulation units. The projecting will be complemented by the search for the optimal conditions for the target processes, the evaluation of the energy efficiency and the ecological safety of the units that produce granules with special properties (porousness) for the mining industry.

CONCLUSION

The use of the computer simulation during the study of the hydrodynamics of the flows in the devices with active regimes and a variable cross-sectional area allows selection of the optimal configuration of the workspace and a construction of the gas flow distributor. Herewith, the necessary quality of the finished product is provided, depending on the requirements to the thermal treatment and the strength.

The results of the computer simulation allow us to distinguish specific zones of the workspace, where the dispersed phase motion may potentially be increased or decreased. The velocity field of the gas flow and the "velocity bursts" (the direction vectors of the resultant velocity in a random area of the device) also depict the possible trajectory of the dispersed phase motion. The second conclusion is based on the fact that the gas flow transfers up to 90% of its kinetic momentum. However, the results of the computer simulation of the gas flow hydrodynamics provide only the qualitative evaluation of the possible trajectory of the solid phase motion. The quantitative evaluation of the possible trajectory of the solid phase motion and its residence time requires an additional computer simulation. It should be based on the presented results, a force analysis of the dispersed phase motion, as well as the conditions for the particles' interaction in the constraint motion (a quantity of the dispersed phase as related to the volume of the device).

The information, presented in this work, allows us to develop the algorithm for conducting the optimization calculations for the granulation and drying equipment, used in the production of porous ammonium nitrate. This algorithm includes the following stages:

1. A stage of moisturizing and granulating (a vortex granulator).
    a. Select an optimal construction of the gas-distributing unit (a vortex).
    b. Select an optimal construction of the workspace of the apparatus.
    c. Calculate the trajectory of the free motion of particles.
    d. Calculate the trajectory of the constraint motion of particles.
    e. Schedule the residence time of the particles in the unit.
2. A stage of complete drying (a multistage gravitational shelf dryer).
    a. Select the construction of the shelf contact.
    b. Calculate the trajectory of the free motion of particles.
    c. Calculate the trajectory of the constraint motion of particles.
    d. Determine the required number of stages (shelf contacts).
    e. Schedule the residence time of the particles in the unit.

The computer simulation results and optimization calculations allowed us to offer new constructions of the vortex granulators (Figures 20 and 21) (Artyukhov & Sklabinskyi, 2013) and multistage gravitational shelf dryers (Figure 22) (Artyukhova, Shandyba, & Artyukhov, 2014).

Figure 21. Vortex granulators: a – with vibrating spray of melt; b – with a two-stage contact of granules and a heat-transfer agent; c – with a multistage zone for a secondary contact of granules and a heat-transfer agent

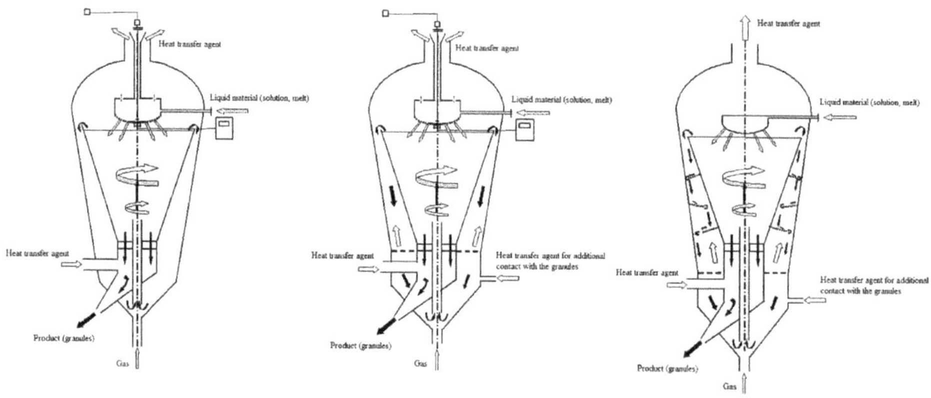

Figure 22. Vortex granulators that produce porous granules: a – with preliminary damping; b – with preliminary damping and drying

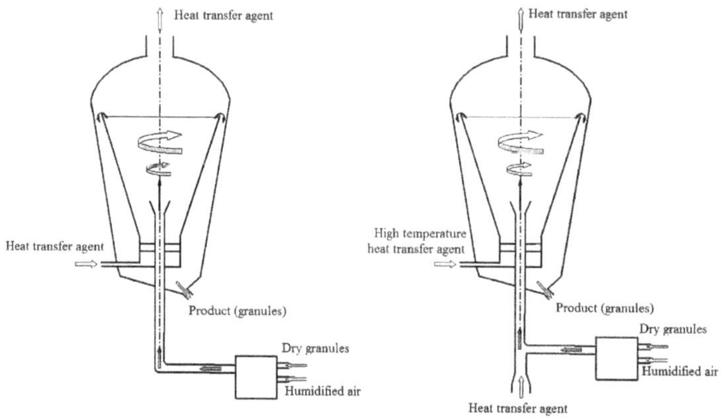

Figure 23. Constructions of the contact shelves in a multistage gravitational dryer: a – a contact shelf with perforations of different diameters; b – a built-up contact shelf with different angles of inclination to the horizon of each part

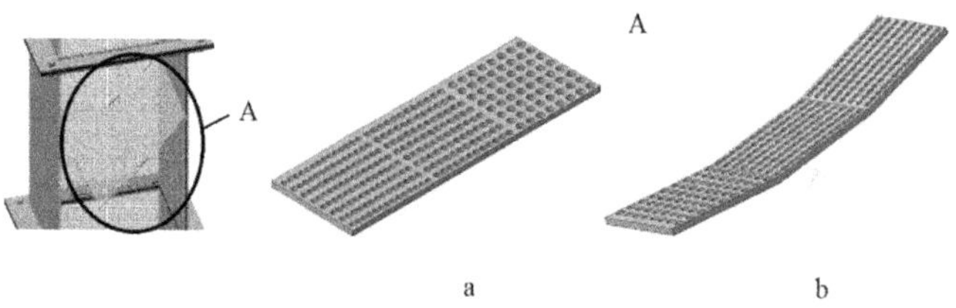

ACKNOWLEDGMENT

This research was supported by the Ministry of Education and Science of Ukraine [grant number 0116U006812].

REFERENCES

Artyukhov, A. (2016). Application software products for calculation trajectories of granules movement in vortex granulator. *CEUR Workshop Proceedings, 1761*, 363–373.

Artyukhov, A., Sklabinskiy, V., Ivaniia, A., & Moskalenko, K. (2016). Software for calculation of vortex type granulation devices. *CEUR Workshop Proceedings, 1761*, 374–385.

Artyukhov, A., & Sklabinskyi, V. (2015). Hydrodynamics of gas flow in small-sized vortex granulators in the production of nitrogen fertilizers. *Chemistry & Chemical Technology, 9*(3), 337–342. doi:10.23939/chcht09.03.337

Artyukhov, A. E., Fursa, A. S., & Moskalenko, K. V. (2015). Classification and separation of granules in vortex granulators. *Chemical and Petroleum Engineering, 51*(5-6), 311–318. doi:10.100710556-015-0044-x

Artyukhov, A. E., Obodiak, V. K., Boiko, P. G., & Rossi, P. C. (2017). Computer modeling of hydrodynamic and heat-mass transfer processes in the vortex type granulation devices. *CEUR Workshop Proceedings, 1844*, 33–47.

Artyukhov, A. E., & Sklabinskyi, V. I. (2013). Experimental and industrial implementation of porous ammonium nitrate producing process in vortex granulators. *Naukovyi Visnyk Natsionalnoho Hirnychoho Universytetu, 6*, 42–48.

Artyukhova, N. A., Shandyba, A. B., & Artyukhov, A. E. (2014). Energy efficiency assessment of multi-stage convective drying of concentrates and mineral raw materials. *Naukovyi Visnyk Natsionalnoho Hirnychoho Universytetu, 1*, 92–98.

Boyadjiev, C. B., & Babak, V. N. (2000). *Non-linear mass transfer and hydrodynamic stability*. Amsterdam: Elsevier Science B.V. doi:10.1016/B978-0-444-50428-9.50012-0

Chen, Q., Zhai, Z., & Wang, L. (2007). Computer modeling of multiscale fluid flow and heat and mass transfer in engineered spaces. *Chemical Engineering Science, 62*(13), 3580–3588. doi:10.1016/j.ces.2006.11.059

Crowe, C. T. (2006). *Multiphase flow handbook*. New York: Taylor & Francis Group, LLC.

Dvornikov, N. A., & Belousov, P. P. (2011). Investigation of a fluidized bed in a vortex chamber. *Journal of Applied Mechanics and Technical Physics, 52*(2), 206–211. doi:10.1134/S0021894411020076

Goldschmidt, M. (2001). *Hydrodynamic Modelling of Fluidised Bed Spray Granulation* (Ph.D. thesis). University of Twente.

Kaewklum, R., Kuprianov, V. I., & Douglas, P. L. (2009). Hydrodynamics of air–sand flow in a conical swirling fluidized bed: A comparative study between tangential and axial air entries. *Energy Conversion and Management, 50*(12), 2999–3006. doi:10.1016/j.enconman.2009.07.019

Khalatov, A. A. (2010). Heat transfer and hydrodynamics in the fields of mass forces. *Journal of Engineering Physics and Thermophysics, 83*(4), 794–808. doi:10.100710891-010-0397-0

Li, X., & Li, G. (2016). Modeling and Simulation of Hydrodynamic Bubble-Liquid Turbulent Flows in Bubble-Column Reactors. *Chemical Engineering & Technology, 39*(12), 2380–2388. doi:10.1002/ceat.201600074

Liu, K., Gu, X. G., Ba, D. C., Li, P. Y., Du, G. Y., Yue, X. J., & Yang, N. (2012). Numerical Research on Flow Characteristics of Vortex Stage in Dry High Vacuum Pump. *Physics Procedia, 32*, 127–134. doi:10.1016/j.phpro.2012.03.529

Parikh, D. (2009). *Handbook of Pharmaceutical Granulation Technology*. London: Informa Healthcare.

Pat. No. 112393 UA (2016). *Vortex granulator with utilization of waste gases.*

Pat. No. 112394 UA (2016). *Vortex granulator.*

Pat. No. 112622 UA (2016). *Vortex granulator.*

Pat. No. 113141 UA (2017). *Vortex granulator.*

Pat. No. 74070 UA (2012). *A device for disperse materials drying.*

Pat. No. 81720 UA (2013). *A device for disperse materials drying.*

Patankar, S. V., & Spalding, D. B. (1972). A calculation procedure for heat, mass and momentum transfer in three-dimensional parabolic flows. *International Journal of Heat and Mass Transfer, 15*(10), 1787–1806. doi:10.1016/0017-9310(72)90054-3

Patel, P., Telange, D., & Sharma, N. (2011). Comparison of Different Granulation Techniques for Lactose Monohydrate. *International Journal of Pharmaceutical Sciences and Drug Research*, *3*, 222–225.

Polascik, T. J., Cottet, G.-H., & Huberson, S. (2013). *Vortex Flows and Related Numerical Methods*. New York: Springer Science & Business Media.

Saikh, M. (2013). A technical note on granulation technology: A way to optimise granules. *International Journal of Pharmaceutical Sciences Review and Research*, *4*, 55–67.

Saqr, K. M., Aly, H. S., Wahid, M. A., & Sies, M. M. (2009). Numerical simulation of confined vortex flow using a modified k−ε turbulence model. *CFD Letters*, *1*(2), 87–94.

Solanki, H., Basuri, T., Thakkar, J., & Patel, C. (2010). Recent advances in granulation technology. *International Journal of Pharmaceutical Sciences Review and Research*, 48–54.

Tahvildari, K., Tavakoli, H., Mashayekhi, A., & Golmohammadzadeh, R. (2016). Modeling and simulation of membrane separation process using computational fluid dynamics. *Arabian Journal of Chemistry*, *9*(1), 72–78. doi:10.1016/j.arabjc.2015.02.022

Trojan, M. (2015). Computer modeling of a convective steam superheater. *Archives of Thermodynamics*, *36*(1), 125–137. doi:10.1515/aoter-2015-0009

Volchkov, E. P., Dvornikov, N. A., Lukashov, V. V., & Abdrakhmanov, R. Kh. (2013). Investigation of the flow in the vortex chamber with centrifugal fluidizing bed with and without combustion. *Thermophysics and Aeromechanics*, *20*(6), 663–668. doi:10.1134/S0869864313060024

Voynov, N., Nikolaev, N., Kustod, F., Nikolaev, A. & Tarovatyu, D. (2008).Vortex contact stages for rectification. *Chemistry of Plant Raw Materials, 3,* 173-184.

Zimmermann, S., & Taghipour, F. (2005). CFD Modeling of the Hydrodynamics and Reaction Kinetics of FCC Fluidized-Bed Reactors. *Industrial & Engineering Chemistry Research*, *44*(26), 9818–9827. doi:10.1021/ie050490+

ADDITIONAL READING

Ashcraft, R. W., Heynderickx, G. J., & Marin, G. B. (2012). Modeling fast biomass pyrolysis in a gas-solid vortex reactor. *Chemical Engineering Journal, 207-208*, 195–208. doi:10.1016/j.cej.2012.06.048

Chen, Q., Zhai, Z., & Wang, L. (2007). Computer modeling of multiscale fluid flow and heat and mass transfer in engineered spaces. *Chemical Engineering Science, 62*(13), 3580–3588. doi:10.1016/j.ces.2006.11.059

Hiltunen, K., Jäsberg, A., & Kallio, S. (2009). *Multiphase Flow Dynamics. Theory and Numerics*. Helsinki: Edita Prima Oy.

Kaewklum, R., Kuprianov, V. I., & Douglas, P. L. (2009). Hydrodynamics of air-sand flow in a conical swirling fluidized bed: A comparative study between tangential and axial air entries. *Energy Conversion and Management, 50*(12), 2999–3006. doi:10.1016/j.enconman.2009.07.019

Kudra, T., & Mujumdar, A. S. (2002). *Advanced Drying Technologies*. New York: Marcel Dekker.

Litster, J. & Ennis, B. (2004). *The science and engineering of granulation processes*. Springer-Science+Business Media.

Salman, A. D., Hounslow, M. J., & Seville, J. P. K. (2006). *Granulation*. Amsterdam: Elsevier Science Ltd.

Stahl, H. (2010). *Comparing Granulation Methods*. Hürth: GEA Pharma Systems.

Yang, W. C. (2003). *Handbook of fluidizaition and fluid-particle systems*. New York: Marcel Dekker. doi:10.1201/9780203912744

Zhaofeng, T. (2006). Numerical modelling of turbulent gas-particle flow and its applications. Ph.D. thesis. RMIT University.

KEY TERMS AND DEFINITIONS

Computational Model: A mathematical model in computational science that requires extensive computational resources to study the behavior of a complex system by computer simulation.

Drying: A mass transfer process consisting of the removal of water or another solvent by evaporation from a solid.

Fluidized Bed: A physical phenomenon occurring when a quantity of a solid particulate substance (usually present in a holding vessel) is placed under appropriate conditions to cause a solid/fluid mixture to behave as a fluid.

Granulation: The act or process of forming into grains or granules.

Hydrodynamics: A branch of physics that deals with the motion of fluids (gases) and the forces acting on solid bodies immersed in fluids (gases) and in motion relative to them.

Navier-Stokes Equations: The equations that describe the motion of viscous fluid substance.

Reynolds-Averaged Navier-Stokes Equations: The time-averaged equations of motion for fluid flow.

Nomenclature

NOMENCLATURE

C_p Specific heat, J.kg^{-1}.K^{-1}
d Diameter, m
D Mass diffusivity, m^2.s^{-1}
h_{fg} latent heat of vaporisation, J.kg^{-1}
$\dot{m}$ Evaporating mass flux, kg.s^{-1}m^{-2}
$\bar{\dot{m}}_{ev}$ Total Evaporation mass flux, kg.s^{-1}
P Pressure, Pa
q_W Flux imposed at evaporator wall, W.m^{-2}
q_{fc} Liquid film convective heat, W.m^{-2}
q_L Latent heat flux, W.m^{-2}
q_S Sensible heat flux, W.m^{-2}
T Temperature, K
u Velocity component in x direction, m.s^{-1}
v Velocity component in y direction, m.s^{-1}
$w_{G,i}$ Vapour mass fraction of constituent i
$w_{L,I}$ Liquid mass fraction of constituent i

Greek Symbols

δ_x Local liquid film thickness, m
λ Thermal conductivity, W.m^{-2}.K^{-1}
μ Dynamic viscosity, Pa.s
ρ Density, kg.m^{-3}
Γ liquid mass flow rate per unit periphery length, kg.s^{-1}.m^{-1}

Subscripts

a Air
b Bulk
G Gas
I Interface
L Liquid
0 Inlet condition
w Water

Compilation of References

Abadie, T., Xuereb, C., Legendre, D., & Aubin, J. (2013). Mixing and recirculation characteristics of gas-liquid Taylor flow in microreactors. *Chemical Engineering Research & Design*, *91*(11), 2225–2234. doi:10.1016/j.cherd.2013.03.003

Abed, W. M., Whalley, R. D., Dennis, D. J. C., & Poole, R. J. (2015). Numerical and experimental investigation of heat transfer and fluid flow characteristics in a micro-scale serpentine channel. *International Journal of Heat and Mass Transfer*, *88*, 790–802. doi:10.1016/j.ijheatmasstransfer.2015.04.062

Abiev, R.S., & Svetlov, S.D. (2014). Two-phase flow regimes in the channels of microfluidic devices by the use of coaxial and coaxial-spherical micro dispersers. *Scientific Instrument Engineering (Nauchnoye Priborostroeniye)*, *4*, 59–66. (in Russian).

Abiev, R. S. (2008). Simulation of the Slug Flow of a Gas–Liquid System in Capillaries. *Theoretical Foundations of Chemical Engineering*, *42*(2), 105–117. doi:10.1134/S0040579508020012

Abiev, R. S. (2009). Circulation and Bypass Modes of the Slug Flow of a Gas–Liquid Mixture in Capillaries. *Theoretical Foundations of Chemical Engineering*, *43*(3), 298–306. doi:10.1134/S0040579509030099

Abiev, R. S. (2010). Method for Calculating the Void Fraction and Relative Length of Bubbles under Slug Flow Conditions in Capillaries. *Theoretical Foundations of Chemical Engineering*, *44*(1), 86–101. doi:10.1134/S0040579510010112

Abiev, R. S. (2011). Modeling of Pressure Losses for the Slug Flow of a Gas–Liquid Mixture in Mini- and Microchannels. *Theoretical Foundations of Chemical Engineering*, *45*(2), 156–163. doi:10.1134/S0040579511020011

Abiev, R. S. (2012). Modern State and Perspectives of Micro Technique Application in Chemical Industry. *Russian Journal of General Chemistry*, *82*(12), 2019–2024. doi:10.1134/S1070363212120237

Abiev, R. S. (2013). Bubbles velocity, Taylor circulation rate and mass transfer model for slug flow in milli and micro channels. *Chemical Engineering Journal*, *227*, 66–79. doi:10.1016/j.cej.2012.10.009

Abiev, R. S. (2015). Effect of Contact Angle Hysteresis on the Pressure Drop under Slug Flow Conditions in Minichannels and Microchannels. *Theoretical Foundations of Chemical Engineering*, *49*(4), 414–421. doi:10.1134/S0040579515040223

Abiev, R. S., & Dymov, A. V. (2013). Modeling the Hydrodynamics of Slug Flow in a Minichannel for Liquid–Liquid Two Phase System. *Theoretical Foundations of Chemical Engineering*, *47*(4), 299–305. doi:10.1134/S0040579513040180

Abiev, R. S., & Lavretsov, I. V. (2011). Hydrodynamics and Mass transfer of Taylor flow of Gas-Liquid Systems in Micro Channels: Theory and Experiment. *Chemical Engineering Journal*, *176-177*, 57–64. doi:10.1016/j.cej.2011.03.107

Abiev, R. S., & Lavretsov, I. V. (2011). Hydrodynamics of Gas–Liquid Slug Flow in Capillaries: Comparing Theory and Experiment. *Theoretical Foundations of Chemical Engineering*, *45*(3), 235–247. doi:10.1134/S0040579511030018

Abiev, R. S., & Lavretsov, I. V. (2012). Intensification of mass transfer from liquid to capillary wall by Taylor vortices in minichannels, bubble velocity and pressure drop. *Chemical Engineering Science*, *74*, 59–68. doi:10.1016/j.ces.2012.02.024

Abraham, R., & Mani, A. (2015). Heat transfer characteristics in horizontal tube bundles for falling film evaporation in multi-effect desalination system. *Desalination*, *375*, 129–137. doi:10.1016/j.desal.2015.06.018

Adamson, A. W., & Gast, A. P. (1997). *Physical Chemistry of Surfaces* (6th ed.). New York: John Wiley & Sons, Inc.

Adib, M., Eckstein, R., Hernandez-Sosa, G., Sommer, M., & Lemmer, U. (2018). SnO 2 Nanowire-Based Aerosol Jet Printed Electronic Nose as Fire Detector. *IEEE Sensors Journal*, *18*(2), 494–500. doi:10.1109/JSEN.2017.2777178

Agarwal, T. (2015). *Heat Sensor Circuit and Working Operation*. Retrieved from: https://www.elprocus.com/heat-sensor-circuit-and-working-operation/

AIAA. (1998). *Guide for the Verification and Validation of Computational Fluid Dynamics Simulations*. AIAA Guide G-077-1998.

Al-Ansari, D., & Owen, I. (1999). Thermal and hydrodynamic analysis of the condensation and evaporation processes in horizontal tube desalination plant. *International Journal of Heat and Mass Transfer*, *42*(9), 1633–1644. doi:10.1016/S0017-9310(98)00240-3

Albadawi, A., Donoghue, D. B., Robinson, A. J., Murray, D. B., & Delauré, Y. M. C. (2013). On the analysis of bubble growth and detachment at low Capillary and Bond numbers using Volume of Fluid and Level Set methods. *Chemical Engineering Science*, *90*, 77–91. doi:10.1016/j.ces.2012.12.004

Al-Neama, A. F., Kapur, N., Summers, J., & Thompson, H. M. (2017). An experimental and numerical investigation of the use of liquid flow in serpentine microchannels for microelectronics cooling. *Applied Thermal Engineering*, *116*, 709–723. doi:10.1016/j.applthermaleng.2017.02.001

Al-Shemmeri, T. (2012). *Engineering Fluid Mechanics*. Ventus Publishing ApS.

American Institute of Chemical Engineers (AIChE) Constitution. (2016). Available in: https://www.aiche.org/about/governance/constitution

Andreani, M., Haller, K., Heitsch, M., Hemstromc, B., Karppinen, I., Maceke, J., ... Toth, I. (2008). A benchmark exercise on the use of CFD codes for containment issues using best practice guidelines: A computational challenge. *Nuclear Engineering and Design*, *238*(3), 502–513. doi:10.1016/j.nucengdes.2007.01.021

Angeli, P., & Gavriilidis, A. (2008). Hydrodynamics of Taylor flow in small channels: A review. Proc IMechE, Part C. *Journal of Mechanical Engineering Science*, *222*(5), 737–751. doi:10.1243/09544062JMES776

Aota, A., Hibara, A., & Kitamori, T. (2007). Pressure Balance at the Liquid–Liquid Interface of Micro Countercurrent Flows in Microchips. *Analytical Chemistry*, *79*(10), 3919–3924. doi:10.1021/ac070031d PMID:17439241

Aota, A., Hibara, A., Sugii, Y., & Kitamori, T. (2012). Shape of the Liquid–Liquid Interface in Micro Counter-Current Flows. *Analytical Sciences*, *28*(1), 9. doi:10.2116/analsci.28.9 PMID:22232217

Aota, A., Mawatari, K., & Kitamori, T. (2009). Parallel multiphase microflows: Fundamental physics, stabilization methods and applications. *Lab on a Chip*, *9*(17), 2470–2476. doi:10.1039/b904430m PMID:19680572

Argüelles-Vivas, F. J., & Babadagli, T. (2016). Gas (air)–heavy oil displacement in capillary media at high temperatures: A CFD approach to model microfluidics experiments. *Chemical Engineering Science*, *140*, 299–308. doi:10.1016/j.ces.2015.10.028

Artyukhov, A. (2016). Application software products for calculation trajectories of granules movement in vortex granulator. *CEUR Workshop Proceedings*, *1761*, 363–373.

Artyukhov, A. E., Fursa, A. S., & Moskalenko, K. V. (2015). Classification and separation of granules in vortex granulators. *Chemical and Petroleum Engineering*, *51*(5-6), 311–318. doi:10.100710556-015-0044-x

Artyukhov, A. E., Obodiak, V. K., Boiko, P. G., & Rossi, P. C. (2017). Computer modeling of hydrodynamic and heat-mass transfer processes in the vortex type granulation devices. *CEUR Workshop Proceedings*, *1844*, 33–47.

Artyukhov, A. E., & Sklabinskyi, V. I. (2013). Experimental and industrial implementation of porous ammonium nitrate producing process in vortex granulators. *Naukovyi Visnyk Natsionalnoho Hirnychoho Universytetu*, *6*, 42–48.

Artyukhov, A., Sklabinskiy, V., Ivaniia, A., & Moskalenko, K. (2016). Software for calculation of vortex type granulation devices. *CEUR Workshop Proceedings*, *1761*, 374–385.

Artyukhov, A., & Sklabinskyi, V. (2015). Hydrodynamics of gas flow in small-sized vortex granulators in the production of nitrogen fertilizers. *Chemistry & Chemical Technology*, *9*(3), 337–342. doi:10.23939/chcht09.03.337

Artyukhova, N. A., Shandyba, A. B., & Artyukhov, A. E. (2014). Energy efficiency assessment of multi-stage convective drying of concentrates and mineral raw materials. *Naukovyi Visnyk Natsionalnoho Hirnychoho Universytetu*, *1*, 92–98.

Atencia, J., & Beebe, D. J. (2005). Controlled microfluidic interfaces. *Nature*, *437*(7059), 648–655. doi:10.1038/nature04163 PMID:16193039

Auroux, P. A., Iossifidis, D., Reyes, D. R., & Manz, A. (2002). Micro total analysis systems. 2. Analytical standard operations and applications. *Analytical Chemistry*, *74*(12), 2637–2652. doi:10.1021/ac020239t PMID:12090654

Aussillous, P., & Quéré, D. (2000). Deposition of a Fluid on the Wall of a Tube. *Physics of Fluids*, *15*(10), 2367–2371. doi:10.1063/1.1289396

Azarmanesh, M., & Farhadi, M. (2016). The effect of weak-inertia on droplet formation phenomena in T-junction microchannel. *Meccanica*, *51*(4), 819–834. doi:10.100711012-015-0245-6

azom.com. (2010). *RTDs: Principle of Operation, Materials, Configuration and Benefits*. Retrieved from: https://www.azom.com/article.aspx?ArticleID=5573

Badilescu, S., & Packirisamy, M. (2012). Microfluidics-Nano-Integration for Synthesis and Sensing. *Polymers*, *4*(2), 1278–1310. doi:10.3390/polym4021278

Bahga, A., & Madisetti, V. (2014). *Internet of Things (A Hands-on-Approach)*. VPT.

Bakhtiary-Davijany, H., Hayer, F., Phanam, X. K., Myrstad, R., Venvik, H. J., Pfeifer, P., & Holmen, A. (2011). Characteristics of an Integrated Micro Packed Bed Reactor-Heat Exchanger for methanol synthesis from syngas. *Chemical Engineering Journal*, *167*(2-3), 496–503. doi:10.1016/j.cej.2010.08.074

Ball, D. W. (2014). *Physical Chemistry* (2nd ed.). Stamford, CT: Cengage Learning.

Barlow, F., Wood, J., Elshabini, A., Stephens, E. F., Feeler, R., Kemner, G., & Junghans, J. (2009). Fabrication of precise fluidic structures in LTCC. *International Journal of Applied Ceramic Technology*, *6*(1), 18–23. doi:10.1111/j.1744-7402.2008.02315.x

Bauer, T., Schubert, M., Lange, R., & Abiev, R. (2006). Intensification of Heterogeneous Catalytic Gas–Fluid Interactions in Reactors with a Multichannel Monolithic Catalyst. *Russian Journal of Applied Chemistry*, *79*(7), 1047–1056. doi:10.1134/S107042720607019

Becker, H., & Locascio, L. E. (2002). Polymer microfluidic devices. *Talanta*, *56*(2), 267–287. doi:10.1016/S0039-9140(01)00594-X PMID:18968500

Becker, R., & Sauter, F. (1964). *Electromagnetic fields and interactions* (Vol. 1). Courier Corporation.

Ben Radhia, R., Harmand, S., Corriou, J. P., & Ben Jabrallah, S. (2014). Liquid film evaporation enhancement in a vertical annulus with preheated air flow. *Desalination and Water Treatment*, *52*(7–9), 1654–1661. doi:10.1080/19443994.2013.806999

Bercic, G., & Pintar, A. (1997). The role of gas bubbles and liquid slug lengths on mass transport in the Taylor flow through capillaries. *Chemical Engineering Science*, *52*(21-22), 3709–3719. doi:10.1016/S0009-2509(97)00217-0

Bertsch, S. S., Groll, E. A., & Garimella, S. V. (2009). Effects of heat flux, mass flux, vapor quality, and saturation temperature on flow boiling heat transfer in microchannels. *International Journal of Multiphase Flow*, *35*(2), 142–154. doi:10.1016/j.ijmultiphaseflow.2008.10.004

Bhagat, R. D., & Watt, K. M. (2015). Effect Of Bond Number On Thermophysical Properties Of Working Fluid Used In Closed Loop. Pulsating Heat Pipe: A Review. *International Journal Of Science And Research*, *4*(9), 544–551.

Bhoi, R., Sen, N., Singh, K. K., Mahajani, S. M., Shenoy, K. T., Rao, H., & Ghosh, S. K. (2014). Transesterification of Sunflower oil in microreactors. *International Journal of Chemical Reactor Engineering*, *12*(1), 47–62. doi:10.1515/ijcre-2013-0105

Blake, T. D. (2006). The physics of moving wetting lines. *Journal of Colloid and Interface Science*, *299*(1), 1–13. doi:10.1016/j.jcis.2006.03.051 PMID:16631781

Blazek, J. (2001). *Computational Fluid Dynamics: Principles and Applications*. Oxford, UK: Elsevier Science Ltd.

Bolivar, J. M., & Nidetzky, B. (2013). Multiphase biotransformations in microstructured reactors: Opportunities for biocatalytic process intensification and smart flow processing. *Green Processes Synthesis*, *2*, 541–559.

Boyadjiev, C. B., & Babak, V. N. (2000). *Non-linear mass transfer and hydrodynamic stability*. Amsterdam: Elsevier Science B.V. doi:10.1016/B978-0-444-50428-9.50012-0

Boyer, S. (2009). SCADA: Supervisory Control and Data Acquisition (4th ed.). ISA: The Instrumentation, Systems, and Automation Society.

Brackbill, J. U., Kothe, D. B., & Zemach, C. (1992). A continuum method for modelling surface tension. *Journal of Computational Physics*, *100*(2), 335–354. doi:10.1016/0021-9991(92)90240-Y

Bretherton, F. P. (1961). The Motion of Long Bubbles in Tubes. *Journal of Fluid Mechanics*, *10*(02), 166–188. doi:10.1017/S0022112061000160

Burkle-Vitzthum, V., Moulis, F., Zhang, J., Commenge, J. M., Schaer, E., & Marquaire, P. M. (2015). Annular flow microreactor: An efficient tool for kinetic studies in gas phase at very short residence times. *Chemical Engineering Research & Design*, *94*, 611–623. doi:10.1016/j.cherd.2014.10.003

Burns, J. R., & Ramshaw, C. (1999). Development of a microreactor for chemical production. *Chemical Engineering Research & Design*, *77*(3), 206–211. doi:10.1205/026387699526106

Capretto, L., Cheng, W., Hill, M., & Zhang, X. (2011). Micromixing within Microfluidic Devices. *Microfluidics*, *304*, 27–68. doi:10.1007/128_2011_150 PMID:21526435

Catelani, M., Ciani, L., Scotto, R., & Zanobini, A. (2016). Measurements and characterization of air temperature sensors for weather stations. In *Proceedings of the IEEE International Conference on Instrumentation and Measurement Technology* (pp. 269-273). Taipei, Taiwan: IEEE.

Celik, I. B., Ghia, U., Roache, P. J., Freitas, C. J., Coleman, H., & Raad, P. E. (2008). Procedure for Estimation and Reporting of Uncertainty Due to Discretization in CFD Applications. *Journal of Fluids Engineering*, *130*, 078001-1–4.

Chalfi, T. Y., & Ghiaasiaan, S. M. (2008). Pressure drop caused by flow area changes in capillaries under low flow conditions. *International Journal of Multiphase Flow*, *34*(1), 2–12. doi:10.1016/j.ijmultiphaseflow.2007.09.004

Chamoun, G. C., Schilder, F., & Brns, M. (2011). Stabilization of vortices in the wake of a circular cylinder using harmonic forcing. *Physical Review. E*, *83*(6), 066308. doi:10.1103/PhysRevE.83.066308 PMID:21797478

Chau, L. K., Osborn, T., Wu, C. C., & Yager, P. (1999). Microfabricated silicon flow-cell for optical monitoring of biological fluids. *Analytical Sciences*, *15*(8), 721–724. doi:10.2116/analsci.15.721

Cheek, B. J., Steel, A. B., Torres, M. P., Yu, Y. Y., & Yang, H. (2001). Chemiluminescence detection for hybridization assays on the flow-thru chip, a three-dimensional microchannel biochip. *Analytical Chemistry*, *73*(24), 5777–5783. doi:10.1021/ac0108616 PMID:11791544

Cheng, P., & Wu, H. Y. (2006). Mesoscale and microscale phase-change heat transfer. *Advances in Heat Transfer*, *39*, 461–563. doi:10.1016/S0065-2717(06)39005-3

Chen, P. C., Wu, M. H., & Wang, Y. N. (2014). Microchannel geometry design for rapid and uniform reagent distribution. *Microfluidics and Nanofluidics*, *17*(2), 275–285. doi:10.100710404-013-1312-8

Chen, Q., Zhai, Z., & Wang, L. (2007). Computer modeling of multiscale fluid flow and heat and mass transfer in engineered spaces. *Chemical Engineering Science*, *62*(13), 3580–3588. doi:10.1016/j.ces.2006.11.059

Chen, S., & Doolen, G. D. (1998). Lattice Boltzmann Method For Fluid Flows. *Annual Review of Fluid Mechanics*, *30*(1), 329–364. doi:10.1146/annurev.fluid.30.1.329

Chen, Y., Lu, B., Chen, Y., & Feng, X. (2015). Breathable and Stretchable Temperature Sensors Inspired by Skin. *Scientific Reports*, *5*(11505), 1–11. PMID:26095941

Cherif, A. A., & Daïf, A. (1999). Etude numérique du transfert dechaleur et de masse entre deux plaques planes verticalesen présence dun film de liquide binaireruisselant sur lune des plaques chauffée. *International Journal of Heat and Mass Transfer*, *42*(13), 2399–2418. doi:10.1016/S0017-9310(98)00339-1

Cheri, M. S., Latifi, H., Moghaddam, M. S., & Shahraki, H. (2013). Simulation and experimental investigation of planar micromixers with short-mixing-length. *Chemical Engineering Journal, 234*, 247–255. doi:10.1016/j.cej.2013.08.067

Choi, S., Park, I., Hao, Z., Holman, H.-Y. N., & Pisano, A. P. (2012). Quantitative studies of long-term stable, top-down fabricated silicon nanowire pH sensors. *Applied Physics. A, Materials Science & Processing, 107*(2), 421–428. doi:10.100700339-011-6754-9

Chueluecha, N., Amaraporn, K., & Attasak, J. (2017). Biodiesel synthesis using heterogeneous catalyst in a packed-microchannel. *Energy Conversion and Management, 141*, 145–154. doi:10.1016/j.enconman.2016.07.020

Chukwudozie, C., & Tyagi, M. (2013). Pore scale inertial flow simulations in 3-D smooth and rough sphere packs using lattice Boltzmann method. *AIChE Journal. American Institute of Chemical Engineers, 59*(12), 4858–4870. doi:10.1002/aic.14232

Chung, C. K., & Shih, T. R. (2007). A rhombic micromixer with asymmetrical flow for enhancing mixing. *Journal of Micromechanics and Microengineering, 17*(12), 2495–2504. doi:10.1088/0960-1317/17/12/016

Crowe, C. T. (2006). *Multiphase flow handbook*. New York: Taylor & Francis Group, LLC.

Cui, B., Diamant, H., Lin, B., & Rice, S. A. (2004). Anomalous Hydrodynamic Interaction in a Quasi-Two-Dimensional Suspension. *Physical Review Letters, 92*(25), 258301. doi:10.1103/PhysRevLett.92.258301 PMID:15245065

Dames, C. (2008). RTDs. In D. Li (Ed.), *Encyclopedia of Microfluidics and Nanofludics* (pp. 1782–1790). Springer. doi:10.1007/978-0-387-48998-8 1354

Das, S. P., Nikolayev, V. S., Lefèvre, F., Pottier, B., Khandekar, S., & Bonjour, J. (2010). Thermally Induced Two-phase Oscillating Flow inside a Capillary Tube. *International Journal of Heat and Mass Transfer, 53*(19-20), 3905–3913. doi:10.1016/j.ijheatmasstransfer.2010.05.009

Daw, R., & Finkelstein, J. (2006). Lab on a chip. *Nature, 442*(7101), 367. doi:10.1038/442367a

de Menech, M., Garstecki, P., Jousse, F., & Stone, H. A. (2008). Transition from squeezing to dripping in a microfluidic T-shaped junction. *Journal of Fluid Mechanics, 595*. doi:10.1017/S002211200700910X

Delaunay, Y., & Kaiktsis, L. (2001). Control of circular cylinder wakes using base mass transpiration. *Physics of Fluids, 13*(11), 3285–3302. doi:10.1063/1.1409968

Delsman, E. R., De Croon, M. H. J. M., Pierik, A., Kramer, G. J., Cobden, P. D., Hofmann, Ch., ... Schouten, J. C. (2004). Design and operation of a preferential oxidation microdevice for a portable fuel processor. *Chemical Engineering Science, 59*(22-23), 4795–4802. doi:10.1016/j.ces.2004.07.099

Deng, D., Wan, W., Zhang, J., & Chu, X. (2017). Flow boiling enhancement of structured microchannels with micro pin fins. *International Journal of Heat and Mass Transfer, 105*, 338–349. doi:10.1016/j.ijheatmasstransfer.2016.09.086

Derksen, J. J. (2015). Simulations of liquid-to-solid mass transfer in a fluidized microchannel. *Microfluidics and Nanofluidics, 18*(5-6), 829–839. doi:10.100710404-014-1475-y

Dernovsek, O., Eberstein, M., Schiller, W. A., Naeini, A., Preu, G., & Wersing, W. (2001). LTCC glass-ceramic composites for microwave application. *Journal of the European Ceramic Society, 21*(10-11), 1693–1697. doi:10.1016/S0955-2219(01)00096-6

Dessimoz, A.-L., Cavin, L., Renken, A., & Kiwi-Minsker, L. (2008). Liquid–liquid two-phase flow patterns and mass transfer characteristics in rectangular glass microreactors. *Chemical Engineering Science, 63*(16), 4035–4044. doi:10.1016/j.ces.2008.05.005

Deutsch, S. (1979). A Preliminary study of the fluid mechanics of liquid penetrant testing. *Journal of Research of the Notionol Bureou of Standords., 84*(4), 287. doi:10.6028/jres.084.012

Dong, S., Triantafyllou, G. S., & Karniadakis, G. E. (2008). Elimination of vortex streets in bluff-body flows. *Physical Review Letters, 100*(20), 204501. doi:10.1103/PhysRevLett.100.204501 PMID:18518541

Dong, Z., Yao, C., Zhang, Y., Chen, G., Yuan, Q., & Xu, J. (2015). Hydrodynamics and Mass Transfer of Oscillating Gas-Liquid Flow in Ultrasonic Microreactors. *American Institute of Chemical Engineers Journal, 61*, 3964–3972.

Dung, S., & Yang, S. (2006). *Second law based optimization of free convection film-wise condensation on a horizontal tube*. Academic Press. doi:10.1016/j.icheatmasstransfer.2006.01.013

Duran Martinez, F. L., Billet, A.-M., Julcour-Lebigue, C., & Larachi, F. (2015). Hydrodynamics And Mass Transfer In Taylor Flow. COMSOL conference, Grenoble, France.

Dvornikov, N. A., & Belousov, P. P. (2011). Investigation of a fluidized bed in a vortex chamber. *Journal of Applied Mechanics and Technical Physics, 52*(2), 206–211. doi:10.1134/S0021894411020076

Ergun, S. (1952). Fluid flow through packed columns. *Chemical Engineering Progress, 48*, 89–94.

Esteban, B., Riba, J. R., Baquero, G., Rius, A., & Puig, R. (2012). Temperature dependence of density and viscosity of vegetable oils. *Biomass and Bioenergy, 42*, 164–171. doi:10.1016/j.biombioe.2012.03.007

Falconi, C. J., Lehrenfeld, C., Marschall, H., Meyer, C., Abiev, R., Bothe, D., ... Wörner, M. (2016). Numerical and experimental analysis of local flow phenomena in laminar Taylor flow in a square mini-channel. *Physics of Fluids, 28*(1), 012109. doi:10.1063/1.4939498

Fasina, O. O., & Colley, Z. (2008). Viscosity and specific heat of vegetable oils as a function of temperature: 35 °C to 180 °C. *International Journal of Food Properties, 11*(4), 738–746. doi:10.1080/10942910701586273

Fazeli, A., & Benham, M. (2010). Hydrogen production in a zigzag and straight catalytic wall coated micro channel reactor by CFD modeling. *International Journal of Hydrogen Energy*, *35*(17), 9496–9503. doi:10.1016/j.ijhydene.2010.05.052

Fazio, R., & Iacono, S. (2009). Entrapped Gas Action for One-Dimensional Models of Capillary Dynamics. *World Congress on Engineering*, 2, 1-3.

Feddaoui, M., Belahmidi, E. M., Mir, A., & Bendou, A. (2001). Numerical study of the evaporative cooling of liquid film in laminar mixed convection tube flows. *International Journal of Thermal Sciences*, *40*(11), 1011–1020. doi:10.1016/S1290-0729(01)01286-8

Feddaoui, M., Meftah, H., & Mir, A. (2006). The numerical computation of the evaporative cooling of falling water film in turbulent mixed convection inside a vertical tube. *International Communications in Heat and Mass Transfer*, *33*(7), 917–927. doi:10.1016/j.icheatmasstransfer.2006.04.004

Ferziger, J. H., & Peric, M. (2002). *Computational Methods for Fluid Dynamics* (3rd. ed.). Berlin, Germany: Springer-Verlag. doi:10.1007/978-3-642-56026-2

Fraden, J. (2010). Temperature Sensors. In J. Fraden (Ed.), *Handbook of Modern Sensors Physics, Designs, and Applications* (pp. 519–567). Springer. doi:10.1007/978-1-4419-6466-3_16

Friedman, S. P. (1999). Dynamic contact angle explanation of flow rate-dependent saturation-pressure relationships during transient liquid flow in unsaturated porous medium. *Journal of Adhesion Science and Technology*, *13*(12), 1495–1518. doi:10.1163/156856199X00613

Fries, N., & Dreye, M. (2008a). The transition from inertial to viscous flow in capillary rise. *Journal of Colloid and Interface Science*, *327*(1), 125–128. doi:10.1016/j.jcis.2008.08.018 PMID:18774142

Fries, N., & Dreye, M. (2008b). An analytic solution of capillary rise restrained by gravity. *Journal of Colloid and Interface Science*, *320*(1), 259–263. doi:10.1016/j.jcis.2008.01.009 PMID:18255086

Fries, N., & Dreye, M. (2009). Dimensionless scaling methods for capillary rise. *Journal of Colloid and Interface Science*, *338*(2), 514–518. doi:10.1016/j.jcis.2009.06.036 PMID:19628213

Galletti, C., Brunazzi, E., Siconolfi, L., Spaltro, D., & Mauri, R. (2017). Mixing Performance of Arrow-Shaped Micro-Devices. *Chemical Engineering Transactions*, *57*, 1309–1314.

Ganapathy, H., Shooshtari, A., Dessiatoun, S., Alshehhi, M., & Ohadi, M. (2014). Fluid flow and mass transfer characteristics of enhanced CO2 capture in a minichannel reactor. *Applied Energy*, *119*, 43–56. doi:10.1016/j.apenergy.2013.12.047

Gao, J., Yin, X. F., & Fang, Z. L. (2004). Integration of single cell injection, cell lysis, separation and detection of intracellular constituents on a microfluidic chip. *Lab on a Chip*, *4*(1), 47–52. doi:10.1039/b310552k PMID:15007440

Gao, P. Z., Rebrov, E. V., Verhoeven, T. M., Schouten, J. C., Kleismit, R., Kozlowski, G., ... Subramanyam, G. (2010). Structural Investigations and Magnetic Properties of Sol-Gel Ni0.5Zn0.5Fe2O4 Thin Films for Microwave Heating. *Journal of Applied Physics*, *107*(4), 44317. doi:10.1063/1.3309767

Gascon, J., van Ommen, J. R., Moulijn, J. A., & Kapteijn, F. (2015). Structuring catalyst and reactor – an inviting avenue to process intensification. *Catalysis Science & Technology*, *5*(2), 807–817. doi:10.1039/C4CY01406E

Gascoyne, P. R. C., Vykoukal, J. V., Schwartz, J. A., Anderson, T. J., Vykoukal, D. M., Current, K. W., ... Andrews, C. (2004). Dielectrophoresis-based programmable fluidic processors. *Lab on a Chip*, *4*(4), 299–309. doi:10.1039/b404130e PMID:15269795

Gilles, T. (2012). *Automotive Science: Inspection, Maintenance, Repair*. Cengage Learning.

Gillet, J. E. (2001). Chemical Engineering Education in the Next Century. *Chemical Engineering & Technology*, *24*(6), 561–570. doi:10.1002/1521-4125(200106)24:6<561::AID-CEAT561>3.0.CO;2-X

Goldschmidt, M. (2001). *Hydrodynamic Modelling of Fluidised Bed Spray Granulation* (Ph.D. thesis). University of Twente.

Golonka, L. J. (2003, August). Low Temperature Co-fired Ceramic (LTCC) technology in microelectronics. In *Proc. Symp. Processing of Electroceramics* (pp. 313-329). Academic Press.

Gongora-Rubio, M. R., Espinoza-Vallejos, P., Sola-Laguna, L., & Santiago-Aviles, J. J. (2001). Overview of low temperature co-fired ceramics tape technology for meso-system technology (MsST). *Sensors and Actuators. A, Physical*, *89*(3), 222–241. doi:10.1016/S0924-4247(00)00554-9

Gonzalez-Mut, P., & Ortega-Casanova, J. (2016). CFD Study on the Drag Reduction of a Bluff Body by a Combined Blowing/Suction Technique. In G. Powell (Ed.), *Computational Fluid Dynamics (CFD): Characteristics, Applications and Analysis* (pp. 117–142). New York: Nova Science Publishers.

Goodrum, J. W. (2002). Volatility and boiling points of biodiesel from vegetable oils and tallow. *Biomass and Bioenergy*, *22*(3), 205–211. doi:10.1016/S0961-9534(01)00074-5

Greco, A., & Vanoli, G. P. (2005). Flow boiling heat transfer with HFC mixtures in a smooth horizontal tube. Part I: Experimental investigations. *Experimental Thermal and Fluid Science*, *29*(2), 189–198. doi:10.1016/j.expthermflusci.2004.03.003

Groß, G. A., Thelemann, T., Schneider, S., Boskovic, D., & Köhler, J. M. (2008). Fabrication and fluidic characterization of static micromixers made of low temperature cofired ceramic (LTCC). *Chemical Engineering Science*, *63*(10), 2773–2784. doi:10.1016/j.ces.2008.02.030

Gruber, R., & Melin, T. (2003). Radial mass-transfer enhancement in bubble-train flow. *International Journal of Heat and Mass Transfer*, *46*(15), 2799–2808. doi:10.1016/S0017-9310(03)00053-X

Guillot, P., & Colin, A. (2005). Stability of parallel flows in a microchannel after a T junction. *Physical Review. E*, *72*(6), 66301. doi:10.1103/PhysRevE.72.066301 PMID:16486053

Günther, A., Khan, S. A., Thalmann, M., Trachsel, F., & Jensen, K. F. (2004). Transport and reaction in microscale segmented gas–liquid flow. *Lab on a Chip*, *4*(4), 278–286. doi:10.1039/B403982C PMID:15269792

Guo, Y., Zhong, J., Xing, Y., Li, D., & Lin, R. (2007). Volatility of Blended Fuel of Biodiesel and Ethanol. *Energy & Fuels*, *21*(2), 1188–1192. doi:10.1021/ef0605807

Gupta, R., Fletcher, D. F., & Haynes, B. S. (2010). CFD modelling of flow and heat transfer in the Taylor flow regime. *Chemical Engineering Science*, *65*(6), 2094–2107. doi:10.1016/j.ces.2009.12.008

Gürsel, I. V., Kockmann, N., & Hessel, V. (2017). Fluidic separation in microstructured devices – Concepts and their Integration into process flow networks. *Chemical Engineering Science*, *169*, 3–17. doi:10.1016/j.ces.2017.03.023

Han, Y., & Shikazono, N. (2009). The effect of bubble acceleration on the liquid film thickness in micro tubes. *International Journal of Heat and Fluid Flow*, *30*(5), 842–853. doi:10.1016/j.ijheatfluidflow.2009.02.019

Harirchian, T., & Garimella, S. V. (2010). A comprehensive flow regime map for microchannel flow boiling with quantitative transition criteria. *International Journal of Heat and Mass Transfer*, *53*(13-14), 2694–2702. doi:10.1016/j.ijheatmasstransfer.2010.02.039

Hazel, A. L., & Heil, M. (2002). The steady propagation of a semi-infinite bubble into a tube of elliptical or rectangular cross-section. *Journal of Fluid Mechanics*, *470*, 91. doi:10.1017/S0022112002001830

He, P., Haswell, S. J., & Fletcher, P. D. I. (2004). Microwave Heating of Heterogeneously Catalysed Suzuki Reactions in a Micro Reactor. *Lab on a Chip*, *4*(1), 38–41. doi:10.1039/b313057f PMID:15007438

He, S., An, P., Li, J., & Jackson, J. D. (1998). Combined heat and mass transfer in a uniformly heated vertical tube with water film cooling. *International Journal of Heat and Fluid Flow*, *19*(5), 401–417. doi:10.1016/S0142-727X(98)10021-8

Hessel, V., Angeli, P., Gavriilidis, A., & Löwe, H. (2005). Gas–Liquid and Gas–Liquid–Solid Microstructured Reactors: Contacting Principles and Applications. *Industrial & Engineering Chemistry Research*, *44*(25), 9750–9769. doi:10.1021/ie0503139

Hessel, V., Müller, A., & Kolb, G. (2005). *Chemical Micro Process Engineering. Processing and Plants*. Weinheim, Germany: Wiley–VCH Verlag. doi:10.1002/3527603581

Hoffman, R. L. (1975). A study of the advancing interface. I. Interface shape in liquid—gas systems. *Journal of Colloid and Interface Science*, *50*(2), 228–241. doi:10.1016/0021-9797(75)90225-8

Hong, C. C., Choi, J. W., & Ahn, C. H. (2004). A novel in-plane passive microfluidic mixer with modified Tesla structures. *Lab on a Chip*, *4*(2), 109–113. doi:10.1039/b305892a PMID:15052349

Hossain, S., Husain, A., & Kim, K. Y. (2011). Optimization of micromixer with staggered herringbone grooves on top and bottom walls. *Engineering Applications of Computational Fluid Mechanics*, *5*(4), 506–516. doi:10.1080/19942060.2011.11015390

Howard, J. A., & Walsh, P. A. (2013). Review and extensions to film thickness and relative bubble drift velocity prediction methods in laminar Taylor or slug flows. *International Journal of Multiphase Flow*, *55*, 32–42. doi:10.1016/j.ijmultiphaseflow.2013.04.005

Hsiao, K. Y., Wu, C. Y., & Huang, Y. T. (2014). Fluid mixing in a microchannel with longitudinal vortex generators. *Chemical Engineering Journal*, *235*, 27–36. doi:10.1016/j.cej.2013.09.010

huber-online.com. (2017). *Temperature Control for Chemical Research and Production*. Retrieved from: http://www.huber-online.com/download/whitepapers_bulletins/Huber_TechDoc8.pdf

Huo, X., Chen, L., Tian, Y. S., & Karayiannis, T. G. (2004). Flow boiling and flow regimes in small diameter tubes. *Applied Thermal Engineering*, *24*(8-9), 1225–1239. doi:10.1016/j.applthermaleng.2003.11.027

Huynh, T. (2015). Fundamentals of Thermal Sensors. In C. Jha (Ed.), *Thermal Sensors: Principles and Applications for Semiconductor Industries* (pp. 5–54). Springer.

Ibanez-Garcia, N., Mercader, M. B., Mendes da Rocha, Z., Seabra, C. A., Góngora-Rubio, M. R., & Chamarro, J. A. (2006). Continuous flow analytical microsystems based on low-temperature co-fired ceramic technology. Integrated potentiometric detection based on solvent polymeric ion-selective electrodes. *Analytical Chemistry*, *78*(9), 2985–2992. doi:10.1021/ac051994k PMID:16642984

Inamuro, T., Yoshino, M., & Ogino, F. (1999). Lattice Boltzmann simulation of flows in a three-dimensional porous structure. *International Journal for Numerical Methods in Fluids*, *29*(7), 737–748. doi:10.1002/(SICI)1097-0363(19990415)29:7<737::AID-FLD813>3.0.CO;2-H

Iudicello, F. (1998). *Guide for the verification and validation of computational fluid dynamics simulations*. AIAA G-077-1998.

Jain, M., Rao, A., & Nandakumar, K. (2013). Numerical study on shape optimization of groove micromixers. *Microfluidics and Nanofluidics*, *15*(5), 689–699. doi:10.100710404-013-1169-x

Janata, J. (2009). Thermal Sensors. In J. Janata (Ed.), *Principles of Chemical Sensors* (pp. 51–62). Springer. doi:10.1007/b136378_3

Jin, C. Y., Li, Z., Williams, R. S., Lee, K.-C., & Park, I. (2011). Localized Temperature and Chemical Reaction Control in Nanoscale Space by Nanowire Array. *Nano Letters*, *11*(11), 4818–4825. doi:10.1021/nl2026585 PMID:21967343

Jindal, R., & Cramer, S. M. (2004). On-chip electrochromatography using sol–gel immobilized stationary phase with UV absorbance detection. *Journal of Chromatography. A*, *1044*(1-2), 277–285. doi:10.1016/j.chroma.2004.05.065 PMID:15354449

Jones, M. (2012). Accurate Temperature Sensing with an External P-N Junction. *Linear technology Corporation – Application Notes, 137*, 1-12.

Jovanović, J., Zhou, W., Rebrov, E. V., Nijhuis, T., Hessel, V., & Schouten, J. C. (2011). Liquid–liquid slug flow: Hydrodynamics and pressure drop. *Chemical Engineering Science*, *66*(1), 42–54. doi:10.1016/j.ces.2010.09.040

Jóźwik, P., Karcz, M., & Badur, J. (2011). Numerical Modelling of a Microreactor for Thermocatalytic Decompostion of Toxic Compounds. *Chemical & Process Engineering*, *32*(3), 215–227. doi:10.2478/v10176-011-0017-3

Kaewklum, R., Kuprianov, V. I., & Douglas, P. L. (2009). Hydrodynamics of air–sand flow in a conical swirling fluidized bed: A comparative study between tangential and axial air entries. *Energy Conversion and Management*, *50*(12), 2999–3006. doi:10.1016/j.enconman.2009.07.019

Kalaivanan, R., & Rathnasamy, R. (2011). Experimental Investigation of Forced Convective Heat Transfer in Rectangular Micro-channels. *Jordan Journal of Mechanical and Industrial Engineering*, *5*, 383–387.

Kandlikar, S. G., & Balasubramanian, P. (2004). An extension of the flow boiling correlation to transition, laminar, and deep laminar flows in minichannels and microchannels. *Heat Transfer Engineering*, *25*(3), 86–93. doi:10.1080/01457630490280425

Kashid, M. N., Harshe, Y. M., & Agar, D. W. (2007). Liquid-Liquid Slug Flow in a Capillary: An Alternative to Suspended Drop or Film Contactors. *Industrial & Engineering Chemistry Research*, *46*(25), 8420–8430. doi:10.1021/ie070077x

Kashid, M. N., Renken, A., & Kiwi-Minsker, L. (2011). Gas–liquid and liquid–liquid mass transfer in microstructured reactors. *Chemical Engineering Science*, *66*(17), 3876–3897. doi:10.1016/j.ces.2011.05.015

Kashid, M., & Kiwi-Minsker, L. (2011). Quantitative prediction of flow patterns in liquid–liquid flow in micro-capillaries. *Chemical Engineering and Processing: Process Intensification*, *50*(10), 972–978. doi:10.1016/j.cep.2011.07.003

Kashmiri, S., & Makinwa, K. (2009). Measuring the thermal diffusivity of CMOS chips. In *Proceedings of IEEE Sensors Conference* (pp. 45-48). Christchurch, New Zealand: IEEE.

Kashmiri, S., Xia, S., & Makinwa, K. (2009). A Temperature-to-Digital Converter Based on an Optimized Electrothermal Filter. *IEEE Journal of Solid-State Circuits*, *44*(7), 2026–2035. doi:10.1109/JSSC.2009.2020248

Kawahara, A., Chung, P. M.-Y., & Kawaji, M. (2002). Investigation of two-phase flow pattern, void fraction and pressure drop in a microchannel. *International Journal of Multiphase Flow*, *28*(9), 1411–1435. doi:10.1016/S0301-9322(02)00037-X

Kececi, S., Wörner, M., Onea, A., & Soyhan, H. S. (2009). Recirculation time and liquid slug mass transfer in co-current upward and downward Taylor flow. *Catalysis Today*, *147*, S125–S131. doi:10.1016/j.cattod.2009.07.054

Kenis, P., Ismagilov, R., & Whitesides, G. (1999, July 2). Microfabrication inside capillaries using multiphase laminar flow patterning. *Science (New York, N.Y.)*. Retrieved from http://www.ncbi.nlm.nih.gov/pubmed/10390366

Kenry, J., Yeo, J., & Lim, C. (2016). Emerging flexible and wearable physical sensing platforms for healthcare and biomedical applications. *Microsystems & Nanoengineering*, *2*(16043), 1–19.

Kew, P. A., & Cornwell, K. (1997). Correlations for the prediction of boiling heat transfer in small-diameter channels. *Applied Thermal Energy.*, *17*(8-10), 705–715. doi:10.1016/S1359-4311(96)00071-3

Khalatov, A. A. (2010). Heat transfer and hydrodynamics in the fields of mass forces. *Journal of Engineering Physics and Thermophysics*, *83*(4), 794–808. doi:10.100710891-010-0397-0

Khandekar, S., Dollinger, N., & Groll, M. (2003). Understanding Operational Regimes of Pulsating Heat Pipes: An Experimental Study. *Applied Thermal Engineering*, *23*(6), 707–719. doi:10.1016/S1359-4311(02)00237-5

Khandekar, S., Gautam, A. P., & Sharma, P. K. (2009). Multiple quasi-steady states in a closed loop pulsating heat pipe. *International Journal of Thermal Sciences*, *48*(3), 535–546. doi:10.1016/j.ijthermalsci.2008.04.004

Kim, B. S., Kwak, B. S., Shin, S., Lee, S., Kim, K. M., Jung, H. I., & Cho, H. H. (2011). Optimization of microscale vortex generators in a microchannel using advanced response surface method. *International Journal of Heat and Mass Transfer*, *54*(1), 118–125. doi:10.1016/j.ijheatmasstransfer.2010.09.061

Kim, D. S., Lee, S. H., Kwon, T. H., & Ahn, C. H. (2005). A serpentine laminating micromixer combining splitting/recombination and advection. *Lab on a Chip*, *5*(7), 739–747. doi:10.1039/b418314b PMID:15970967

Kim, J., Li, Z., & Park, I. (2011). Direct synthesis and integration of functional nanostructures in microfluidic devices. *Lab on a Chip*, *11*(11), 1946. doi:10.1039/c1lc20079h PMID:21499615

Kim, S. M., & Mudawar, I. (2013a). Universal approach to predicting saturated flow boiling heat transfer in mini/micro-channels – Part I. Dryout incipience quality. *International Journal of Heat and Mass Transfer*, *64*, 1226–1238. doi:10.1016/j.ijheatmasstransfer.2013.04.016

Kim, S. M., & Mudawar, I. (2013b). Universal approach to predicting saturated flow boiling heat transfer in mini/micro-channels – Part II. Two-phase heat transfer coefficient. *International Journal of Heat and Mass Transfer*, *64*, 1239–1256. doi:10.1016/j.ijheatmasstransfer.2013.04.014

Kim, S. M., & Mudawar, I. (2014). Review of databases and predictive methods for heat transfer in condensing and boiling mini/micro-channel flows. *International Journal of Heat and Mass Transfer*, *77*, 627–652. doi:10.1016/j.ijheatmasstransfer.2014.05.036

Kirby, B. (2010). *Micro- and Nanoscale Fluid Mechanics: Transport in Microfluidic Devices*. Cambridge, UK: Cambridge University Press. doi:10.1017/CBO9780511760723

Kiwi-Minsker, L., & Renken, A. (2005). Microstructured reactors for catalytic reactions. *Catalysis Today*, *110*(1-2), 2–14. doi:10.1016/j.cattod.2005.09.011

Kolb, W. B., & Cerro, R. L. (1991). Coating the inside of a capillary of square cross section. *Chemical Engineering Science*, *46*(9), 2181–2195. doi:10.1016/0009-2509(91)85119-I

Kovacs, G. T. (1998). *Micromachined transducers sourcebook*. New York: WCB/McGraw-Hill.

Kreutzer, M. T., Du, P., Heiszwolf, J. J., Kapteijn, F., & Moulijn, J. A. (2001). Mass transfer characteristics of three-phase monolith reactors. *Chemical Engineering Science*, *56*(21-22), 6015–6023. doi:10.1016/S0009-2509(01)00271-8

Kreutzer, M. T., Kapteijn, F., Moulijn, J. A., & Heiszwolf, J. J. (2005a). Multiphase monolith reactors: Chemical reaction engineering of segmented flow in microchannels. *Chemical Engineering Science*, *60*(22), 5895–5916. doi:10.1016/j.ces.2005.03.022

Kreutzer, M. T., Kapteijn, F., Moulijn, J. A., Kleijn, C. R., & Heiszwolf, J. J. (2005b). Inertial and interfacial effects on pressure drop of Taylor flow in capillaries. *AIChE Journal. American Institute of Chemical Engineers*, *51*(9), 2428–2440. doi:10.1002/aic.10495

Kumar, A. (2017). Micro-channel fabrication using Cu-sheets as electrode in EDM process and optimization of process parameters using Taguchi Method. *International Journal of Current Engineering and Technology*, *7*(1), 15–18.

Kuo, J. N., & Li, Y. S. (2017). Centrifuge-based micromixer with three-dimensional square-wave microchannel for blood plasma mixing. *Microsystem Technologies*, *23*(7), 2343–2354. doi:10.100700542-016-2950-z

Kuswandi, B., Huskens, J., & Verboom, W. (2007). Optical sensing systems for microfluidic devices: A review. *Analytica Chimica Acta*, *601*(2), 141–155. doi:10.1016/j.aca.2007.08.046 PMID:17920386

labjack.com. (2016). *EI-1022 sensor.* Retrieved from: https://labjack.com/support/datasheets/accessories/ei-1022

Lee, C. Y., Chang, C. L., Wang, Y. N., & Fu, L. M. (2011). Microfluidic mixing: A review. *International Journal of Molecular Sciences*, *12*(5), 3263–3287. doi:10.3390/ijms12053263 PMID:21686184

Lee, H. J., & Lee, S. Y. (2001). Pressure Drop Correlations for Two-Phase Flow within Horizontal Rectangular Channels with Small Height. *International Journal of Multiphase Flow*, *27*(5), 783–796. doi:10.1016/S0301-9322(00)00050-1

Le, T. H., Le, T. H., Dong, T., Ta, Q. B., Tran, M. N., & Karlsen, F. (2015). An effective passive micromixer with shifted trapezoidal blades using wide Reynolds number range. *Chemical Engineering Research & Design*, *93*, 1–11. doi:10.1016/j.cherd.2014.12.003

Li, G., Chen, Q., Li, J., Hu, X., & Zhao, J. (2010). A Compact Disk-Like Centrifugal Microfluidic System for High-Throughput Nanoliter-Scale Protein Crystallization Screening. *Analytical Chemistry*, *82*(11), 4362–4369. doi:10.1021/ac902904m PMID:20459060

Liu, B. F., Ozaki, M., Utsumi, Y., Hattori, T., & Terabe, S. (2003). Chemiluminescence detection for a microchip capillary electrophoresis system fabricated in poly (dimethylsiloxane). *Analytical Chemistry*, *75*(1), 36–41. doi:10.1021/ac026096s PMID:12530816

Liu, H., Li, P., & Lew, J. V. (2010). CFD study on flow distribution uniformity in fuel distributors having multiple structural bifurcations of flow channels. *International Journal of Hydrogen Energy*, *35*(17), 9186–9198. doi:10.1016/j.ijhydene.2010.06.043

Liu, H., Vandu, C. O., & Krishna, R. (2005). Hydrodynamics of Taylor Flow in Vertical Capillaries: Flow Regimes, Bubble Rise Velocity, Liquid Slug Length, and Pressure Drop. *Industrial & Engineering Chemistry Research*, *44*(14), 4884–4897. doi:10.1021/ie049307n

Liu, K., Gu, X. G., Ba, D. C., Li, P. Y., Du, G. Y., Yue, X. J., & Yang, N. (2012). Numerical Research on Flow Characteristics of Vortex Stage in Dry High Vacuum Pump. *Physics Procedia*, *32*, 127–134. doi:10.1016/j.phpro.2012.03.529

Liu, Z.-H., & Yi, J. (2002). Falling film evaporation heat transfer of water/salt mixtures from roll-worked enhanced tubes and tube bundle. *Applied Thermal Engineering*, *22*(1), 83–95. doi:10.1016/S1359-4311(01)00061-8

Liu, Z., & Winterton, R. H. S. (1991). A general correlation for saturated and subcooled flow boiling in tubes and annuli based on a nucleate pool boiling equation. *International Journal of Heat and Mass Transfer*, *34*(11), 2759–2766. doi:10.1016/0017-9310(91)90234-6

Li, W., & Wu, Z. (2010). A general correlation for evaporative heat transfer in micro/mini-channels. *International Journal of Heat and Mass Transfer*, *53*(9-10), 1778–1787. doi:10.1016/j.ijheatmasstransfer.2010.01.012

Li, X., & Li, G. (2016). Modeling and Simulation of Hydrodynamic Bubble-Liquid Turbulent Flows in Bubble-Column Reactors. *Chemical Engineering & Technology*, *39*(12), 2380–2388. doi:10.1002/ceat.201600074

Li, Y., Wang, J., Li, D., & Liu, R. (2018). Flow pattern diagrams of oil-water two-phase microflows and stable parallel flows obtained at low Reynolds numbers. *International Journal of Multiphase Flow*, *98*, 139–146. doi:10.1016/j.ijmultiphaseflow.2017.09.008

Lohse, S. E., Eller, J. R., Sivapalan, S. T., Plews, M. R., & Murphy, C. J. (2013). A Simple Millifluidic Benchtop Reactor System for the High-Throughput Synthesis and Functionalization of Gold Nanoparticles with Different Sizes and Shapes. *ACS Nano*, *7*(5), 4135–4150. doi:10.1021/nn4005022 PMID:23634842

Lo, R. C. (2013). Application of Microfluidics in Chemical Engineering. *Chemical Engineering & Process Techniques*, *1*, 1002.

Love, J. (2007). Temperature Measurement. In J. Love (Ed.), *Process Automation Handbook – A Guide to Theory and Practice* (pp. 99–106). Springer.

Lutze, P., Gani, R., & Woodley, J. M. (2010). Process intensification: A perspective on process synthesis. *Chemical Engineering and Processing: Process Intensification.*, *49*(6), 547–558. doi:10.1016/j.cep.2010.05.002

Maaref, S., Rokhforouz, M.-R., & Ayatollahi, S. (2017). Numerical Investigation of Two Phase Flow in Micromodel Porous Media: Effects of Wettability, Heterogeneity, and Viscosity. *Canadian Journal of Chemical Engineering*, *95*(6), 1213–1223. doi:10.1002/cjce.22762

Madou, M. J. (2002). *Fundamentals of microfabrication: the science of miniaturization.* CRC Press. doi:10.1201/9781482274004

Mahmouda, M. M., & Karayiannis, T. G. (2013). Heat transfer correlation for flow boiling in small to micro tubes. *International Journal of Heat and Mass Transfer*, *66*, 553–574. doi:10.1016/j.ijheatmasstransfer.2013.07.042

Maims, C., Hulme, J., Fielden, P. R., & Goddard, N. J. (2001). Grating coupled leaky waveguide micro channel sensor chips for optical analysis. *Sensors and Actuators. B, Chemical*, *77*(3), 671–678. doi:10.1016/S0925-4005(01)00773-0

Makinwa, K. (2010). Smart temperature sensors in standard CMOS. *Procedia Engineering*, *5*, 930–939. doi:10.1016/j.proeng.2010.09.262

Malecha, K., Pijanowska, D. G., Golonka, L. J., & Torbicz, W. (2009). LTCC microreactor for urea determination in biological fluids. *Sensors and Actuators. B, Chemical*, *141*(1), 301–308. doi:10.1016/j.snb.2009.06.026

Malecha, Z. M., & Malecha, K. (2014). Numerical Analysis of Mixing Under Low and High Frequency Pulsations at Serpentine Micromixers. *Chemical & Process Engineering*, *35*(3), 369–385. doi:10.2478/cpe-2014-0028

Mansoor, M., Haneef, I., Akhtar, S., DeLuca, A., & Udrea, F. (2015). Silicon diode temperature sensors – A review of applications. *Sensors and Actuators. A, Physical*, *232*, 63–74. doi:10.1016/j.sna.2015.04.022

Marre, S., & Jensen, K. F. (2010). Synthesis of micro and nanostructures in microfluidic systems. *Chemical Society Reviews*, *39*(3), 1183. doi:10.1039/b821324k PMID:20179831

Marschall, H., Falconi, C. J., Lehrenfeld, C., Abiev, R., Wörner, M., Reusken, A., & Bothe, D. (2017). *Direct Numerical Simulations of Taylor bubbles in a square mini-channel: detailed shape and flow analysis with experimental validation. In Transport Processes at Fluidic Interfaces* (pp. 663–679). Berlin, Germany: Springer-Verlag.

Martínez-Cisneros, C. S., da Rocha, Z., Ferreira, M., Valdés, F., Seabra, A., Góngora-Rubio, M., & Alonso-Chamarro, J. (2009). A monolithic continuous-flow microanalyzer with amperometric detection based on the green tape technology. *Analytical Chemistry*, *81*(17), 7448–7453. doi:10.1021/ac9012418 PMID:19715366

Martin, R. S., Gawron, A. J., Lunte, S. M., & Henry, C. S. (2000). Dual-electrode electrochemical detection for poly (dimethylsiloxane)-fabricated capillary electrophoresis microchips. *Analytical Chemistry*, *72*(14), 3196–3202. doi:10.1021/ac000160t PMID:10939387

McQuade, D. T., & Seeberger, P. H. (2013). Applying Flow Chemistry: Methods, Materials, and Multistep Synthesis. *The Journal of Organic Chemistry*, *78*(13), 6384–6389. doi:10.1021/jo400583m PMID:23750988

Mehta, B., & Khandekar, S. (2014). Taylor Bubble-train Flow and Heat Transfer in the Context of Pulsating Heat Pipes. *International Journal of Heat and Mass Transfer*, *79*, 279–290. doi:10.1016/j.ijheatmasstransfer.2014.08.005

Mikaelian, D., Haut, B., & Scheid, B. (2015). Bubbly flow and gas–liquid mass transfer in square and circular microchannels for stress-free and rigid interfaces: CFD analysis. *Microfluidics and Nanofluidics*, *19*(3), 523–545. doi:10.100710404-015-1578-0

Mishima, K., & Hibiki, T. (1996). Some characteristics of air–water two-phase flow in small diameter vertical tubes. *International Journal of Multiphase Flow*, *22*(4), 703–712. doi:10.1016/0301-9322(96)00010-9

Mohamad, A. A. (2011). *Lattice Boltzmann Method: Fundamentals and Engineering Applications with Computer Codes*. London: Springer-Verlag London Limited. doi:10.1007/978-0-85729-455-5

Moharana, M. K., Nemade, R. M., & Khandekar, S. (2013). Phase-Change Heat Transfer of Ethanol-Water Mixtures: Towards Development of a Distributed Hydrogen Generator. *Proceedings of the ASME 2013 Summer Heat Transfer Conference.*

Mondal, D., Pal, D., & Chaudhuri, C. R. (2015). Real-time sensing of epithelial cell-cell and cell-substrate interactions by impedance spectroscopy on porous substrates. *Journal of Applied Physics*, *118*(4), 044701. doi:10.1063/1.4927215

Moulijn, J. A., Kreutzer, M. T., Nijhuis, T. A., & Kapteijn, F. (2011). Monolithic Catalysts. and Reactors: High Precision with Low Energy Consumption. In B. C. Gates & H. Knozinger (Eds.), *Advances in Catalysis* (Vol. 54, pp. 249–327). San Diego, CA: Elsevier. Academic Press Inc.

Mourik, S. V., Veldman, A. E. P., & Dreyer, M. E. (2005). Simulation of Capillary Flow with a Dynamic Contact Angle. *Microgravity Science and Technology*, *17*(3), 87–93. doi:10.1007/BF02872093

Mourzina, Y., Kalyagin, D., Steffen, A., & Offenhausser, A. (2005). Electrophoretic separations of neuromediators on microfluidic devices. *Electrophoresis*, *26*, 1849–1860. doi:10.1002/elps.200410295 PMID:15719361

Murphy, C. J., & Jana, N. R. (2002). Controlling the Aspect Ratio of Inorganic Nanorods and Nanowires. *Advanced Materials*, *14*(1), 80–82. doi:10.1002/1521-4095(20020104)14:1<80::AID-ADMA80>3.0.CO;2-#

Nahum, A., & Seifert, A. (2006). On the application of confined twin-jet instability to micro-mixing enhancement. *Physics of Fluids*, *18*(6), 064107. doi:10.1063/1.2214882

Nait Alla, A., Feddaoui, M., & Meftah, H. (2015a). Numerical Simulation of Heat and Mass Transfer in the Evaporation of Glycols Liquid Film along an Insulated Vertical Channel. *American Journal of Heat and Mass Transfer*, *2*(2), 59–75. doi:10.7726/ajhmt.2015.1005

Nait Alla, A., Feddaoui, M., & Meftah, H. (2015b). Numerical study of the evaporation by mixed convection of ethanol in partially heated plate along a vertical channel. *International Journal of Heat and Mass Transfer*, *89*, 206–215. doi:10.1016/j.ijheatmasstransfer.2015.04.076

Najim, M., Feddaoui, M., Nait Alla, A., Charef, A., & Labsi, N. (2017). Gas flow humidification enhancement by evaporating seawater liquid film. *23ème Congrès Français de Mécanique*.

Najim, M., Feddaoui, M., Charef, A., & Meftah, H. (2017). Computational study of saline water film evaporation in a vertical tube. *Desalination*, *408*, 81–91. doi:10.1016/j.desal.2017.01.008

Narayanan, S., Fedorov, A. G., & Joshi, Y. K. (2013). Heat and mass transfer during evaporation of thin liquid films confined by nanoporous membranes subjected to air jet impingement. *International Journal of Heat and Mass Transfer*, *58*(1–2), 300–311. doi:10.1016/j.ijheatmasstransfer.2012.11.015

Nguyen, N.-T., & Wereley, S. T. (2006). *Fundamentals and applications of microfluidics*. Boston: Publishing House Artech House.

Nichita, B. A., Zun, I., & Thome, J. R. (2010). *A VOF method coupled with a dynamic contact angle model for simulation of two-phase flows with partial wetting.* Paper presented at the 7th International Conference on Multiphase Flow, Tampa, FL.

Niu, X., Liu, L., Wen, W., & Sheng, P. (2006). Active microfluidic mixer chip. *Applied Physics Letters*, *88*(15), 153508. doi:10.1063/1.2195567

Noroozi, Z., Kido, H., Micic, M., Pan, H., Bartolome, C., Princevac, M., ... Madou, M. (2009). Reciprocating flow-based centrifugal microfluidics mixer. *The Review of Scientific Instruments*, *80*(7), 075102. doi:10.1063/1.3169508 PMID:19655976

Noureddini, H., Teoh, B. C., & Davis Clements, L. (1992). Viscosities of vegetable oils and fatty acids. *Journal of the American Oil Chemists' Society*, *69*(12), 1189–1191. doi:10.1007/BF02637678

Oberkampf, W. L., & Trucano, T. G. (2002). Verification and Validation in Computational Fluid Dynamics. *Progress in Aerospace Sciences*, *38*(3), 209–272. doi:10.1016/S0376-0421(02)00005-2

Ober, T. J., Foresti, D., & Lewis, J. A. (2015). Active mixing of complex fluids at the microscale. *Proceedings of the National Academy of Sciences of the United States of America*, *112*(40), 12293–12298. doi:10.1073/pnas.1509224112 PMID:26396254

Oh, K. W., & Ahn, C. H. (2006). A review of microvalves. *Journal of Micromechanics and Microengineering*, *16*(5), R13–R39. doi:10.1088/0960-1317/16/5/R01

omega.com. (2016). *Thermocouples: Using Thermocouples in Temperature Measurement.* Retrieved from: https://www.omega.com/prodinfo/thermocouples.html

omega.com. (2017). *What are RTD Sensors? Why Use Them? How Do They Work?* Retrieved from: https://www.omega.com/temperature/pdf/rtd_gen_specs_ref.pdf

omega.com. (2018). *Technical Learning: Theory and application of the infrared temperature sensor.* Retrieved from: https://www.omega.co.uk/technical-learning/infrared-temperature-measurement-theory-application.html

Onea, A., Wörner, M., & Cacuci, D. G. (2009). A Qualitative Computational Study of Mass Transfer in Upward Bubble Train Flow through Square and Rectangular Mini-Channels. *Chemical Engineering Science*, *64*(7), 1416–1435. doi:10.1016/j.ces.2008.11.008

Orfi, J., Debbissi, C., Mohamed, A. B., & Nasrallah, S. B. (2004). Air humidification by free convection in a vertical channel. *Desalination*, *168*, 161–168. doi:10.1016/j.desal.2004.06.182

Ortega-Casanova, J. (2016). Enhancing mixing at a very low Reynolds number by a heaving square cylinder. *Journal of Fluids and Structures*, *65*, 1–20. doi:10.1016/j.jfluidstructs.2016.05.002

Ortega-Casanova, J. (2017a). CFD study on mixing enhancement in a channel at a low Reynolds number by pitching a square cylinder. *Computers & Fluids*, *145*, 141–152. doi:10.1016/j.compfluid.2016.12.022

Ortega-Casanova, J. (2017b). On the onset of vortex shedding from 2D confined rectangular cylinders having different aspect ratios: Application to promote mixing fluids. *Chemical Engineering and Processing: Process Intensification*, *120*, 81–92. doi:10.1016/j.cep.2017.06.014

ourpcb.com. (2016). *Temperature Sensor Thermocouple design considerations using LT ICs.* Retrieved from: https://www.ourpcb.com/thermocouple.html

Öztaskin, M. C., Wörner, M., & Soyhan, H. S. (2009). Numerical investigation of the stability of bubble train flow in a square minichannel. *Physics of Fluids*, *21*(4), 042108. doi:10.1063/1.3101146

Pangarkar, K., Schildhauer, T. J., van Ommen, J. R., Nijenhuis, J., Kapteijn, F., & Moulijn, J. A. (2008). Structured Packings for Multiphase Catalytic Reactors. *Industrial & Engineering Chemistry Research*, *47*(10), 3720–3751. doi:10.1021/ie800067r

Parikh, D. (2009). *Handbook of Pharmaceutical Granulation Technology*. London: Informa Healthcare.

Parsa, M. K., & Hormozi, F. (2014). Experimental and CFD modeling of fluid mixing in sinusoidal microchannels with different phase shift between side walls. *Journal of Micromechanics and Microengineering*, *24*(6), 065018. doi:10.1088/0960-1317/24/6/065018

Pat. No. 112393 UA (2016). *Vortex granulator with utilization of waste gases.*

Pat. No. 112394 UA (2016). *Vortex granulator.*

Pat. No. 112622 UA (2016). *Vortex granulator.*

Pat. No. 113141 UA (2017). *Vortex granulator.*

Pat. No. 74070 UA (2012). *A device for disperse materials drying.*

Pat. No. 81720 UA (2013). *A device for disperse materials drying.*

Patankar, S. V. (1980). *Numerical heat transfer and fluid flow*. New York: Hemisphere/McGraw Hill. doi:10.1201/9781482234213

Patankar, S. V., & Spalding, D. B. (1972). A calculation procedure for heat, mass and momentum transfer in three-dimensional parabolic flows. *International Journal of Heat and Mass Transfer*, *15*(10), 1787–1806. doi:10.1016/0017-9310(72)90054-3

Patel, P., Telange, D., & Sharma, N. (2011). Comparison of Different Granulation Techniques for Lactose Monohydrate. *International Journal of Pharmaceutical Sciences and Drug Research*, *3*, 222–225.

Pauzauskie, P. J., Radenovic, A., Trepagnier, E., Shroff, H., Yang, P., & Liphardt, J. (2006). Optical trapping and integration of semiconductor nanowire assemblies in water. *Nature Materials*, *5*(2), 97–101. doi:10.1038/nmat1563 PMID:16429143

Peng, X. F., Hu, H. Y., & Wang, B. X. (1998). Boiling nucleation during liquid flow in Microchannels. *International Journal of Heat and Mass Transfer*, *41*(1), 101–106. doi:10.1016/S0017-9310(97)00096-3

Petersen, K. E., & Petersen, K. E. (1979). Fabrication of an integrated, planar silicon ink-jet structure. *IEEE Transactions on Electron Devices*, *26*(12), 1918–1920. doi:10.1109/T-ED.1979.19796

Peterson, K. A., Knudson, R. T., Garcia, E. J., Patel, K. D., Okandan, M., Ho, C. K., . . . Zawicki, L. R. (2008, June). LTCC in microelectronics, microsystems, and sensors. In *Mixed Design of Integrated Circuits and Systems, 2008. MIXDES 2008. 15th International Conference on* (pp. 23-37). IEEE.

Pistoresi, C., Fan, Y., & Luo, L. (2015). Numerical study on the improvement of flow distribution uniformity among parallel mini-channels. *Chemical Engineering and Processing*, *95*, 63–71. doi:10.1016/j.cep.2015.05.014

Pohar, A., & Plazl, I. (2009). Process Intensification through Microreactor Application. *Chemical and Biochemical Engineering Quarterly*, *23*, 537–544.

Pohorecki, R., & Kula, K. (2008). A simple mechanism of bubble and slug formation in Taylor flow in Microchannels. *Chemical Engineering Research & Design*, *86*(9), 997–1001. doi:10.1016/j.cherd.2008.03.013

Pokusaev, B. G., Zaitsev, A. A., & Zaitsev, V. A. (1999). Transfer Processes under Slug Flow Conditions in Three-Phase Media. *Theoretical Foundations of Chemical Engineering*. *33*(6), 539 -549.

Polascik, T. J., Cottet, G.-H., & Huberson, S. (2013). *Vortex Flows and Related Numerical Methods*. New York: Springer Science & Business Media.

Poling, B. E., Thomson, G. H., Friend, D. G., Rowley, R. L., & Wilding, W. V. (2008). Physical and Chemical Data. In *Perry's Handbook of Chemical Engineering* (8th ed.). New York: McGraw-Hill Education.

Popescu, M. N., Ralston, J., & Sedev, R. (2008). Capillary Rise with Velocity-Dependent Dynamic Contact Angle. *Langmuir*, *24*(21), 12710–12716. doi:10.1021/la801753t PMID:18834162

Popova, E. A., Abiev, R. Sh., Lappalainen, L. A., Svetlov, S. D., Andreeva, T. V., Trifonov, R. E., & Ostrovskii, V. A. (2014). Synthesis of 5-phenyltetrazole and its N-methyl Derivatives in a Microreactor. *Chemical and Biochemical Engineering Quarterly*, *28*(2), 241–246. doi:10.15255/CABEQ.2013.1936

Pradhan, T. K., & Panigrahi, P. K. (2016). Evaporation-induced natural convection of a liquid slug of binary mixture inside a microchannel: Effect of confinement. *Microfluidics and Nanofluidics*, *20*(8), 115. doi:10.100710404-016-1782-6

Raach, H., & Mitrovic, J. (2005). Seawater falling film evaporation on vertical plates with turbulence wires. *Desalination*, *183*(1–3), 307–316. doi:10.1016/j.desal.2005.02.053

Radiom, M., Chan, W. K., & Yang, C. (2010). Capillary filling with the effect of pneumatic pressure of trapped air. *Microfluidics and Nanofluidics*, *9*(1), 65–75. doi:10.100710404-009-0527-1

Rafferty, I., & Kaminski, D. (2014). Non-dimensionalization and mixing quantification of laminar twin semi-confined jets. *International Journal of Heat and Fluid Flow*, *48*, 15–23. doi:10.1016/j.ijheatfluidflow.2014.03.007

Rahimi, M., Aghel, B., Alitabar, M., Sepahvand, A., & Ghasempour, H. R. (2014). Optimization of biodiesel production from soybean oil in a microreactor. *Energy Conversion and Management*, *79*, 599–605. doi:10.1016/j.enconman.2013.12.065

Rahimi, M., Azimi, N., & Parvizian, F. (2013). Using microparticles to enhance micromixing in a high frequency continuous flow sonoreactor. *Chemical Engineering and Processing: Process Intensification*, *70*, 250–258. doi:10.1016/j.cep.2013.03.013

Rahmat, M., & Hubert, P. (2010). Two-phase simulations of micro heat pipes. *Computers & Fluids*, *39*(3), 451–460. doi:10.1016/j.compfluid.2009.09.014

Raithby, G. D., & Schneider, G. E. (1979). Numerical solution of problems in incompressible fluid flow: Treatment of the velocity-pressure coupling. *Numerical Heat Transfer*, *2*(4), 417–440. doi:10.1080/10407787908913423

Ramon, G., & Oron, A. (2008). Capillary rise of a meniscus with phase change. *Journal of Colloid and Interface Science*, *327*(1), 145–151. doi:10.1016/j.jcis.2008.08.016 PMID:18722625

Ramshaw, C. (1995). *The Incentive for Process Intensification*. International Conference on Process Intensification for the Chemical Industry, London, UK.

Ranade, V. V. (2002). Multiphase Flow Processes. In Computational Flow Modeling for Chemical Reactor Engineering (pp. 85-122). Academic Press.

Rebenklau, L., Wolter, K. J., & Howitz, S. (2000). Realization of hybrid microfluidic systems using standard LTCC process. In Electronic Components & Technology Conference, 2000. 2000 Proceedings (pp. 1696-1700). IEEE.

Rebrov, E. V. (2010). TwoPhase Flow Regimes in Microchannels. *Theoretical Foundations of Chemical Engineering, 44*(4), 355–367. doi:10.1134/S0040579510040019

Reinelt, D. A. (1987). The Rate at Which a Long Bubble Rises in a Vertical Tube. *Journal of Fluid Mechanics, 175*(-1), 557–565. doi:10.1017/S0022112087000521

Renken, A., Hessel, V., Löb, P., Uerdingen, M., & Kiwi-Minsker, L. (2007). Chemical Engineering and Processing. *Process Intensification, 46*, 840–845. doi:10.1016/j.cep.2007.05.020

resistorguide.com. (2018). *Thermistors*. Retrieved from: http://www.resistorguide.com/thermistor/

Revellin, R., & Thome, J. R. (2007). Adiabatic two-phase frictional pressure drops in microchannels. *Experimental Thermal and Fluid Science, 31*(7), 673–685. doi:10.1016/j.expthermflusci.2006.07.001

Revertera, F., Perpiñàb, X., León b, J., Vellvehib, M., Jordàb, X., & Altet, J. (2015). Characterization of MOSFET temperature sensors for on-chip dynamic thermal measurements. *Procedia Engineering, 120*, 836–839. doi:10.1016/j.proeng.2015.08.699

Reyes, D. R., Iossifidis, D., Auroux, P. A., & Manz, A. (2002). Micro total analysis systems. 1. Introduction, theory, and technology. *Analytical Chemistry, 74*(12), 2623–2636. doi:10.1021/ac0202435 PMID:12090653

Roache, P. J. (1998). *Verification and Validation in Computational Science and Engineering*. Albuquerque, NM: Hermosa Publishers.

Rocha, F. R., & Reis, B. F. (2000). A flow system exploiting multicommutation for speciation of inorganic nitrogen in waters. *Analytica Chimica Acta, 409*(1-2), 227–235. doi:10.1016/S0003-2670(99)00883-1

Ro, K. W., Lim, K., Shim, B. C., & Hahn, J. H. (2005). Integrated light collimating system for extended optical-path-length absorbance detection in microchip-based capillary electrophoresis. *Analytical Chemistry, 77*(16), 5160–5166. doi:10.1021/ac050420c PMID:16097754

Roy, S., Bauer, T., Al-Dahhan, M., Lehner, P., & Turek, T. (2004). Monoliths as Multiphase Reactors: A Review. *AIChE Journal. American Institute of Chemical Engineers, 50*(11), 2918–2938. doi:10.1002/aic.10268

Řužička, J., & Hansen, E. H. (1975). Flow injection analyses: Part I. A new concept of fast continuous flow analysis. *Analytica Chimica Acta, 78*(1), 145–157. doi:10.1016/S0003-2670(01)84761-9

Saikh, M. (2013). A technical note on granulation technology: A way to optimise granules. *International Journal of Pharmaceutical Sciences Review and Research, 4*, 55–67.

Salim, A., Fourar, M., Pironon, J., & Sausse, J. (2008). Oil-water two-phase flow in microchannels: Flow patterns and pressure drop measurements. *Canadian Journal of Chemical Engineering*, *86*(6), 978–988. doi:10.1002/cjce.20108

Samiec, D. (2012). Distributed fiber-optic temperature and strain measurement with extremely high spatial resolution. *Optical Metrology*, *6*, 10–13.

Santacesaria, E., Di Serio, M., Tesser, R., Turco, R., Tortorelli, M., & Russo, V. (2012). Biodiesel process intensification in a very simple microchannel device. *Chemical Engineering and Processing*, *52*, 47–54. doi:10.1016/j.cep.2011.12.001

Santana, H. S., Sanchez, G. B., & Taranto, O. P. (2017a). Evaporation of excess alcohol in biodiesel in a microchannel heat exchanger with Peltier module. *Chemical Engineering Research & Design*, *124*, 20–28. doi:10.1016/j.cherd.2017.05.022

Santana, H. S., Silva, J. L. Jr, & Taranto, O. P. (2015a). Numerical simulation of mixing and reaction of *Jatropha curcas* oil and ethanol for synthesis of biodiesel in micromixers. *Chemical Engineering Science*, *132*, 159–168. doi:10.1016/j.ces.2015.04.014

Santana, H. S., Silva, J. L. Jr, & Taranto, O. P. (2015b). Numerical simulations of biodiesel synthesis in microchannels with circular obstructions. *Chemical Engineering and Processing*, *98*, 137–146. doi:10.1016/j.cep.2015.10.011

Santana, H. S., Tortola, D. S., Reis, E. M., Silva, J. L. Jr, & Taranto, O. P. (2016). Transesterification reaction of sunflower oil and ethanol for biodiesel synthesis in microchannel reactor: Experimental and simulation studies. *Chemical Engineering Journal*, *302*, 752–762. doi:10.1016/j.cej.2016.05.122

Santana, H. S., Tortola, D. S., Silva, J. L. Jr, & Taranto, O. P. (2017). Biodiesel synthesis in micromixer with static elements. *Energy Conversion and Management*, *141*, 28–39. doi:10.1016/j.enconman.2016.03.089

Saqr, K. M., Aly, H. S., Wahid, M. A., & Sies, M. M. (2009). Numerical simulation of confined vortex flow using a modified k–ε turbulence model. *CFD Letters*, *1*(2), 87–94.

Sarkar, S., Singh, K. K., Shankar, V., & Shenoy, K. T. (2014). Numerical simulation of mixing at 1–1 and 1–2 microfluidic junctions. *Chemical Engineering and Processing: Process Intensification*, *85*, 227–240. doi:10.1016/j.cep.2014.08.010

Sarpkaya, T. (2004). A critical review of the intrinsic nature of vortex-induced vibrations. *Journal of Fluids and Structures*, *19*(4), 389–447. doi:10.1016/j.jfluidstructs.2004.02.005

Scholz, C., Wirner, F., Li, Y., & Bechinger, C. (2012). Measurement of permeability of microfluidic porous media with finite-sized colloidal tracers. *Experiments in Fluids*, *53*(5), 1327–1333. doi:10.100700348-012-1362-9

Schubert, K., Brandner, J., Fichtner, M., Linder, G., Schygulla, U., & Wenka, A. (2001). Microstructure devices for applications in thermal and chemical process engineering. *Microscale Thermophysics Engineering*, *5*(1), 17–39. doi:10.1080/108939501300005358

Schwaller, A. (2005). *Total Automotive Technology*. Thomas Delmar Learning.

Schwartz, L. W., Princen, H. M., & Kiss, A. D. (1986). On the motion of bubbles in capillary tubes. *Journal of Fluid Mechanics, 172*(-1), 259–275. doi:10.1017/S0022112086001738

Sempértegui-Tapia, D. F., & Ribatski, G. (2017). Two-phase frictional pressure drop in horizontal micro-scale channels: Experimental data analysis and prediction method development. *International Journal of Refrigeration, 79*, 143–163. doi:10.1016/j.ijrefrig.2017.03.024

Shamloo, A., Madadelahi, M., & Akbari, A. (2016). Numerical simulation of centrifugal serpentine micromixers and analyzing mixing quality parameters. *Chemical Engineering and Processing, 104*, 243–252. doi:10.1016/j.cep.2016.03.017

Shao, N., Gavriilidis, A., & Angeli, P. (2009). Flow regimes for adiabatic gas-liquid flow in microchannels. *Chemical Engineering Science, 64*(11), 2749–2761. doi:10.1016/j.ces.2009.01.067

Sharma, A., & Eswaran, V. (2004). Heat and fluid flow across a square cylinder in the two-dimensional laminar flow regime. *Numerical Heat Transfer Part A, 47*, 79–107. doi:10.1080/10407780490520760

Shikhmurzaev, Y. D., & Sprittles, J. E. (2013). Dynamic contact angle of a liquid spreading on an unsaturated wettable porous substrate. *Journal of Fluid Mechanics, 715*, 273–282. doi:10.1017/jfm.2012.517

Shojaeian, M., & Kosar, A. (2016). Convective heat transfer of non-Newtonian power-law slip flows and plug flows with variable thermophysical properties in parallel-plate and circular microchannels. *International Journal of Thermal Sciences, 100*, 155–168. doi:10.1016/j.ijthermalsci.2015.09.024

Silva, J. L. Jr, Santana, H. S., Sanchez, G. B., & Taranto, O. P. (2017). Numerical Simulation of Excess Ethanol Evaporation from Biodiesel in a Micro Heat-Exchanger. *Chemical Engineering Transactions, 57*, 1123–1128.

Sobieszuk, P., Aubin, J., & Pohorecki, R. (2012). Hydrodynamics and Mass Transfer in GasLiquid Flows in Micro-reactors. *Chemical Engineering & Technology, 35*(8), 1346–1358. doi:10.1002/ceat.201100643

Sobieszuk, P., Ilnicki, F., & Pohorecki, R. (2014). Contribution of Liquid- and Gas-Side Mass Transfer Coefficients to Overall Mass Transfer Coefficient in Tayloer Flow in a Microreactor. *Chemical & Process Engineering, 35*(1), 35–45. doi:10.2478/cpe-2014-0003

Sohankar, A., Khodadadi, M., & Rangraz, E. (2015). Control of fluid flow and heat transfer around a square cylinder by uniform suction and blowing at low Reynolds numbers. *Computers & Fluids, 109*, 155–167. doi:10.1016/j.compfluid.2014.12.020

Solanki, H., Basuri, T., Thakkar, J., & Patel, C. (2010). Recent advances in granulation technology. *International Journal of Pharmaceutical Sciences Review and Research*, 48–54.

Solehati, N., Bae, J., & Sasmito, A. P. (2014). Numerical investigation of mixing performance in microchannel T-junction with wavy structure. *Computers & Fluids*, *96*, 10–19. doi:10.1016/j.compfluid.2014.03.003

Soleymani, A., Kolehmainen, E., & Turunen, I. (2008). Numerical and experimental investigations of liquid mixing in T-type micromixers. *Chemical Engineering Journal*, *135S*, S219–S228. doi:10.1016/j.cej.2007.07.048

Som, S. K., Biswas, G., & Chakraborty, S. (2017). *Introduction to Fluid Mechanics and Fluid Machines* (3rd ed.). McGraw Hill Education.

Song, H., Chen, D. L., & Ismagilov, R. F. (2006). Reactions in Droplets in Microfluidic Channels. *Angewandte Chemie International Edition*, *45*(44), 7336–7356. doi:10.1002/anie.200601554 PMID:17086584

Sotowa, K. I. (2014). Fluid Behavior and Mass Transport Characteristics of Gas^|^#8211;Liquid and Liquid^|^#8211;Liquid Flows in Microchannels. *Journal of Chemical Engineering of Japan*, *47*(3), 213–224. doi:10.1252/jcej.13we141

Souri, K. (2016). *Energy-Efficient Smart Temperature Sensors in CMOS Technology* (Unpublished doctoral dissertation). Electronic Instrumentation Laboratory of Delft University of Technology.

Squires, T. M., & Quake, S. R. (2005). Microfluidics: Fluid physics at the nanoliter scale. *Reviews of Modern Physics*, *77*(3), 977–1026. doi:10.1103/RevModPhys.77.977

Stankiewicz, A. I., & Moulijn, J. A. (2000). Process Intensification: Transforming Chemical Engineering. *Chemical Engineering Progress*, *96*(1), 22–33.

Stern, F., Wilson, R. V., Coleman, H. W., & Paterson, E. G. (1999). *Verification and Validation of CFD Simulations*. Iowa Institute of Hydraulic Research Report No. 407.

Stern, F., Wilson, R. V., Coleman, H. W., & Paterson, E. G. (2001a). Comprehensive Approach to Verification and Validation of CFD Simulations - Part 1: Methodology and Procedures. *Journal of Fluids Engineering*, *123*(4), 793–802. doi:10.1115/1.1412235

Stroock, A., Weck, M., Chiu, D., Huck, W., Kenis, P., Ismagilov, R., & Whitesides, G. (2000). Patterning Electro-osmotic Flow with Patterned Surface Charge. *Physical Review Letters*, *84*(15), 3314–3317. doi:10.1103/PhysRevLett.84.3314 PMID:11019078

Sung, J., Kim, Y. B., & Lee, M. H. (2012). *Transient Phenomena of Dynamic Contact Angle in Micro Capillary Flow*. Paper presented at the 15th International Symposium on Flow Visualization, Minsk, Belarus.

Sun, L., & Mishima, K. (2009). An evaluation of prediction methods for saturated flow boiling heat transfer in mini-channels. *International Journal of Heat and Mass Transfer*, *52*(23-24), 5323–5329. doi:10.1016/j.ijheatmasstransfer.2009.06.041

Suzuki, H., Inoue, Y., Nishimura, T., Fukutani, K., & Suzuki, K. (1993). Unsteady flow in a channel obstructed by a square rod (crisscross motion of vortex). *International Journal of Heat and Fluid Flow*, *14*(1), 2–9. doi:10.1016/0142-727X(93)90034-K

Svetlov, S. D., & Abiev, R. S. (2016). Modeling Mass Transfer in a Taylor Flow Regime through Microchannels Using a Three-Layer Model. *Theoretical Foundations of Chemical Engineering*, *50*(6), 975–989. doi:10.1134/S0040579516060166

Szekely, J., Neumann, A. W., & Chuang, Y. K. (1971). The rate of capillary penetration and the applicability of the washburn equation. *Journal of Colloid and Interface Science*, *35*(2), 273–278. doi:10.1016/0021-9797(71)90120-2

Tabeling, P. (2010). *Introduction to Microfluidics*. Oxford University Press.

Taha, T., & Cui, Z. F. (2004). Hydrodynamics of Slug Flow Inside Capillaries. *Chemical Engineering Science*, *59*(6), 1181–1190. doi:10.1016/j.ces.2003.10.025

Tahvildari, K., Tavakoli, H., Mashayekhi, A., & Golmohammadzadeh, R. (2016). Modeling and simulation of membrane separation process using computational fluid dynamics. *Arabian Journal of Chemistry*, *9*(1), 72–78. doi:10.1016/j.arabjc.2015.02.022

Tantra, R., & Manz, A. (2000). Integrated potentiometric detector for use in chip-based flow cells. *Analytical Chemistry*, *72*(13), 2875–2878. doi:10.1021/ac000036+ PMID:10905321

Taylor, G. I. (1961). Deposition of a Viscous Fluid on the Wall of a Tube. *Journal of Fluid Mechanics*, *10*(02), 161–165. doi:10.1017/S0022112061000159

Terry, S. C., Jerman, J. H., & Angell, J. B. (1979). A gas chromatographic air analyzer fabricated on a silicon wafer. *IEEE Transactions on Electron Devices*, *26*(12), 1880–1886. doi:10.1109/T-ED.1979.19791

Thelander, C., Agarwal, P., Brongersma, S., Eymery, J., Feiner, L. F., Forchel, A., ... Samuelson, L. (2006). Nanowire-based one-dimensional electronics. *Materials Today*, *9*(10), 28–35. doi:10.1016/S1369-7021(06)71651-0

Thulasidas, T. C., Abraham, M. A., & Cerro, R. L. (1995). Bubble Train Flow in Capillaries of Circular and Square Cross Section. *Chemical Engineering Science*, *50*(2), 183–199. doi:10.1016/0009-2509(94)00225-G

Thulasidas, T. C., Abraham, M. A., & Cerro, R. L. (1997). Flow Patterns in Liquid Slugs during Bubble Train Flow Inside Capillaries. *Chemical Engineering Science*, *52*(17), 2947–2962. doi:10.1016/S0009-2509(97)00114-0

Tibiriçá, C. B., & Ribatski, G. (2013). Flow boiling in micro-scale channels - Synthesized literature review. *International Journal of Refrigeration*, *36*(2), 301–324. doi:10.1016/j.ijrefrig.2012.11.019

Tibiriçá, C. B., Ribatski, G., & Thome, J. R. (2012). Saturated flow boiling heat transfer and critical heat flux in small horizontal flattened tubes. *International Journal of Heat and Mass Transfer*, *55*(25-26), 7873–7883. doi:10.1016/j.ijheatmasstransfer.2012.08.017

Tonkovich, A., Kuhlmann, D., Rogers, A., McDanie, J., Fitzgerald, S., Arora, R., Tu, J., Yeoh, G. H., & Liu, C. (2008). *Computational Fluid Dynamics: A Practical approach*. Oxford, UK: Butterwoth-Heinemann/Elsevier.

Triplett, K. A., Ghiaasiaan, S. M., Abdel-Khalik, S. I., & Sadowski, D. L. (1999). Gas–Liquid Two-Phase Flow in Microchannels. Part I: Two-Phase Flow Patterns. *International Journal of Multiphase Flow*, *25*(3), 377–394. doi:10.1016/S0301-9322(98)00054-8

Trojan, M. (2015). Computer modeling of a convective steam superheater. *Archives of Thermodynamics*, *36*(1), 125–137. doi:10.1515/aoter-2015-0009

Tsaoulidis, D., & Angeli, P. (2014). Effect of channel size on mass transfer during liquid-liquid plug flow in small scale extractors. *Chemical Engineering Journal*, *262*, 785–793. doi:10.1016/j.cej.2014.10.012

Tseng, L. Y., Yang, A. S., Lee, C. Y., & Hsieh, C. Y. (2011). CFD-based optimization of a diamond-obstacles inserted micromixer with boundary protrusions. *Engineering Applications of Computational Fluid Mechanics*, *5*(2), 210–222. doi:10.1080/19942060.2011.11015365

Tsoligkas, A. N., Simmons, M. J. H., & Wood, J. (2007).The Effect of Hydrodynamics on Reaction Rates in Capillary Reactor. *Proc. 6th Int. Conf. on Multiphase Flow*, 1.

Tsoligkas, A. N., Simmons, M. J. H., Wood, J., & Frost, C. G. (2007). Kinetic and selectivity studies of gas-liquid reaction under Taylor flow in a circular capillary. *Catalysis Today*, *128*(1-2), 36–46. doi:10.1016/j.cattod.2007.07.001

Tuckerman, D. B., & Pease, R. F. W. (1981). High-performance heat sinking for VLSI. Electron Device Lett., 2, 126-129.

Ueda, T., & Tanaka, H. (1975). Measurements of velocity, temperature and velocity fluctuation distributions in falling liquid films. *International Journal of Multiphase Flow*, *2*(3), 261–272. doi:10.1016/0301-9322(75)90014-2

Ukil, A., Braendle, H., & Krippner, P. (2012). Distributed Temperature Sensing: Review of Technology and Applications. *IEEE Sensors Journal*, *12*(5), 885–892. doi:10.1109/JSEN.2011.2162060

Van Baten, J. M., & Krishna, R. (2005). CFD Simulations of Wall Mass Transfer for Taylor Flow in Circular Capillaries. *Chemical Engineering Science*, *60*(4), 1117–1126. doi:10.1016/j.ces.2004.10.001

van der Linden, H. J., Jellema, L. C., Holwerda, M., & Verpoorte, E. (2006). Stabilization of two-phase octanol/water flows inside poly(dimethylsiloxane) microchannels using polymer coatings. *Analytical and Bioanalytical Chemistry*, *385*(8), 1376–1383. doi:10.100700216-006-0526-y PMID:16773299

Varjo, S. J., Ludwig, M., Belder, D., & Riekkola, M. L. (2004). Separation of fluorescein isothiocyanate-labeled amines by microchip electrophoresis in uncoated and polyvinyl alcohol-coated glass chips using water and dimethyl sulfoxide as solvents of background electrolyte. *Electrophoresis*, *25*(12), 1901–1906. doi:10.1002/elps.200405914 PMID:15213991

Versteeg, H. K., & Malalasekera, W. (2007). *An Introduction to Computational Fluid Dynamics: The Finite Volume Method* (2nd ed.). Edinburgh Gate, UK: Pearson Education Limited.

Volchkov, E. P., Dvornikov, N. A., Lukashov, V. V., & Abdrakhmanov, R. Kh. (2013). Investigation of the flow in the vortex chamber with centrifugal fluidizing bed with and without combustion. *Thermophysics and Aeromechanics*, *20*(6), 663–668. doi:10.1134/S0869864313060024

Voynov, N., Nikolaev, N., Kustod, F., Nikolaev, A. & Tarovatyu, D. (2008).Vortex contact stages for rectification. *Chemistry of Plant Raw Materials, 3,* 173-184.

Wang, J., Wang, S., Zhang, P., & Li, Y. (2017). Easy-disassembly bonding of PDMS used for leak-tight encapsulation of microfluidic devices. In *2017 18th International Conference on Electronic Packaging Technology (ICEPT)* (pp. 1051–1055). IEEE. 10.1109/ICEPT.2017.8046623

Wang, G. R., Yang, F., & Zhai, W. (2014). There can be turbulence in microfluidics at low Reynolds number. *Lab on a Chip*, *14*(8), 1452–1458. doi:10.1039/C3LC51403J PMID:24599543

Wang, H., Iovenitti, P., Harvey, E., & Masood, S. (2002). Optimizing layout of obstacles for enhanced mixing in microchannels. *Smart Materials and Structures*, *11*(5), 662–667. doi:10.1088/0964-1726/11/5/306

Wang, J., Jiao, N., Tung, S., & Liu, L. (2014). Magnetic microrobot and its application in a microfluidic system. *Robotics and Biomimetics*, *1*(1), 1–8. doi:10.118640638-014-0018-z PMID:26613074

Wang, L., Ma, S., Wang, X., Bi, H., & Han, X. (2014). Mixing enhancement of a passive microfluidic mixer containing triangle baffles. *Asia-Pacific Journal of Chemical Engineering*, *9*(6), 877–885. doi:10.1002/apj.1837

Wang, Y., & Wang, Z. G. (2014). An overview of liquidevapor phase change, flow and heat transfer in mini- and micro-channels. *International Journal of Thermal Sciences*, *86*, 227–245. doi:10.1016/j.ijthermalsci.2014.07.005

Wang, Y., Zhang, G., & Ma, J. (2002). Research of LTCC/Cu, Ag multilayer substrate in microelectronic packaging. *Materials Science and Engineering B*, *94*(1), 48–53. doi:10.1016/S0921-5107(02)00073-9

Warnier, M. J. F., de Croon, M. H. J. M., Rebrov, E. V., & Schouten, J. C. (2010). Pressure Drop of Gas–Liquid Taylor Flow in Round Micro Capillaries for Low to Intermediate Reynolds Numbers. *Microfluidics and Nanofluidics*, *8*(1), 33–45. doi:10.100710404-009-0448-z

Washburn, E. W. (1921). The dynamics of capillary flow. *Physical Review*, *17*(3), 273–283. doi:10.1103/PhysRev.17.273

Watts, P., & Haswell, S. J. (2005). The application of micro reactors for organic synthesis. *Chemical Society Reviews*, *34*(3), 235–246. doi:10.1039/b313866f PMID:15726160

Whitesides, G. M. (2006). The origins and the future of microfluidics. *Nature*, *442*(7101), 368–373. doi:10.1038/nature05058 PMID:16871203

Wicks, D. A., & Li, P. C. (2004). Separation of fluorescent derivatives of hydroxyl-containing small molecules on a microfluidic chip. *Analytica Chimica Acta*, *507*(1), 107–114. doi:10.1016/j.aca.2003.09.067

Wilcox, D. C. (1998). *Turbulence Modeling for CFD*. La Cañada, CA: DCW Industries.

Williamson, C. H. K., & Govardhan, R. (2004). Vortex-induced vibrations. *Annual Review of Fluid Mechanics*, *36*(1), 413–455. doi:10.1146/annurev.fluid.36.050802.122128

Woldemariam, M., Filimonov, R., Purtonen, T., Sorvari, J., Koiranen, T., & Eskelinen, H. (2016). Mixing performance evaluation of additive manufactured milli-scale reactors. *Chemical Engineering Science*, *152*, 26–34. doi:10.1016/j.ces.2016.05.030

Wörner, M. (2012). Numerical modeling of multiphase flows in microfluidics and micro process engineering: A review of methods and applications. *Microfluidics and Nanofluidics*, *12*(6), 841–886. doi:10.100710404-012-0940-8

Wörner, M. (2013). Computational study on scaling of co-current downward Taylor flow in small square channels of three different sizes. *Journal of Chemical Engineering of Japan*, *46*, 335–341. doi:10.1252/jcej.12we068

Wörner, M., Ghidersa, B., & Onea, A. (2007). A Model for the Residence Time Distribution of Bubble Train Flow in a Square Mini Channel Based on Direct Numerical Simulation Results. *International Journal of Heat and Fluid Flow*, *28*(1), 83–94. doi:10.1016/j.ijheatfluidflow.2006.04.011

Wu, C. J., Wang, L., & Wu, J. Z. (2007). Suppression of the von Karman vortex street behind a circular cylinder by a travelling wave generated by a flexible surface. *Journal of Fluid Mechanics*, *574*, 365–391. doi:10.1017/S0022112006004150

Wu, Z., & Chen, X. (2018). A novel design for 3D passive micromixer based on Cantor fractal structure. *Microsystem Technologies*, 1–12.

Xia, Y., & Whitesides, G. M. (1998a). Soft Lithography. *Angewandte Chemie International Edition*, *37*(5), 550–575. doi:10.1002/(SICI)1521-3773(19980316)37:5<550::AID-ANIE550>3.0.CO;2-G PMID:29711088

Xia, Y., & Whitesides, G. M. (1998b). Soft lithography. *Annual Review of Materials Science*, *28*(1), 153–184. doi:10.1146/annurev.matsci.28.1.153 PMID:29711088

Xing, Y., & Dittrich, P. (2018). One-Dimensional Nanostructures: Microfluidic-Based Synthesis, Alignment and Integration towards Functional Sensing Devices. *Sensors (Basel)*, *18*(2), 134. doi:10.339018010134 PMID:29303990

Xin, W., De Rosa, I. M., Cao, Y., Yin, X., Yu, H., & Ye, P. … Yang, J.-M. (2018). Ultrasonication-assisted Synthesis of High Aspect Ratio Gold Nanowires on Graphene Template and Investigation of Their Growth Mechanism. *Chemical Communications*. doi:10.1039/C8CC01727A

Yagodnitsyna, A., Kovalev, A., & Bilsky, A. (2015). Experimental investigation of immiscible liquids flow in a T-shaped microchannel. *10th Pacific Symposium on Flow Visualization and Image Processing*.

Yanagisawa, K., Kuwano, H., & Tago, A. (1995). Electromagetically driven microvalve. *Microsystem Technologies*, *2*(1), 22–25. doi:10.1007/BF02739524

Yang, A. S., Chuang, F. C., Chen, C. K., Lee, M. H., Chen, S. W., Su, T. L., & Yang, Y. C. (2015). A high-performance micromixer using three-dimensional Tesla structures for bio-applications. *Chemical Engineering Journal*, *263*, 444–451. doi:10.1016/j.cej.2014.11.034

Yang, D., Wang, Y., Ding, G., Jin, Z., Zhao, J., & Wang, G. (2017). Numerical and experimental analysis of cooling performance of single-phase array microchannel heat sinks with different pin-fin configurations. *Applied Thermal Engineering*, *112*, 1547–1556. doi:10.1016/j.applthermaleng.2016.08.211

Yang, F., Shi, J., Zhang, X., Hao, S., Liu, Y., Feng, C., & Cui, L. (2018). The microstructure and magnetic properties of Cu/CuO/Ni core/multi-shell nanowire arrays. *Chemical Physics Letters*, *697*, 43–47. doi:10.1016/j.cplett.2018.02.064

Yang, L., Loubière, K., Dietrich, N., Le Men, C., Gourdon, C., & Hébrard, G. (2017). Local investigations on the gas-liquid mass transfer around Taylor bubbles flowing in a meandering millimetric square channel. *Chemical Engineering Science*, *165*, 192–203. doi:10.1016/j.ces.2017.03.007

Yan, W. M., & Lin, T. F. (1991). Evaporative cooling of liquid film through interfacial heat and mass transfer in a vertical channel—II. Numerical study. *International Journal of Heat and Mass Transfer*, *34*(4–5), 1113–1124. doi:10.1016/0017-9310(91)90021-6

Yan, W. M., Lin, T. F., & Tsay, Y. L. (1991). Evaporative cooling of liquid film through interfacial heat and mass transfer in a vertical channel—I. Experimental study. *International Journal of Heat and Mass Transfer*, *34*(4–5), 1105–1111. doi:10.1016/0017-9310(91)90020-F

Yao, C. Q., Dong, Z. Y., Zhao, Y. C., & Chen, G. W. (2014a). An online method to measure mass transfer of slug flow in a microchannel. *Chemical Engineering Science*, *112*, 15–24. doi:10.1016/j.ces.2014.03.016

Yao, C. Q., Dong, Z. Y., Zhao, Y. C., & Chen, G. W. (2014b). The effect of system pressure on gas-liquid slug flow in a microchannel. *American Institute of Chemical Engineering Journal, 60*(3), 1132–1142. doi:10.1002/aic.14306

Yoshino, M., & Inamuro, T. (2003). Lattice Boltzmann simulations for flow and heat/mass transfer problems in a three-dimensional porous structure. *International Journal for Numerical Methods in Fluids, 43*(2), 183–198. doi:10.1002/fld.607

Yuan, F., & Isaac, K. M. (2017). A study of MHD-based chaotic advection to enhance mixing in microfluidics using transient three dimensional CFD simulations. *Sensors and Actuators. B, Chemical, 238*, 226–238. doi:10.1016/j.snb.2016.07.063

Yuschak, T. (2005). Microchannel Technology Scale-Up to Commercial Capacity. *Chemical Engineering Research & Design, 83*(6), 634–639. doi:10.1205/cherd.04354

Zhang, J., Fletcher, D. F., & Li, W. (2016). Heat transfer and pressure drop characteristics of gas–liquid Taylor flow in mini ducts of square and rectangular cross-sections. *International Journal of Heat and Mass Transfer, 103*, 45–56.

Zhang, S., Yang, Y., & Nie, Z. (2018). Controllable Self-Assembly of One-Dimensional Nanocrystals. In Functional Organic and Hybrid Nanostructured Materials (pp. 1–38). Weinheim, Germany: Wiley-VCH Verlag GmbH & Co. KGaA. doi:10.1002/9783527807369.ch1

Zhang, Y., Li, Z., Xu, K., Chencheng, G., Hao, Y., Meng, F., & Gui, Y. (2017). A low-cost flexible electrochemical accelerometer using graphene-integrated microchannel. In *Micro Electro Mechanical Systems (MEMS), 2017 IEEE 30th International Conference on* (pp. 1268–1271). Las Vegas, NV: IEEE. 10.1109/MEMSYS.2017.7863648

Zhang, X., Hu, N., Chen, X., Fan, T., Wang, Z., Zheng, X., ... Qian, S. (2017). Controllable cell electroporation using microcavity electrodes. *Sensors and Actuators. B, Chemical, 240*, 434–442. doi:10.1016/j.snb.2016.08.172

Zhao, B., Viernes, N. O. L., Moore, J. S., & Beebe, D. J. (2002). Control and Applications of Immiscible Liquids in Microchannels. *Journal of the American Chemical Society, 124*(19), 5284–5285. doi:10.1021/ja025835j PMID:11996566

Zhao, T. S., & Bi, Q. C. (2001). Pressure drop characteristics of gas-liquid two-phase flow in vertical miniature triangular channels. *International Journal of Heat and Mass Transfer, 44*(13), 2523–2534. doi:10.1016/S0017-9310(00)00282-9

Zhao, Y., Chen, G., & Yuan, Q. (2006). Liquid-liquid two-phase flow patterns in a rectangular microchannel. *AIChE Journal. American Institute of Chemical Engineers, 52*(12), 4052–4060. doi:10.1002/aic.11029

Zhmud, B. V., Tiberg, F., & Hallstensson, K. (2000). Dynamics of Capillary Rise. *Journal of Colloid and Interface Science*, *228*(2), 263–269. doi:10.1006/jcis.2000.6951 PMID:10926465

Zhou, T., Liu, T., Deng, Y., Chen, L., Qian, S., & Liu, Z. (2017). Design of microfluidic channel networks with specified output flow rates using the CFD-based optimization method. *Microfluidics and Nanofluidics*, *21*(1), 11. doi:10.100710404-016-1842-y

Zhu, C., Fu, T., Gao, X., Liu, W., & Ma, Y. (2011). Study on the liquid-liquid two-phase flow patterns in microchannels. *Chemical Industry and Engineering Progress*, *30*(s1), 65–69.

Zimmermann, S., & Taghipour, F. (2005). CFD Modeling of the Hydrodynamics and Reaction Kinetics of FCC Fluidized-Bed Reactors. *Industrial & Engineering Chemistry Research*, *44*(26), 9818–9827. doi:10.1021/ie050490+

Ziqian, C., Hongfei, Z., Kaiyan, H., & Chaochen, M. (2007). Steady-state experimental studies on a multi-effect thermal regeneration solar desalination unit with horizontal tube falling film evaporation. *Desalination*, *207*(1–3), 59–70. doi:10.1016/j.desal.2006.08.003

About the Contributors

Harrson Santana earned his Ph.D in Chemical Engineering from University of Campinas, Brazil in 2016. His Ph.D thesis was the investigation of biodiesel synthesis in microchannels. In 2015 he spent several months in the group of Prof. Lee Cronin at the University of Glasgow (UK) studying the application of microwaves in inorganic reactions and manufacture of microdevices using 3D printers. He is currently an associate researcher working in the field of microchemical plants and the use of 3D printers in chemical processes. He has published several articles exploring from numerical simulations in the development of microdevices to the use of Microfluidics in chemical reactions and unit operations. His scientific interest focuses on transport phenomena in microfluidic systems and 3D printers applied to chemical processes in microscale.

João Lameu Silva Jr., Chemical Engineer graduated in 2009, is currently Professor at the Federal Institute of Education, Science and Technology of the South of Minas Gerais, Campus Pouso Alegre and collaborator in research projects in the area of Microfluidics. Master and PhD degrees by the School of Chemical Engineering at the University of Campinas (FEQ/UNICAMP). He performed researches in UNICAMP/Petrobras Cooperation projects in the Laboratory of Chemical Processes and Corporate Management (PQGe), Department of Process Engineering of the School of Chemical Engineering/UNICAMP, between 2010 and 2015. He has experience in the area of Transport Phenomena and Particulate Systems, Microfluidics and Computational Fluid Dynamics (CFD).

Osvadir Pereira Taranto is currently a Full Professor at School of Chemical Engineering in the University of Campinas (UNICAMP). He has experience in Chemical Engineering, acting on the following subjects: drying, jetting bed, fluidized bed, coating and fluid dynamics.

* * *

Rufat Abiev has been, since 2008 till the present time, Full Professor, Chair of Optimization of Chemical and Biotechnological Equipment, St. Petersburg State Institute of Technology (TU). In October-November 2017 has worked as invited professor at ENSIACET, Institute National Politechnique Toulouse, at the Laboratory of Chemical Engineering (LGC). In December 2016 delivered a course of lectures 'Microreactors: mathematical modelling of hydrodynamics, heat and mass transfer, design and fabrication' as an invited professor in Indian Institute of Technology Indore. In October 2016 has worked as invited researcher in Ecole des Mines d'Ales (Mining School of Ales, France) in the Laboratory of Building Material Science, project was done in the field of foam concrete production and investigation of properties. In 2015 was a Chairman of 15th International Conference on Mixing (Mixing-15) in St.Petersburg (more than 100 participants from 34 countries). In 2014 has received a research grant from German Academy Exchange Service (DAAD) for 1 month to perform investigations in Karlsruhe Institut für Technologie (KIT, Germany) in the Institut fuer Mikroverfahrenstechnik (Direktor Prof. Dr.-Ing. Roland Dittmeyer, Head of the laboratory PD Dr.-Ing. Juergen Brandner). In 2006 for 2 months has received a research grant from Altana-Quandt Foundation and later in the same year for 3 months a research grant from German Academy Exchange Service (DAAD). The research was made at the Technical University Dresden in the Department of Chemical Engineering (Head – Prof. Ruediger Lange). In 1998 has received the grant of the Swiss Academy of engineering Science (SATW- Branco Weiss, Schweizerische Akademie der Technischen Wissenschaften). Since August, 1998 till March, 1999 worked as a trainee in laboratory of polymers branch of concern "CIBA Specialitaeten Chemie" in Basel (Switzerland). April 2015: Professional retraining course for economics and management for enterprise. June 2000: a defense of second thesis, degree of the Doctor of Science (habilitation). 1987 – 1990 - post-graduate student of Department of Optimization of Chemical and Biotechnological Equipment, St. Petersburg State Institute of Technology (Technical University), defense of a thesis (March 1991), degree of the Candidate (equivalent to PhD). 1982 – 1987 – St. Petersburg State Institute of Technology (Technical University); the Chemical Engineering diploma (honors degree). Fields of interest: Process Intensification, Microreactors, Process Simulation, Bubbles and Droplets break up, Heat- and Mass Transfer intensification, Pulsations, Bio fuels production. Twice (in 2005 and 2006) has obtained a honorary research grant from Russian Ministry of Research and Education for one of the best projects for young habilitated Doctors. In 2015 R. Abiev as a Chairman and a Head of Organizing Committee has organized 15th European Conference on Mixing (St. Petersburg, Russia), which attracted more than 100 participants from 34 countries. In 2009 R. Abiev was decorated with Mendeleyev's medal for achievements in the field of chemical engineering science and education. R. Abiev has more than 350 publications, including 6 books, 5 chapters in a "New Handbook of chemist and technologist" (in Russian), 2 chapters in "Engineering Ecology Handbook" (in Russian), more than 100 papers in peer-reviewed Russian and international journals and more than 60 patented inventions.

Abderrahman Nait Alla is a Professor of Process Engineering at the Technology High School, Hassan II University. He has published more than 6 research articles in international journals. He has worked in several fields, such as multiphasic flow, phase change, porous media, heat and mass transfer.

Artem Artyukhov is a member of Processes and Equipment of Chemical and Petroleum-Refineries Department. Research areas: 1. Mathematical modeling of the main processes and devices of chemical technology. 2. Optimization calculations of equipment for chemical production. 3. Creation of three-dimensional models of technological units and equipment. 4. Computer simulation of hydromechanical and heat-mass transfer processes in chemical technology. 5. Development of new types of heat-mass transfer equipment with the use of swirling flows.

Nadiia Artyukhova is a member of Processes and Equipment of Chemical and Petroleum-Refineries Department. Research areas: 1. Mathematical modeling of the main processes and devices of chemical technology. 2. Optimization calculations of equipment for chemical production. 3. Computer simulation of hydromechanical and heat-mass transfer processes in chemical technology. 4. Development of technological and design documentation for the manufacturing of equipment for chemical production. 5. Development of new types drying equipment.

Goutam Kumar Bose is currently working as Professor and Head in Mechanical Engineering Department, Haldia Institute of Technology, Haldia, India. He obtained his PhD in Production Engineering from the Jadavpur University, Kolkata, India. He has obtained a Master's in Engineering in Mechanical Engineering from the Bengal Engineering & Science University, Shibpur, India. He has worked as an Assistant Professor in the Department of Mechanical Engineering at the College of Engineering & Management, Kolaghat for ten years. He was an Engineer in R & D Centre of M/s Hindustan Motors Ltd. West Bengal, India. His active areas of interest are Metal Cutting, Non- conventional machining and Industrial and Production Management. He has published research papers in journals of international repute. He has attended several international conferences in India and abroad.

Adil Charef received his M. Sc. In energy and process engineering 2014 and his Ph.D. in energy and fluid mechanics, both at National School of Applied Sciences of Agadir, at Ibn Zohr University in April 2018. His research interests have encompassed multiphase flow, heat and heat transfer, laminar and turbulent flows modelling in addition to binary liquid film condensation.

M'barek Feddaoui is a professor of mechanical engineering at National School of applied Sciences, Ibn Zohr University. He is head of Energy Engineering, Materials and Systems. He received his professor (HDR) degree from Ibn Zohr University in 2011. He has published more than 40 research papers in international journals and more than 80 papers in edited proceedings and actes. He has been working in several fields, such as multiphase flow, phase change, porous media, heat and mass transfer, nanofluids.

Francisco-Jose Galindo-Rosales finished the MS in Engineering and the PhD in Rheology of Colloidal Suspensions at the University of Málaga (Spain). Since October 2008, he started his professional career as a scientist at the Katholieke University Leuven (Belgium) and, since may 2010, he works at the University of Porto (Portugal). His research interests lie on the field of Rheology and Microfluidics. He started to be interested in complex fluids since his Master Thesis, working with aqueous clay suspensions. In his PhD Thesis, he worked on the rheological characterization of fumed silica in PPG. Afterward, he kept on working in Rheology and Colloids, joining the Laboratory of Applied Rheology in the K.U. Leuven (Belgium) to study about the dispersion quality of fillers in nanocomposites. He also worked as a postdoctoral researcher at the Transport Phenomena Research Center (CEFT) in the University of Porto (Portugal) and he had the opportunity of furthering his expertise towards Microfluidics and Extensional Flows, working mainly in the development a micro-extensional rheometer by means of experimental and numerical techniques. Recently, in 2013 he was awarded one of the FCT Investigator Starting Grants to develop advanced energy absorbing composites based on natural cork. This latter research project has led to a technology consisting of reinforcing any scaffold material by the addition of complex fluids through microfluidic channels (PCT/IB2015/057399), which will allow the development of composites with optimal damping performance under an external mechanical load, either an impact or a vibration

Pritam Ghosh completed B.Tech in Mechanical Engineering from West Bengal University of Technology in 2014 and M.Tech in Mechanics of Fluids from Indian Institute of Engineering Science and Technology, Shibpur in 2016. Presently working as an Assistant Professor at Haldia Institute of Technology, Haldia.

Yuan Hai received his Ph.D. in Materials Science from Northwestern Polytechnical University, China in 2013. He joined the Xi'an Microelectronic Technology Institute in the same year and worked as a senior engineer till now. As a supervisor for Master Degree Candidates, his main research interests include: (1) high-density integrated packaging; (2) thin film preparation and performance testing; (3) research and implementation on efficient thermal management in tiny space. He published more than 10 journal articles and applied for 2 Chinese patents.

Jan Krmela is a head of Department of Numerical Methods and Computational Modelling. Research areas: 1. Computational modeling and tests of composites with a rubber (an elastomer) matrix such as tire casings, especially with focus on strain-stress states and modal analysis and computational modeling of composites after the degradation processes, tire-road interaction and vehicle parts; 2. Tests of tires on static and dynamic test machines with the pressure footprint analyses of contact footprints between tires and a plane road or a bump and also prediction of radial stiffness of tires; 3. Low cyclic loading tests of composites with textiles on a test machine with a video extensometer and static tests of composites with planning of tests with a design of geometric parameters of test samples and test conditions for testing of tire casing parts; 4. Determination of material parameters that can be used as input data into computational models; Microscopic observation of interface bonds between steel and textile cords and an elastomer; 5. 3D printing of technical objects.

Di Li received his Master's degree in Material Processing Engineering from Harbin Institute of Technology at Weihai in 2015. The title of his thesis was Liquid-liquid Two-phase Microfluid Flow and Mixing at Low Reynolds Number. In Aug. 2015, he joined Beijing Institute of Remote-Sense Equipment.

Jianguo Li received his bachelor's degree from Changchun Institute of Technology in 1998. He started his work in Xi'an Microelectronic Technology Institute as a senior engineer in the same year. His main research interests include: (1) LTCC substrate fabrication and performance testing; and (2) high-density integration and heat dissipation technology based on LTCC packaging platform.

Yujie Li received her Ph.D. in Materials Science from Northwestern Polytechnical University, China in 2002. She worked as a postdoc first in the School of Physics, Peking University, China and then in the Department of Semiconductor Physics, Ulm University, Germany 2002-2005. In Sep. 2005, she was awarded her scholarship by the Alexander von Humboldt foundation and continued her research work in Ulm University as an Alexander von Humboldt fellow. In 2008, Yujie joined the Max-Planck-Institute for Metal Research in Stuttgart and started her research in microfluidics. After coming back to China, she accepted the professorship in the School of Materials Science and Engineering, Harbin Institute of Technology at Weihai in 2011 and worked in Weihai until now. Her main research interests include: (1) novel microfluidic devices and their applications; (2) new materials and techniques for micro- and nano-joining; (3) integration and packaging of multi-functional microsystems; and (4) printable electronics.

Monssif Najim received his M.Sc. in energy and process engineering 2014 and his Ph.D. in energy and fluid mechanics 2018, both at National School of Applied Sciences of Agadir, Morocco. His research activities focus on phase-change phenomenon, especially on falling liquid films. Besides the research work, he has been actively involved in teaching at National School of Applied Sciences of Agadir, within the courses for renewable energies, refrigeration, mass transfer and energy systems modelling.

Joaquin Ortega-Casanova is native of Tijola (Almeria, Andalusia, Spain), a small city located in the valley of the Almanzora River. received his M.Sc. in Industrial Engineering and the Ph.D. in Fluid Mechanics from the University of Malaga in 1995 and 2000, respectively. He has worked on several Spanish and European competitive projects. He is currently Associate Professor at the University of Malaga where teaches in different Grade and Master's programs in the field of fluid mechanics. His main research interests include CFD, FSI, microfluidics, heat transfer and aero-hydrodynamics.

Debashis Pal earned his Ph.D. in Indian Institute of Technology Kharagpur, his ME in Indian Institute of Science Bangalore and BE in Jadavpur University.

María Sanchez-Claros earned his Master Degree in Industrial Engineering by ETSII from the University of Malaga (Spain) in 2017, and the Degree in Industrial Technologies Engineering with hydraulic and thermal branch specialization by ETSII from the University of Malaga (Spain) in 2015.

Reinaldo Lucas dos Santos Rosa is Engineer (2014) by the Federal University of São João Del Rei in Mechatronics Engineering area. Was substitute professor on the Mechatronics Engineering course at Federal University of São João Del Rei. Participates in research and development projects in the areas of micro and nanofabrication, nanosensors and electronic instrumentation, mainly in water quality monitoring. Master in Electrical Engineering, Microelectronic subarea, by the Department of Electronic Systems in the Laboratory of Integrated Systems, associated with the Polytechnic School of USP. Currently a doctoral student in the area of

Electrical Engineering, Microelectronic subarea, by the Department of Electronic Systems, Polytechnic School of USP.

Antonio Carlos Seabra, Engineer (1984), Masters (1989) and Doctorate (1997) degree, all by the Polytechnic School of University of São Paulo (USP), Brazil, in the Electric Engineering field. He completed basic education, high school and vocational education in public schools. Electronic Technician (1979) by the Technic School Lauro Gomes in São Bernardo do Campo. Professor in the Polytechnic School of University of São Paulo (USP) for more than 25 years, as full dedication Titular Professor at the Department of Electronic Systems Engineering, and Vice-President of graduation commission of EPUSP (2018-2020). Head of the Department of Electronic Systems Engineering from 2014 and 2017. He was coordinator of the basic cycle of the Graduation (first two years of the graduation) in the Polytechnic School of University of São Paulo (USP) from 2012 to 2013. He is ad-hoc consultant of the São Paulo Research Foundation (FAPESP) in projects of the Pesquisa Inovativa na Pequena e Média Empresa (PIPE) and of the CNPq. He is coordinator and collaborator in research and development in the fields of micro and nanofabrication, nanosensors and electronic instrumentation, especially in the monitoring of water quality. He was president (2009-2010) of the (IEEE-CAS) and developed the experimental part of his doctorate degree at the Interuniversity Microelectronics Center (IMEC) in Belgium between 1991 and 1993, considered the greatest independent centre of microelectronics in Europe. He is author of more than 100 scientific publications, 5 books and 2 translations.

Shan Song studied Electronic Packaging Technology in 2011-2015 in Harbin Institute of Technology at Weihai and was awarded her Bachelor of Engineering. In 2016 she went to Leibniz Universität Hannover and studied Nanotechnology there until now.

Jie Wang received his Master's degree in Material Processing Engineering from Harbin Institute of Technology at Weihai in 2017. His main research content was microfluidic devices and their applications in 2014-2017. In Aug. 2017, he joined The 54th Research Institute of China Electronics Technology Group Corporation. He is mainly engaged in the research of packaging processes and thermal stress simulation at the moment.

Shijie Wang received his Bachelor's Degree in Electronics Packaging from Harbin Institute of Technology at Weihai in 2017. Since Sep.2017, he studied for his Master's Degree in Materials Processing Engineering in the same university. His research topic at the moment is novel microfluidic devices and their applications.

Ashraf A. Zaher is the Chair of the Electrical and Computer Engineering Department in the American University of Kuwait. He was awarded both his B.Eng. (Honors) in Electronics and Communication Engineering in 1988 and his M.Eng. in Automatic Control Engineering in 1994 from Mansoura University, Egypt. He received his first Ph.D. in Systems Engineering from Oakland University, Michigan, USA in 2001, and his second Ph.D. in Control Engineering from Oakland /Mansoura Universities (joint supervision) in 2004. He participated in many projects, received many grants, and published several technical reports in the field of virtual engineering applications in the Automotive industry. Dr. Ashraf has plenty of publications in both international journals and conferences and is an associate editor in the American Control Conferences, as well as a reviewer in many periodicals. His research interests include virtual engineering, nonlinear dynamics, hybrid signal processing, engineering physics, and chaos. He is a member of IEEE, AACC and AIP.

Peng Zhang received his Ph.D. in Materials Processing Engineering from Harbin Institute of Technology, China in 2008. In Jul. 2008, he joined the School of Materials Science and Engineering in Harbin Institute of Technology at Weihai and started his research in materials science until now. In 2011 and 2016, he obtained associate professor and professorship successively. His main research interests include: (1) design and fabrication of materials with micro/nano characteristics; (2) relationship between structures and properties of micro/nano-scale materials and (3) micro/nano manufacturing of advanced materials.

Index

F

G

H

I

L

M

N

O

P

R

S

T

U

V

W

www.ingramcontent.com/pod-product-compliance
Lightning Source LLC
LaVergne TN
LVHW080605300926
849162LV00005B/1047